Jürgen Valldorf · Wolfgang Gessner (Eds.)

Advanced Microsystems for Automotive Applications 2003

Springer-Verlag Berlin Heidelberg GmbH

Jürgen Valldorf · Wolfgang Gessner (Eds.)

Advanced Microsystems for Automotive Applications 2003

With 357 Figures

Springer

Dr. Jürgen Valldorf
VDI/VDE-Technologiezentrum Informationstechnik GmbH
Rheinstr. 10B
D-14513 Teltow
e-mail: valldorf@vdivde-it.de

Dr. Wolfgang Gessner
VDI/VDE-Technologiezentrum Informationstechnik GmbH
Rheinstr. 10B
D-14513 Teltow
e-mail: gessner@vdivde-it.de

ISBN 978-3-662-31209-4 ISBN 978-3-540-76988-0 (eBook)
DOI 10.1007/978-3-540-76988-0

Cataloging-in-Publication Data applied for
Bibliographic information published by Die Deutsche Bibliothek
Die Deutsche Bibliothek lists this publication in the Deutsche Nationalbibliografie;
detailed bibliographic data is available in the Internet at <http://dnb.ddb.de>

http://www.springer.de

Originally published by Springer-Verlag Berlin Heidelberg New York in 2003.

Softcover reprint of the hardcover 1st edition 2003

Typesetting: Digital data supplied by authors
Cover-Design: deblik, Berlin
Printed on acid free paper 68/3020/kk - 5 4 3 2 1 0

Preface

Microsystems have become an indispensable constituent of modern automobiles. Many features of the vehicle requested by today's customers can only be offered by applying intelligent and extremely miniaturised components. Improvements in comfort, safety, engine performance and emissions are unimaginable without microsystems and their underlying technologies.

Still there remain a series of obstacles to overcome and challenges to face: cost-effectiveness of microsystems and in this context manufacturing, resistance against environmental parameters such as humidity, temperature, reliability and – may be the most exigent – systems integration. The latter implies the integration of microsystems into new electronic architecture concepts as well as issues related to a possible centralisation and combination of functions. Present and future development strategies related to microsystems for automobiles have to take into account the communication of the vehicle with an intelligent external environment.

The Yearbook 2003 "Advanced Microsystems for Automotive Applications" reflects the present status of discussions within industry and outlines the R&D perspectives related to the aforementioned areas. The second Yearbook after the 2002 edition succeeds the previous and successful annual conference publications dating back to the year 1998. With the Yearbook concept a promising attempt was made to follow systematically the progress in the various areas of microsystems for automotive applications and to provide yearly up-dated information of the state-of-the art in industrial research and technological development.

In the section "Technologies and Materials" the emphasis is on the development of core technologies and materials specifically for automotive microsystems applications. "Functions and Applications" contains contributions expressing the priorities of industrial R&D in view of applications. The contributions published under this heading have been selected by an industrial steering group of representatives of car manufacturers and system suppliers. Some of them are being presented at the 7th International Conference on Advanced Microsystems for Automotive Applications held on May 22-23, 2003 in Berlin.

My explicit thanks goes to the authors for their valuable contributions to this publication and to the members of the Honorary and Steering Committee for their commitment and support. Particular thanks are addressed to Graham Bell and Detlef Ricken, Delphi Delco, Roger Grace, Roger Grace Associates, Berthold Ulmer, DaimlerChrysler and Florian Solzbacher, First Sensor Technology, for inspiring conversations.

I would like to thank the European Commission for their financial support through the Innovation Relay Centre Northern Germany and to the supporting organisations Delphi Delco, First Sensor Technology, Investitionsbank Berlin, mstnews and Victorian Trade Commission.

Last but not least, I would like to express my sincere thanks to the Innovation Relay Centre team at VDI/VDE-IT, especially Uwe Klädtke and Jasmin Mehrgan for their help in preparing the book for publication, and not forgetting Jürgen Valldorf, who is project manager responsible for this initiative.

Teltow/Berlin, May 2003

Wolfgang Gessner

Public Financers

Berlin Senate for Economics and Technology

European Commission

Ministry for Economics Brandenburg

Supporting Organisations

Delphi Automotive Systems

Investitionsbank Berlin (IBB)

First Sensor Technology

mstnews

State Government of Victoria, Australia

Co-Organisators

European Council for Automotive R&D (EUCAR)

European Association of Automotive Suppliers (CLEPA)

Honorary Commitee

Domenico Bordone	President and CEO Magneti Marelli S.p.A., Italy
Günter Hertel	Vice President Research and Technology, DaimlerChrysler AG, Germany
Rémi Kaiser	Director Technology and Quality Delphi Automotive Systems Europe, France
Gian C. Michellone	President and CEO Centro Ricerche FIAT, Italy
Karl-Thomas Neumann	Head of Electronic Strategy VW AG, Germany

Steering Commitee

Giancarlo Alessandretti	Centro Ricerche FIAT, Orbassano, Italy
Wilhelm Bois	Audi AG, Ingolstadt, Germany
Serge Boverie	Siemens VDO Automotive, Toulouse, France
Albert Engelhardt	Conti Temic microelectronic GmbH, Nürnberg, Germany
Roger Grace	Roger Grace Associates, San Francisco, USA
Henrik Jakobsen	SensoNor A.S., Horten, Norway
Peter Lidén	AB Volvo, Göteborg, Sweden
Torsten Mehlhorn	Investitionsbank Berlin, Germany
Ulf Meriheinä	VTI Hamlin, Vantaa, Finland
Roland Müller-Fiedler	Robert Bosch GmbH , Stuttgart, Germany
Paul Mulvanny	QinetiQ, Farnborough, UK
Andy Noble	Ricardo Consulting Engineers Ltd, Shoreham-by-Sea, UK
Gerhard Pellischek	Clepa, Brussels, Belgien
David B. Rich	Delphi Delco Electronics Systems, Kokomo,USA
Detlef E. Ricken	Delphi Delco Electronics Europe GmbH, Wiesbaden, Germany
Jean-Paul Rouet	Sagem SA, Paris, France
Christian Rousseau	Renault S.A. D.I.V., Guyancourt, France
Ernst Schmidt	BMW AG, Munich, Germany
John P. Schuster	Motorola Inc. , Northbrook Illinois, USA
Bob Sulouff	Analog Devices Inc. Cambridge, USA
Gerd Teepe	Motorola GmbH, München, Germany
Jürgen Valldorf	VDI/VDE-IT, Teltow, Germany
Egon Vetter	Ewv Management Consultancy Pty Ltd., Melbourne, Australia
Ralf Voss	DaimlerChrysler Corporation, Detroit, USA
Matthias Werner	Deutsche Bank AG, Berlin, Germany

Conference chair:

Wolfgang Gessner	VDI/VDE-IT, Teltow, Germany

Table of Contents

Technologies and Materials

Prologue – Beyond Microsystems

F. Solzbacher, First Sensor Technology GmbH

Abstract

This almanch is aimed at executive staff and key development engineers in the automotive systems supply chain. Our vision is to establish information exchange and communication between technology driven sensor and actuator development and automotive systems suppliers. We are therefore trying to supply both parties with a handbook of available technologies and potentials. While trying to give a concise overview for the systems designer, sufficient detail is included in order to improve sensitivity towards strengths and weaknesses, opportunities and risks of microsystems technology.

Topics covered include all components of state of the art microsystems technologies, ranging from materials and basic processing technologies to sensors/actuators, packaging and electronics leading to complex micro-mechatronic systems and general issues such as reliability considerations.

This year only marks the second issue of this compendium, which we are planning on improving and expanding every year, eventually leading to a concise and complete handbook of advanced Microsystems for automotive applications. We are therefore looking forward to our readers' input and comments on this years first edition.

1 Introduction

Microsystems technology (MST) is moving towards being the dominating key technology to match the sensing and control tasks faced in automotive applications through the ever increasing requirements for safety, environment and comfort.

At the same time, MST is at a watershed in its development. After about two decades of evolution, driven chiefly by a few key applications, the MST lifecycle has reached a point, where strategic planning will have to fill the technological and structural gaps in research and industry, present today. Structural changes in the industrial and research environment as well as the government economic and research development policy will have to be made. New roadmaps for MST – in particular with respect to the automotive industry – will have to be found.

2 Technology Path

MST has come a long way from the crude beginnings of the first pressure sensors in Manifold Air Pressure applications (MAP) in the late 1970's. Whereas generally speaking, the time for market introduction of new products is still relatively long, it has nevertheless decreased significantly in past years to about 7-10 years from initial product evolution to commercialisation.

The MST technology concept combines new materials with micro processing technology, mostly well suited for low cost mass production purposes and machining technologies to form three basic building blocks of every micro system: sensing/actuation element, package and signal processing.

Some key challenges faced by MST products are:

- **Price** – new technologies usually have to match or outperform existing technologies at 10-20% lower price.
- **Media compatibility** – most of the promising applications today require high media compatibility, since the processes to be controlled frequently involve aggressive liquid or gaseous media.
- **High temperature compatibility** – most of the key applications in high volume markets are found in automotive control systems with high temperature requirements.
- **Integration of multiple functionalities** – apart from the direct cost reduction of a cheap microsystem compared to conventional solutions, frequently, production logistics and quality control can be simplified when integrating several components into one single microsystem rather than using several discrete devices.
- **Reliability** – failure modes and reliability of microsystems are already well investigated in many areas.

Advances in materials science and processing are the corner stone of each MST product. The main groups of materials to be distinguished are: materials for the package, the actual device, the electronics and the mechanical/electrical connection between these. Progress in semiconductor processing has evolved in a number of substrate materials, well suited for the use in micro structured devices, such as Silicon, Silicon-on-Insulator (SOI), Silicon Carbide and Gallium Arsenide. Pricing and reliability considerations have lead to an almost exclusive use of Si-based micro machined devices. Packaging and assembly has focused on ceramics, PCB-technology, and MCM's.

A key process for MST is the 3D-machining technique of semiconductor materials leading to the miniaturised structures constituting the sensing, actuating or other functional parts. The two main processes to be stated here are bulk and surface micro machining. In bulk micro machining a wet etching agent such as KOH, EDP or TMAH or alternatively, a dry etching process using e.g. SF_6 gas is used to etch 3D structures directly into the bulk Si-wafer. Structures of up to wafer thickness can easily be realised. Surface micro machining uses sandwiches of so called sacrificial and functional layers on top of the Si-wafer. Removal of the sacrificial layers results in free standing structures with thickness of about 1-5µm.

The choice of 3D-micromachining technology depends strongly on the application field and design of the MST product to be manufactured. Cost considerations usually strongly favour the cheap KOH wet etching process allowing batch processing of large quantities of wafers. Nevertheless, surface micromachined pressure sensors or gyroscopes as well as a number of acceleration sensors have achieved good commercial success. The decision on which fabrication technology to use however still has to be made individually for each application and design case. No general industrial standards or modular processing blocks are available, except for internal solutions.

3 Industry Structure

When MST technology was first introduced in the 1980's and early 1990's its use was limited to a few applications with the automotive industry being the key driver in promoting technology and device development. Many of the key industrial players, with Europe in the lead, were incorporating MST activities and investing (e.g. DaimlerChrysler, Siemens, Bosch, Temic...) into R&D and production facilities.

Today, the industry structure has changed and a growing number of MST based high-tech MST companies have evolved working in a tightly knit network of OEM's, first tier suppliers and MST spin-offs from university backgrounds. Singular applications have through their economic success triggered the development of application fields, slowly closing the gaps in processing production technology.

4 Markets

MST products have entered the market in far more depth than commonly known: a huge variety of MST-components have become standards, such as acceleration sensors for airbag control, pressure sensors for a variety of automotive, medical applications (e.g. blood pressure sensors) and white goods appliances (e.g. for water and foam level measurement) , inkjet printheads in office printers and many more. The current 7series BMW marks a new milestone in the use of MST-components with a total of 70 MST devices installed in the car. This is but the start of a new generation of products. Nowadays, upper class cars comprise more than 100 sensors of which almost 30 are MST based. The migration of microsystems based sensors is growing. In contrast to the shrinking annual growth rate of microelectronics from 15% in the past to now about 7%, microsensors are expected to grow with up to 23 to 25% annually. The automotive MST market is showing a sound growth rate of about 15% p.a. with a total volume of about 1,000 Mio. Euro. In comparison, the total automotive Sensor market amounts to 4,230 Mio. Euro. Prices for sensors range from about 3.50 Euro per piece for the packaged device (i.e. the smallest unit handled by non-specialist industrial customers) to 200.00 Euro per piece. The MST market remains a very attractive market and is subject to intensive research and analysis. Even though it is still essentially a "killer application" driven market, by now, all major physical and a growing number of chemical properties can be detected by MST devices.

Microsensors will on the one hand replace conventional systems. On the other hand they will as key elements generate many new applications and functions like assistant systems for brakes, suspension, vision and pre-crash-detection. Clear advantages besides price like reliability, self testing capabilities, speed, increased functionality and the ability to communicate yield further momentum to the success story of microsystems in automotive applications.

5 Summary and Outlook

MST can be considered an "established" technology for automotive applications. MST has migrated into various technology and application fields. MST is developing in a growing market environment, thereby partially replacing conventional solutions and adding new and advanced functionalities based on new degrees of freedom and new concepts for automotive systems, like e.g. ambient intelligence, active safety, pre-crash, x-by-wire... MST industry and market will have to re-organise. Whereas initially, the market was driven by a few key applications promoted by automotive OEM's and a few of their key suppliers, today, the entire supply chain has to co-operate in order to generate the next generation of devices and solutions.

The automotive system environment is becoming more and more complex. With the ability of spontaneous communication in different networks (local and wide area), the "system" expands beyond the single car. Whereas formerly, one single supplier delivered all components of an ABS system or all sensors for airbag control, today, the networked architecture allows for merged sensor systems for different functions. Therefore, function and sensor system responsibility might not be on the same partner. The vision of "ambient intelligence" provides an additional degree of freedom for realising new automotive functions. Ambient intelligence, meaning an environment of interacting smart devices, opens up new information sources for the vehicle. A body network could for example monitor the health status of the driver and administrate personal data, cars might inform each other about potentially dangerous traffic situations and road infrastructure might tell the car about its icy surface. The automotive industry is working on car-based solutions to get this information by developing eye tracking systems, scanner, road condition sensors and other solutions. An environment being capable to communicate might however allow to get the same type of information through different channels. The automotive engineer now has the difficult task to decide on the level for a problem solution and not just to design a new sufficient sensor. Because of synergies, one effect of ambient intelligence will be the reduced need for sensor systems fulfilling a specific function. The idea itself, however, asks for a huge number of systems ready to provide all different kind of data. Discussions what kind of sensors might be feasible eventually lead to autonomous and self organising nanosensors (swarms) with single measurement capabilities.

Florian Solzbacher
First Sensor Technology GmbH
Carl-Scheele-Straße 16
12489 Berlin
Germany
solzbacher.consultant@first-sensor.com

Keywords: tendencies, supply chain, sensor, actuator development, automotive suppliers, packaging, micro-mechatronic systems, reliability, mst-lifecycle, media compatibility, high temperature compatibility, silicon, silicon-on-insulator (SOI), silicon carbide, gallium arsenide, PCB-technology, wet etching, 3D-micromachining

Processes for the Fabrication of Three Dimensional MEMS Structures

E. Obermeier, Berlin University of Technology

Abstract

The similarity between MEMS and integrated circuit technology allows for the production of MEMS devices using standard manufacturing techniques developed by the semiconductor industry. Because MEMS are mechanical as well as electrical devices, however, their production requires the manufacture of three-dimensional structures as well as multiwafer stacks, whereby this requirement is lacking for pure microelectronic devices. This important difference has led to the development of special processes as well as the adaptation of existing semiconductor processes to MEMS-specific needs. Examples are wafer bonding, chemical- and mechanical planarization and deep reactive ion etching (DRIE). These processes will be described here, and their application to the manufacture of MEMS sensors, primarily for the automotive industry, will be demonstrated.

1 Introduction

Many MEMS products developed in the last ten years have shown a large market potential. Just as MEMS applications have found commercial success in the automotive industry [1-3], they are increasingly finding similar success in optical, biological and microfluidic areas. This success can be attributed to the close relationship between the MEMS and the microelectronic industry. In fact, the MEMS industry has profited from the existing processing and equipment technologies as well as from the proven experiences of the microelectronics industry. However, microelectronic processes are primarily planar, but MEMS processing technologies must be capable of realizing three-dimensional structures and wafer stacking. Therefore, the scope and diversity of fabrication techniques for manufacturing MEMS has widened remarkably in the past years. Some of them are still closely related to IC-type fabrication, while others are primarily specialized MEMS processes. Examples of these processes include: Wafer bonding, chemical mechanical polishing (CMP) and deep reactive ion etching (DRIE).

Wafer bonding is a key technology for developing three-dimensional structures, and for first level packaging of MEMS devices. CMP plays a significant role in wafer bonding and the planarization of multilevel MEMS made for example by surface micromachining. DRIE enables the fabrication of three-dimensional silicon structures with high etch rates and high aspect ratios. The DRIE process is an example of a specialized MEMS etching process that was developed by the MEMS industry. Recent MEMS (e. g. micro sensors, microfluidic systems, BioMEMS, microoptical MEMS) have demonstrated that the combination of the aforementioned technologies can enable the development of micromechanical devices with an unprecendented degree of versatility and flexibility.

2 Wafer Bonding

Wafer bonding, also called wafer-to-wafer bonding, is a key technology for the manufacture of three-dimensional MEMS structures and microelectronic devices. A multitude of sensors and other interesting devices exist using wafer bonding. This bonding method allows for hermetic sealing and first level packaging with high packaging density. Furthermore, wafer level packaging can protect the structures and circuits on the chips during the wafer dicing process.

The types of wafer bonding processes that are most commonly employed can be classified as [4]:

- Anodic bonding.
- Direct bonding.
- Intermediate-layer bonding.

The next sections will discuss each of these bonding techniques in detail. Representative microsensor devices for automotive applications using wafer-bonding will also be presented.

2.1 Anodic Bonding

Anodic bonding, also known as Field Assisted Bonding or Electrostatic Bonding, is a commonly used process for the packaging of micromechanical devices [5]. The process is often employed when bonding silicon to glass at temperatures below the softening temperature of the glass. However, this technique can also be used for bonding metals or alloys to glass or ceramics provided that the surfaces of the materials to be bonded are polished, smooth and free of particles and that the thermal expansion coefficients of the two materials are closely matched [6]. The anodic bonding process is shown schematically in figure 1 for silicon to glass bonding.

The equipment for the process consists of a hot plate on which the parts to be bonded are heated to temperatures between approximately 300 °C and 450 °C and a DC-power supply which is connected in such a way so that the silicon is positively charged with respect to the glass.

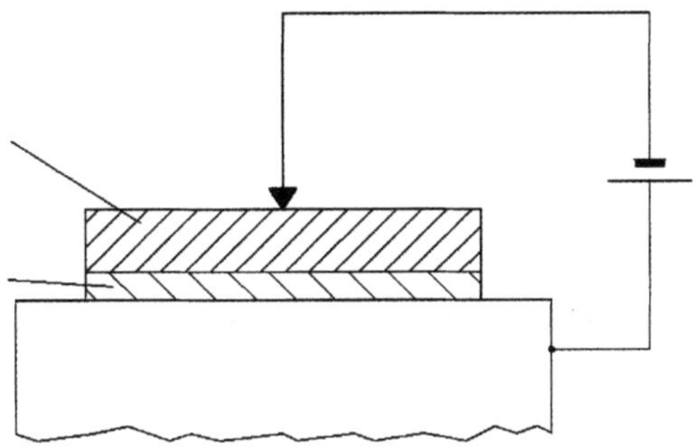

Fig. 1. Schematic setup for anodic bonding of silicon to glass.

Usually borosilicate glass (e. g. Corning PYREX®, code 7740 glass or TEMPAX®, Schott Glaswerke) is used. These glasses have thermal expansion coefficients that approximately equal the thermal expansion coefficient of silicon as well as the necessary electrical conductivity at the temperature at which the bonding process occurs. A necessary requirement for a good seal is that the two parts have a very close contact. For anodic bonding, this is achieved by applying the DC-voltage mentioned above across the parts to be bonded. Without applying an external force, the mirror-like bond surfaces are separated by a gap of several µm. The applied voltage (ca. 400V - 1000V) creates an electrostatic field in the gap, which generates an electrostatic force that pulls the parts into direct contact. At the same time, Na^+ ions, which are mobile at the bonding temperature, move through the electrostatic field from the glass/silicon interface to the negative electrode where they are neutralized. This causes a depletion of free positive ions in the region near the glass/silicon interface. The fixed negative oxygen ions remaining create a space charge region (figure 2). Due to the high electric field, oxygen ions leave the glass and create Si-O-Si bonds, resulting in a thin SiO_2 layer. This provides a hermetic and irreversible bonding of the silicon to the glass wafer.

Bond strengths are characterized by a number of techniques, one of which is the Pull Test. Pull tests provide values of bond strengths between 25MPa and 40MPa.

Anodic bonding can also be used for joining silicon with silicon. In this case, a thin borosilicate glass layer is deposited on one of the silicon wafers. The thermal mismatch is therefore caused only by the thin glass film [7, 8].

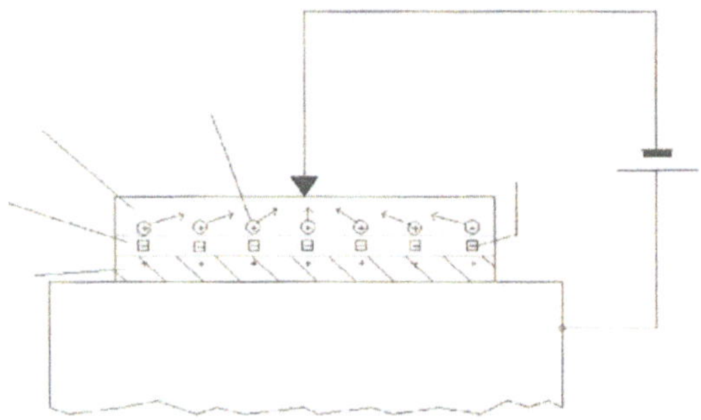

Fig. 2. Borosilicate glass with space charge region at the silicon glass interface during anodic bonding.

Figure 3 shows a piezoresistive pressure sensor for automotive air-conditioning systems. The Si die is bonded to an opposing glass plate. The essential requirements of the bonding process are to provide a hermetic seal and a first-level packaging ability.

a) b)

Fig. 3. Piezoresistive silicon pressure sensor for automotive air- conditioning system (Courtesy of Invensys Sensor Systems). a) Sensor dies anodically bonded to glass backplates; b) Fully packaged sensor.

In figure 4 an accelerometer from VTI Hamlin is illustrated which consists of three layers of silicon. The outer layers are coated with a borosilicate glass. The stack of three Si wafers is anodically bonded in a single step, then diced, and contact pads are then fabricated on the side of the sensor.

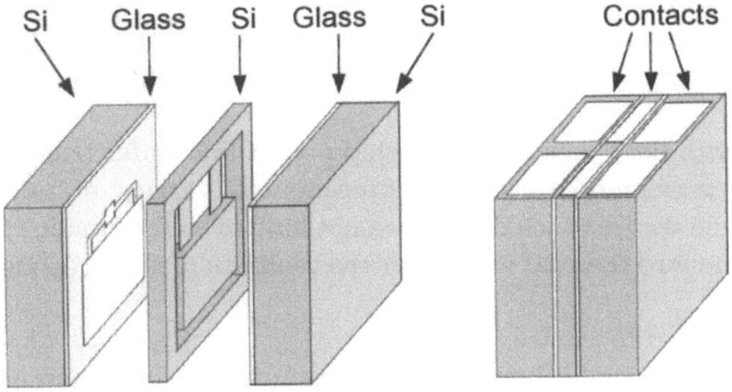

Fig. 4. Capacitive Accelerometer from VTI Hamlin with a five-layer-sandwich of silicon and glass [9].

2.2 Silicon Fusion Bonding

The process relies on the adhesion of two extremely clean, smooth and flat silicon surfaces, brought into close contact at room temperature in a clean room environment. A surface treatment (e.g. hydration, plasma exposure) is performed prior to the contacting, and the chemical species formed on the surfaces serve to promote the actual attraction between and bonding of the surfaces. Typically, the bonding strength is increased by a high-temperature annealing step. The temperature of the annealing process defines the bonding as being low temperature (<450 °C) or high temperature (>800 °C) [4]. The average roughness of the wafer to be bonded has to be on the order of 1 nm. Silicon fusion bonding eliminates thermal mismatch problems and has the potential to achieve bonds with strengths comparable to those of bulk silicon.

In short, the silicon fusion bonding process consists of three basic steps: surface treatment, wafer alignment and contacting, and annealing (Fig. 5), whereby the hydration process creates a hydrophilic wafer surface, i.e. a surface that is covered with OH groups.

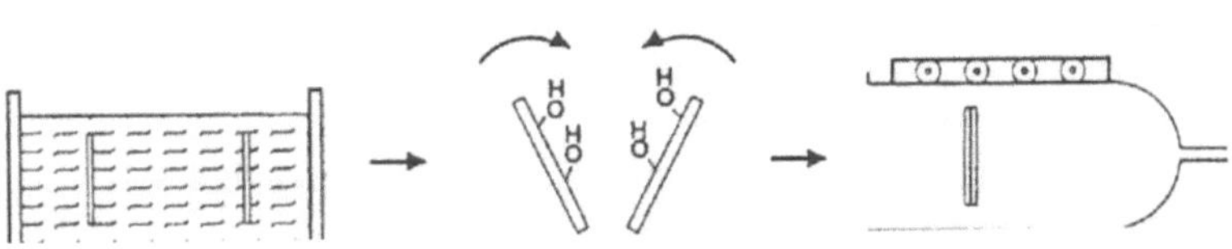

Fig. 5. Basic steps of the wafer bonding with hydrophilic surfaces.

As the temperature increases during the annealing process, three distinct phases in the bonding are observed. First, at the lower temperature, hydrogen bonds develop. Next, at the intermediate temperatures, Si-O-Si bonding accompanied by elastic deformation of the wafer occurs. Finally, at the highest temperatures the oxide flows plastically, promoting further Si-O-Si bond formation.

Together with deep reactive ion etching and CMP, wafer bonding affords the capability to manufacture an unprecedented variety of micromechanical devices. Such devices include but are not limited to five wafer stacks [13] and stacks with more than 20 wafers for the realization of microchemical reactors [14].

The use of silicon wafer bonding in the manufacture of an angular rate sensor is illustrated in figure 6. Here, the sensor is fabricated by a combination of deep reactive ion etching, high temperature wafer bonding, anisotropic wet etching, and anodic bonding [26].

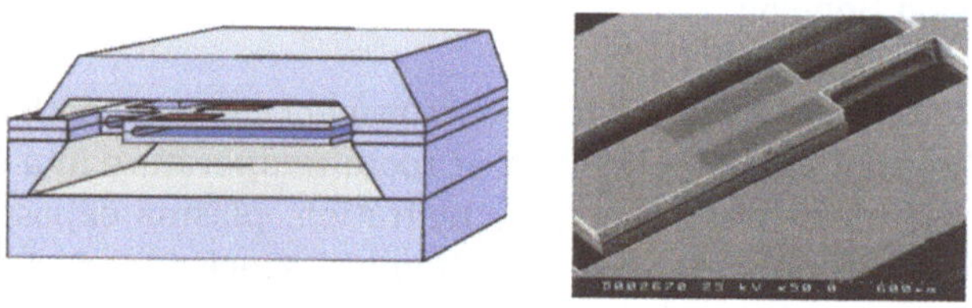

a) b)

Fig. 6. Angular rate sensor with tuning fork for automotive application.
a) Schematic cross section. b) Micromachined tuning fork fabricated by DRIE.

2.3 Intermediate-Layer Bonding

A large number of bonding processes using intermediate layers are available in MEMS technology to bond wafers and chips to each other. Some of the more common are [15]:

- Adhesive bonding (using a great variety of polymers).
- Low temperature glass sealing (using vacuum evaporated, sputtered or screen printed glass films).
- Soldering (using preforms, pastes and thin film solders).
- Eutectic bonding (using Au/Si and Al/Si thin films).

Adhesive bonding has some interesting properties for the fabrication and packaging of MEMS devices. These properties are given below. The other bonding methods and their properties will not be discussed here but can be found in the literature [17, 18].

Distinct advantages of adhesive bonding are [16]:

- A wide array of materials can be bonded with each other.
- Bonding temperatures lower than 100°C and even reaching room temperature can be used.
- High bond strengths can be achieved.
- The thermal mismatch of bonded materials can be reduced as a result of the elastic properties of some of the adhesive polymers used in the bonding.
- Low material and process costs can be achieved.
- Compatibility of most polymers with standard clean room processing is realized.

Disadvantages of polymer bonding are:

- Hermetic sealing is not possible with adhesive bonding.
- Long-term stability is limited.
- High coefficients of thermal expansion exist.
- Bonding strength decreases significantly with temperature.

Adhesive Bonding offers the potential that many different materials can be bonded to each other such as with Si to Si, glass, ceramic, or metal.

3 Chemical Mechanical Planarization (CMP)

In the past decade, chemical-mechanical planarization, also known as chemical-mechanical polishing, has emerged as the fastest growing process in the semiconductor manufacturing industry [19]. Note, however, that polishing and planarization are not synonyms. Polishing generally refers to smoothing of the surface, which is not necessarily planar. Chemical-mechanical planarization is an excellent planarization technique yielding a surface roughness of less than 1nm over large dimensions, but is slow, with removal rates of less than 100nm/min, compared to 1µm/min for standard polishing. Figure 7 illustrates the CMP process [20]:

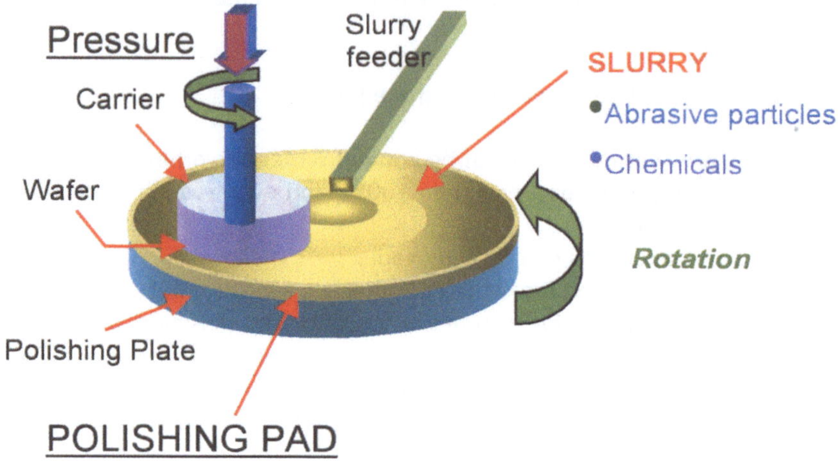

Fig. 7. Schematic of the CMP. The wafer to be planarized is mounted on a carrier and is brought in contact with a polishing pad mounted on a rotating plate.

A liquid slurry is continuously dispensed onto the surface of the polishing pad. A combination of the vertical force between the wafer and the abrasive pad with the chemical action of the slurry is used to polish the surface to a highly planar state [21].

Chemical-mechanical planarization is increasingly being used for the production of MEMS devices, particularly in polysilicon surface micromachining [22].

CMP planarization provides a means of integrating separate process technologies such as micromechanics and microelectronics or surface micromachining and high-aspect ratio micromachining and is also necessary for the realization of multi-wafer devices using anodic bonding or fusion bonding.

4 Deep Reactive Ion Etching (DRIE)

The DRIE process for creating three-dimensional silicon structures has proven to be absolutely essential for MEMS technology. Because DRIE allows for etching capabilities independent of the crystallographic structure of the silicon substrate being etched, it enables the manufacture of structures and components that were previously unrealizable or realizable only by means of more complicated processing steps. In fact the manufacture of structures with heights ranging from a few micrometers to the whole wafer thickness and with high aspect ratios are possible.

4.1 "BOSCH" Process

The "BOSCH" process, developed and patented by Bosch, is the etching process most frequently used today for the creation of three-dimensional silicon structures with large aspect ratios [24]. The process uses a fluorine-based plasma to etch the silicon, combined with a fluorocarbon plasma coating process to provide sidewall passivation and improved selectivity to the masking material. In fact, the plasma is generated outside of the reacting area of the ICP reactor (inductively coupled plasma), and the process cycles between etch and passivation steps to achieve the deep vertical profiles. A particular sidewall roughness, though, is unavoidable with the BOSCH process.

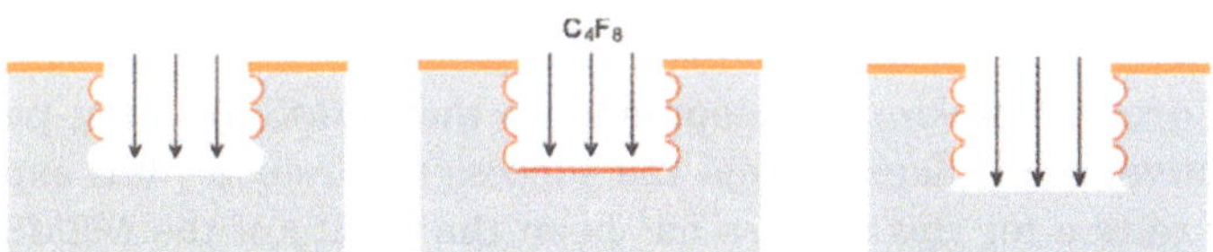

Fig. 8. DRIE employing the "BOSCH" process uses short etch steps using SF_6 combined with sidewall passivation using C_4F_8.

The BOSCH process has been used for the manufacture of a wide variety of MEMS devices, ranging from microsensors such as those for measuring pressure or acceleration, microactuators including valves, pumps, and grippers, to microoptical devices like mirrors and scanners. The example shown in figure 9 demonstrates the possibility of achieving an etch rate of up to 20μm/min and aspect ratio of 60:1 using this process.

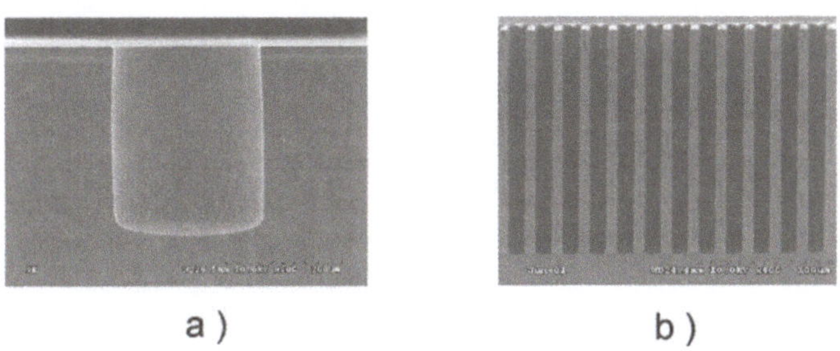

Fig. 9. Silicon microstructures fabricated with an etch rate of 20μm/min (figure 9a), and an aspect ratio of 60:1 (figure 9b) (Courtesy of Alcatel Vacuum Technology).

4.2 "Cryo" Process

An alternative to the "BOSCH" process is the "Cryo" process with the silicon wafers sitting on a liquid nitrogen cooled electrode (-100 °C) [25]. The process also uses SF_6 for the silicon etching. The mechanism for sidewall passivation and mask protection is, however, different. Instead of using a fluorocarbon plasma to create the passivation layer, oxide/fluoride layers are formed on the sidewalls and serve to passivate them. Etch rates of up to approximately 5µm/min and very smooth sidewalls can be achieved with the process. Disadvantages of the process are the need for cryogenic temperatures and good thermal contacts on the backsides of the wafers to achieve uniform profiles.

5 Conclusion

Numerous groundbreaking developments in the MEMS industry, particularly for automotive needs, demonstrate the success the industry has experienced. A primary reason for this success has been the ability of the MEMS industry to build upon existing microelectronic fabrication processes. In fact, by using common semiconductor batch processing techniques, MEMS device throughput has been increased while manufacturing costs have been reduced. The industry will continue to innovate, developing new devices for new applications according to market needs. Furthermore, new fabrication processes, such as the wafer bonding and DRIE processes discussed herein, will be developed, thereby nurturing an industry of equipment vendors and other supporting industries.

References

[1] H.-P. Trah und R. Müller-Fiedler, „Mikrosystemtechnik im Automobil", Physik Journal 1, nr. 11, p. 39, 2002

[2] P. Arminen, „Microsensors in Vehicle Stability and Driving Control", available: http://www.ad.tut.fi/aci/courses/76527/seminars2000/Microsensors_in vehicle stability.pdf

[3] R. Grace, „Market Opportunities for MEMS/MST in Automotive Applications", AMAA Yearbook 2001, Springer-Verlag Berlin Heidelberg New York, p. 3, 2001

[4] M. A. Schmidt, „Wafer-to-Wafer Bonding for Microstructure Formation", Proc. IEEE, vol. 86, no. 8, p. 1575, 1998

[5] G. Wallis and D. I. Pomerantz, „Field assisted glass-metal sealing", J. Amer. Ceramic Soc., vol. 53, no. 10, p. 564, 1969

[6] E. Obermeier, „Anodic wafer bonding", Proc. 3rd Int. Symp. on Semicond. Wafer Bonding: Sience, Technology and Applications, vol. 95-7, The Electrochem. Society, Pennington, NJ, p. 212, 1995

[7] A. D. Brooks, et. al, „Low temperature electrostatic silicon-to-silicon seals using sputtered borosilicate glass", J. Electrochem. Soc., vol. 119, no. 4, p. 545, 1972

[8] P. Krause, et. al, „Silicon to Silicon Anodic Bonding Using Evaporated Glass", Digest of Techn. Papers, Int. Conf. Solid-State Sensors and Actuators, Stockholm, Sweden, p. 228, 1995

[9] H. Kuisma, „Intertial Sensors for Automotive Applications", Digest of Techn. Papers, 11th Int. Conf. Solid-State Sensors and Actuators, Transducers'01, Munich, Germany, p. 430, 2001

[10] J. Lasky, et. al, „Silicon-on-Insulator (SOI) by Bonding and Etch-back", „Digest Electron Devices Meeting, Washington DC, p. 684, 1985

[11] Q.-Y. Tong, U. Gösele, „Semiconductor Wafer Bonding", John Wiley & Sons, Inc., New York, 1999

[12] K. Petersen et. al, „Silicon Fusion Bonding For Pressure Sensors", IEEE Solid-State Sensor and Actuator Workshop, Hilton Head, SC, p. 144, 1988

[13] M. A. Schmidt, „Technologies for Micro-turbomachinery", Digest Techn. Papers, 11th Int. Conf. on Solid-State Sensors and Actuators, Transducers'01, Munich, Germany, p. 2, 2001

[14] J. J. Leon, et. al, „Mikrosystem technology of chemical and biological microreactors", DECHEMA Monographs, vol. 132, no. 51, 1996

[15] J. F. Niklaus, „Low temperature full wafer adhesive bonding", J. Micromech. and Microeng., vol. 11, no. 2, p. 100, 2001

[16] J. F. Niklaus, „Void-free full-wafer adhesive Bonding", Proc. 13th IEEE Int. MEMS Conf., Myjazaki, Japan, 2000, p. 247

[17] W. H. Ko, et. al, „Bonding techniques for micorsensors", in Micromachining and Micropackaging of Transducers", C. D. Fung, P. W. Cheung, W. H. Ko, and D. G. Fleming, Eds., Amsterdam: The Netherlands: Elsevier, 1985, p. 41

[18] L. Ristic (ed.), „Sensortechnology and devices", Artech House, Inc. Boston-London, 1994

[19] J. M. Steigerwald, S. P. Murata, R. J. Gutmann, „Chemical Mechanical Planarization of Microelectronic Materials", John Wiley & Sons, Inc., New York, 1997

[20] S. Aksu, „Chemical Mechanical Planarization (CMP), available: http://www.mse.berkeley.edu/groups/doyle/serdar/CMP/CMP.html

[21] R. K. Singh and R. Bajaj, „Advances in Chemical-Mechanical Planarization", MRS Bulletin, Oct. 2002, p. 743

[22] J. J. Sniegowski, „Chemical-mechanical polishing: enhancing the manufacturability of MEMS", Proc. SPIE Conf.: Micromachining and Microfabrication Process Technology, Austin, Texas, vol. 2879, p. 104, 1996

[23] J. Bhardwaj et. al, „Dry Silicon Etching for MEMS", The Symposium on Microstructures and Microfabricated Systems, Annual Meeting of the Electrochem. Soc., Montreal, Quebec, Canada, p.118, 1997

[24] A. Schilp et. al, „Advanced Etch Tool for High Etch Rate Deep Reactive Ion Etching in Silicon Micromachining Production Environment", Proc. MST 2001, Düsseldorf

[25] M. J. Walker, „Comparison of Bosch and cryogenic processes for patterning high aspect ratio features in silicon", Proc. SPIE Conf.: MEMS Design, Fabrication, Characterization and Packaging, Edingburgh, UK, vol. 4401, p. 89, 2001

[26] R. Voss et. al, „Silicon angular rate sensor for automotive applications with piezoelectric drive and piezoresistive read-out", Techn. Digest 9th Int. Conf. on Sensors and Actuators, Transducers'97, Chicago, USA, p. 879, 1997

Ernst Obermeier
Berlin University of Technology, MAT
Gustav-Meyer-Allee 25
13355 Berlin
Germany
ernst.obermeier@mat.ee.tu-berlin.de

Keywords: fabrication, MEMS manufacturing, integrated circuit technology, multi-wafer stacks, wafer bonding, chemical planarization, mechanical planarization, deep reactive ion etching (DRIE), chemical mechanical polishing (CMP), anodic bonding, silicon fusion bonding, intermediate-layer bonding

Signal Processing for Automotive Applications

T. Riepl, S. Bolz, G. Lugert, Siemens VDO Automotive AG

Abstract

This article gives a brief overview on the major signal forms and the related signal processing in automotive applications with focus on powertrain applications. It describes the principle flow for signal processing and discusses different approaches for the integration of the signal processing hardware. Two examples for analogue signal processing are given: linear oxygen sensor and piezoresistive pressure sensor. Finally status and trends for signal transfer are addressed.

1 Sensor Signals as Input for Automotive Control Units

Engine and transmission control, as well as many other functions of automotive systems (safety, comfort, environmental protection, ...) require effective control concepts in order to fulfil the requirements from legislation and the customers. These control concepts can be basically reduced to a control loop: sensor measures the controlled variable - control unit processes the sensor input and drives the actuator - actuator adjusts the controlled variable.

In modern cars up to 70 control units are working. Most of the implemented control concepts are rather complex and involve a number of sensor signals to handle. As an example the block diagram of an engine control unit (ECU) is shown in figure 1.

A typical ECU provides several interfaces to sensors:

- Analogue input.
- Digital input (on/off).
- Pulse width modulated (PWM) input.
- CAN interface.

In order to protect the electronics from overvoltage all sensor signals have to pass input protection circuits (passive: R, RC circuits; active: specific semiconductors) and filters. These circuits suspend noise and limit the signals to the allowed input range of the microcontroler (0V to 5V). After analogue to digital conversion, the sensor signals are typically processed fully digital by the microcontroler.

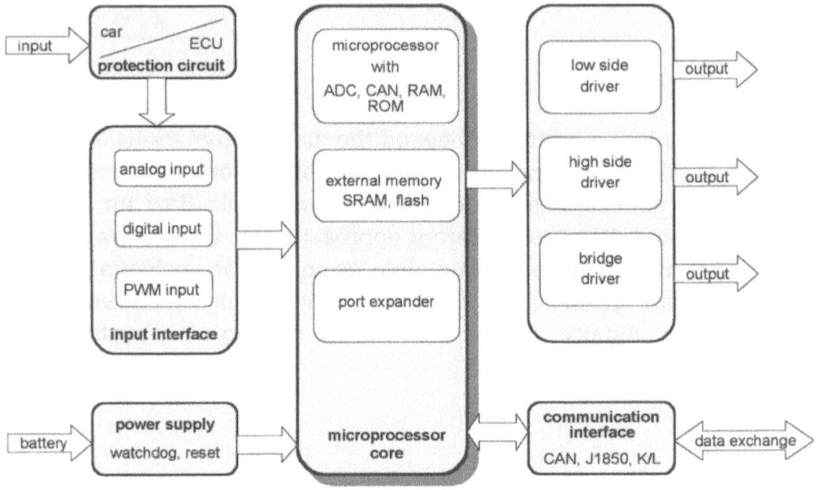

Fig. 1. Block diagram of a typical Engine Control Unit (ECU) [1].

In addition to these wire-based interfaces, also optical and telemetric interfaces are under development (see chapter 4). Today a typical control unit is able to handle the following signals:

Analogue Signals

Analogue signals are the input from signals from analogue sensors like lambda probe, pressure sensor and potentiometer. The signal information is described be the voltage with a typical range of 0V to 5V (typically 0.5V and 4.5V for a defined range of the measurand). This normalisation of the input voltages requires that analogue signal processing has to be included in the sensor.

Pulsed Signals

A typical example for pulsed signals is the output of a inductive rotational speed sensor. Its voltage range is up to 100V. Usually the signal is converted to a digital signal and then be handled by a counter. Typical ECUs have specific input interfaces that perform this conversion. Another example for pulsed signals is the pulse width modulated output of e.g. mass air flow and manifold absolute pressure sensors.

Digital Signals

The digital input interface handles the digital signals (on/off) of switches with a voltage range of 0V to battery voltage. The voltage is limited to 5V by the input protection circuit. In addition to this, also digitalised pulsed signals like that of hall probe with a digital output are handled by the digital input interface. Usually this signal type has a voltage range from 0V to 5V).

For extremely safety relevant purposes like the signal of an airbag satellite, electrical current is used as a digital signal. For this application an idle current is defined, which allows to control the integrity of the connection between airbag ECU and satellite and of the satellite itself. The amplitude of the signal current is typically 25mA.

Data Transferred via Bus

The communication interface of the ECU is designed to exchange digital information with sensors via bus. This requires a smart bus capable sensor with integrated analogue to digital conversion and communication interface. The signal form has to follow the bus protocol, usually CAN.

2 Signal Processing

The full sequence from primary signal to digital information is explained in brief in this chapter. Also the so called smart sensors are introduced. In addition, the advantages and disadvantages of monolithic integration, as well as the trend towards systems integration are discussed.

2.1 From Primary Electrical Signal to Digital Information

The basic sequence of signal processing is illustrated in figure 2. The measurand is transformed to a primary signal by the sensor element. In the analogue signal processing step, the signal is filtered, amplified and normalised - usually to a voltage range of 0V to 5V. After analogue to digital conversion, the signal can be processed fully digital by the microprocessor of the ECU.

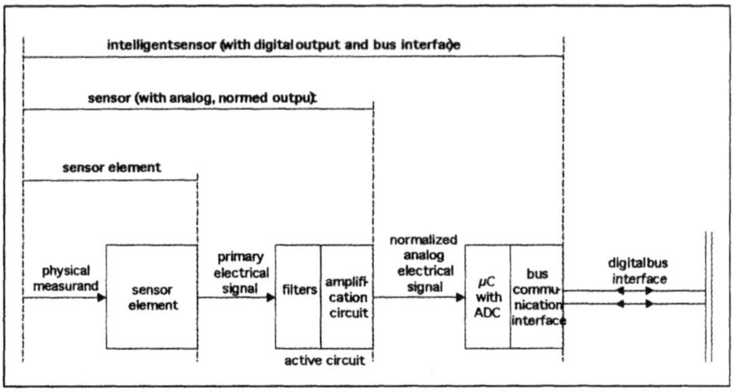

Fig. 2. Basic sequence of signal processing and definition of terms: sensor element - sensor - intelligent/smart sensor [2].

Besides these basic steps, signal processing includes also linearisation and compensation of perturbances, which can be performed in an analogue or digital way. Usually the analogue signal processing unit includes also a calibration circuit, but also calibration can be done digitally (software calibration).

Sensor signals are affected by nonlinearities and interfering quantities, like temperature. In general, the achievable characteristics of a sensor depend on the structure and complexity of the individual device. Mostly the following structures are used for analogue compensation [6]:

- Serial structure (e.g. temperature compensation by a resistor network).
- Parallel structure (e.g. differential structure).
- Loop structure (e.g. closed-loop accelerometer).

The most common is the serial structure, while the loop structure is only used when it is impossible to achieve the requirements by other means. Detailed information on serial structure and loop structure can be taken from the examples, given in chapter 4.

Depending on application, the signal processing is performed partially or fully by the sensor. Corresponding to this, the ECU receives either a partly processed analogue/digital signal, that requires post-processing, or it receives fully processed digital information via bus. In the following the basic steps are explained in more detail:

Primary electrical signal

The primary electrical signal is described by voltage/current (analogue signals), or by frequency (pulsed signals). It is created by the sensor element and has a direct, in some cases linear correlation to the physical or chemical measurand.

Amplification

As the primary analogue signals are typically rather weak, amplification is required. Mostly amplification is accompanied by offset correction and normalisation of the signal to a range between 0V and 5V.

Filtering

Usually analogue signals come along with high frequency noise. Therefore a low pass filter is often part of the analogue signal conditioning. The cutoff frequency of the filter has to respect the required dynamics of the sensor as well as the sampling frequency of the AD converter (anti aliasing filter).

Compensation

Most physical and chemical effects that are used to create the primary electrical signal are affected by perturbing effects like temperature. Especially in automotive applications, where temperature varies over a wide range, compensation of these undesired effects is required. This can be done in different ways:

- Analogue compensation by using the perturbance effect itself as a reference (e.g. compensation of the temperature dependence of a piezoelectric sensor element).
- Analogue compensation by an adapted (trimmed) compensation circuit.
- Digital compensation via look up table using the information of an additional temperature sensor e.g. via a diode as a reference.

Calibration

For high accuracy measurements, calibration is mandatory in order to compensate the manufacturing tolerances of sensor element and analogue signal processing circuit. This can be done on hard ware level, e.g. by laser trimming but also on software level, like by fine tuning of the lookup table. During calibration also the temperature compensation can be adapted to the specimen's characteristics.

Linearisation

Usually a sensor signal is expected to have a linear correlation to the measurand. However, many physical/chemical effects that are used to create the primary signal are of nonlinear nature. Therefore most sensor signals are linearised during signal processing. Linearisation can be part of the analogue signal conditioning, but is often performed in the digital stage using a lookup table.

Analogue to Digital Conversion

This step requires a normalised signal with controlled offset. The analogue to digital conversion can be performed by the (smart) sensor or by the microcontroler of the control unit. In any case it has to be assured that the sampling rate is high enough to detect the highest occurring frequencies of the signal and to avoid aliasing (oversampling).

2.2 Smart Sensors

According to the degree of implemented signal processing, sensing devices are usually referred to as sensor element, sensor or smart sensor. A definition of these terms has been given in [2] and is illustrated in figure 2. The sensor element transforms a measurand (temperature, pressure, ...) into a primary electrical signal. This analogue signal is usually characterised by voltage and often depends linearly from the measurand. Otherwise, linearisation either analogue or digital, is part of the signal processing. An active circuit amplifies and normalises the primary signal, typically to a range of 0V to 5V. In many cases also a compensation of temperature effects is included in this analogue signal processing step. When we talk about a sensor, we usually think of sensor element plus active circuit for analogue signal processing. If the sensing device has included on top also analogue to digital (AD) conversion and a bus interface it is usually called an intelligent or smart sensor.

In some cases even signal interpretation and decision making is implemented in the sensor. As an example the airbag satellite evaluates the sensed acceleration or pressure rise and makes the decision crash / no crash. The result of this decision is then transferred to the airbag ECU.

In general a sensor element is sensitive not only to the intended measurand but also to other effects, often being a perturbation to the primary signal. The most critical perturbing effect in automotive applications is usually temperature. In cases where the relation between output and the perturbing effect is simple compensation can be done by analogue means, i.e. by incorporating a temperature sensing element in the sensor package which is used to apply a measure of correction. In cases where there are a number of perturbing effects, especially if they are nonlinear, digital techniques must be applied to obtain a "true" signal. In these cases it is often useful to implement signal processing and compensation in a smart sensor device.

Figure 3 shows schematically the concept of a smart sensor. The signal to be measured is obtained by a sensor whose output, together with the outputs from sensors for the perturbing variables is passed via analogue to digital converter to a correction unit. This is programmed with a description of the transfer functions for each variable, obtained during initial calibration. Examples for this concept are NOx sensor and hall sensor with digital output.

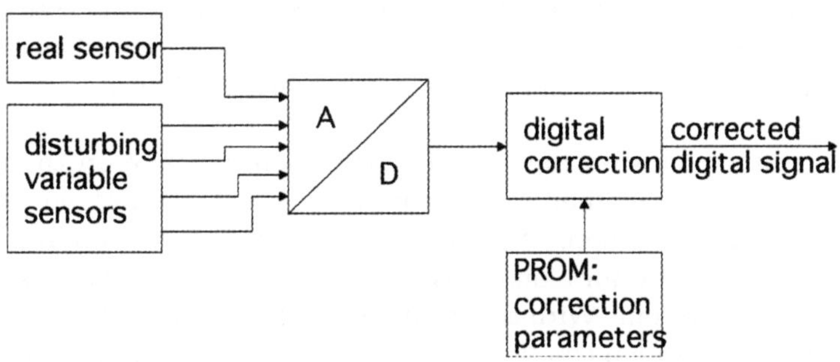

Fig. 3. Concept of a fully smart sensor with digital compensation of all perturbances [3].

2.3 Monolithic Integration

Classical sensor concepts were typically hybrid solutions consisting of a sensor element and a separate analogue signal processing circuit, e.g. realised in surface mount technology. With the advent of micromashined sensors based on silicon substrates it is now in principal possible to integrate the signal processing onto the silicon of the sensor element. This comes along with the following advantages:

- Reduction of interconnects (cost advantage, improved reliability).
- Reduction of total sensor size.
- Signal processing circuit is manufactured as integrated circuit (cost advantage, perfect matching of resistors, ...).

On the other hand, there are some disadvantages of monolithic integration:

- ▶ Poor compatibility of process technologies for micromashining and manufacturing of the analogue/digital circuit.
- ▶ Limited synergy between process for manufacturing the electric structures of sensor element and analogue signal processing.
- ▶ Depending on application and design, the analogue signal processing circuit might be subjected to aggressive media and/or high temperature, when integrated monolithically.
- ▶ In some cases, still some external resistors are needed for calibration purposes (SW calibration can avoid this disadvantage) and on top of that usually an external RC-network is required to match EMC directives.
- ▶ Especially for powertrain applications, some sensors are subjected to high temperature. This temperature often exceeds the application range for standard signal processing circuits. In these cases a remote signal processing or a specific , high temperature enhardened separate signal processing circuit is often the preferred solution.

Hence, it still depends on the specific application, but also on the expected production volume whether monolithic integration provides an advantage compared to hybrid solution. As a compromise, we find very often a two-chip solution, where the sensor element is on one die and the analogue signal processing on the other one. With this modular approach, most of the advantages of monolithic integration can be used, while the technology related disadvantages are avoided. Furthermore, the analogue signal processing for different sensor applications can be performed by one standard signal processing IC. This generates production volume, which decreases cost. Additionally the engineering effort is reduced, as basically only the sensor element itself has to be designed specifically.

Despite these advantages of two chip solutions, the overall trend heads towards monolithic integration of sensor element, analogue signal processing and even analogue to digital conversion followed by digital signal processing and bus communication.

2.4 Systems Integration

With the increasing complexity of vehicle systems and subsystems it becomes apparent that the output of a sensor may be required by several subsystems. It thus becomes attractive to integrate the subsystems in a bus architecture with smart sensors that communicate directly with the bus and therefore avoid sensor duplication. In early electronic engine management systems even the ignition and fuelling systems were completely separate, and had separate sets of sensors. Today manifold pressure sensor, engine speed and position sensors are used in the ECU to determine fuelling, ignition, and exhaust gas recirculation requirements. Similarly it is obvious to integrate the sensing requirements of braking, steering and suspension, because of the interactions of braking and cornering forces with suspension leveling. Pursuing this line of development, one can think of a total vehicle control system combining driver's input with various automatic safety controls and guidance systems for a complete control of the vehicle dynamic. The system shown in figure 4 was already proposed in 1993 [4]. In the recent years the major aspects of this proposal have been implemented.

The general trend towards systems integration of sensors is a main driver for the market penetration of smart sensors.

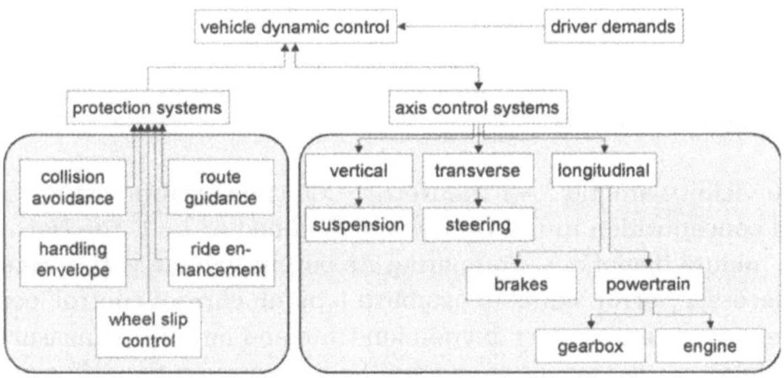

Fig. 4. Hierarchical vehicle control system [4].

3 Examples for Analogue Signal Processing

3.1 Linear Oxygen Sensor

The precise, fast and wide range measurement of exhaust gas oxygen concentration is a key element to the development of low emissions combustion engines. The signal processing for a linear oxygen sensor follows mainly a closed loop concept and is shown in figure 5.

The basic sensing element of a linear oxygen sensor is a Nernst Cell. One electrode is exposed to the measured gas, the other to an oxygen reference. The sensing element is confined to a measurement cavity, which is located inside the sensor and linked to the exhaust gas by a gas diffusion barrier. The sensor is fitted with another pair of electrodes (pump cell), which allows to transport oxygen ions from the exhaust gas into the cavity and vice versa. Additionally it contains a heating element to maintain a constant operating temperature.

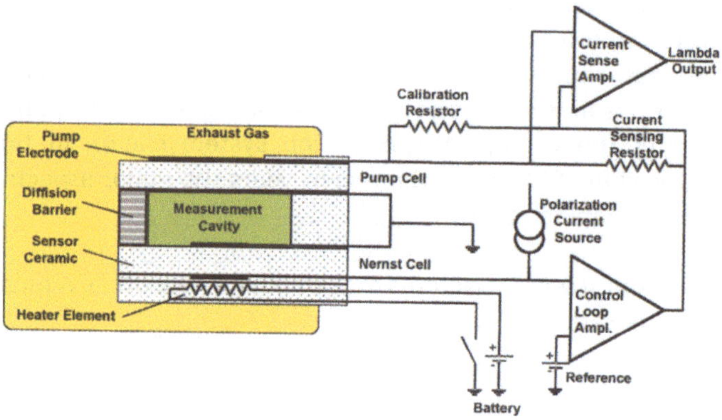

Fig. 5. Linear oxygen sensor: Basic construction and closed loop analogue signal processing.

In the close vicinity around $\lambda = 1$ the Nernst Cell transfer function is linear. If the oxygen concentration in the cavity is maintained at $\lambda = 1$, the Nernst Cell will report minute deviations. Comparing its output voltage with a reference gives the necessary error signal to establish a pump current control loop. The pump current used to transport oxygen ions into and out of the measurement cavity correlates to the amount of oxygen flowing through the diffusion barrier. This in turn is related to the difference of oxygen concentration between exhaust gas and cavity. As the lambda value inside the cavity is maintained at 1 it is now possible to derive the oxygen concentration of the exhaust gas from the pump current.

Tolerances of the diffusion barrier are measured during manufacturing. The sensor is then fitted with an individual calibration resistor. Reading the resistor value allows either a gain adjustment of the current measurement or a software correction of the measured current value and by this a compensation of the manufacturing tolerance.

Due to its construction the diffusion barrier resistance changes with temperature. The temperature - impedance characteristics of the sensor equals an NTC, making it ideal for detecting the sensor temperature. A temperature stabilisation of the sensor can now be established by reading the sensor impedance, comparing it against a target value, and controlling the heater power.

Having applied all these measures, the pump current is proportional to the oxygen partial pressure of the exhaust gas. The precision of its measurement defines the sensor quality. Several actions have to be taken to improve accuracy:

- Suppression of common mode signal.
- Compensation of sensor manufacturing tolerance.
- Low pass filtering to remove undesired high frequency components from the output signal.

Analogue signal conditioning is followed by AD-conversion. Communication of the sensor signal to the microcontroler is performed e.g. by Serial Peripheral Interface (SPI). A more detailed description of sensor, analogue signal processing and also the digital interface to the microcontroler can be taken from [5].

3.2 Piezoresistive Pressure Sensors

One of the first micromechanic sensors to enter the automotive market was the piezoresistive sensor for measuring manifold absolute pressure (MAP). The sensor element itself is basically a piezoresistive film on a silicon membrane. The piezoresistive film forms a Wheatstone bridge with the bridge voltage as a primary electrical signal. Membrane and resistors are designed to achieve a voltage response which is linear to the pressure that is applied to the membrane. In general the analogue signal processing involves components for amplification, offset correction, and compensation of temperature effects on the offset and sensitivity of the sensor.

Figure 6 shows the analogue signal processing unit for the pressure sensor. The sensor element is followed by a resistor network which affects the temperature compensation. Ideally, the correction of the temperature effects on offset and sensitivity should be accomplished independently of each other, but in practice this is very difficult to achieve and depends strongly on the electric circuit design. Here the offset temperature coefficient is corrected by trimming the resistors R1 and R2, while temperature influences to the pressure sensitivity of the sensor are compensated by the temperature independent resistor R3. The necessary values of the resistors are calculated for each individual sensor according to the calibration data at different temperatures and pressures. Finally, the temperature corrected sensor signal is amplified by a symmetrical amplifier stage with an amplification factor of up to 50.

Alternative to the described analogue signal processing, meanwhile also ICs performing digital signal processing on the output of piezoresistive pressure sensors are commercially available. These ICs perform an analogue to digital conversion of the primary signal, followed by compensation of thermal effects, calibration and linearisation of the digital data. This also includes a permanent memory to store the sensor specific data. Usually digital to analogue conversion is performed as a last step in order to be compatible to the standard analogue voltage signal.

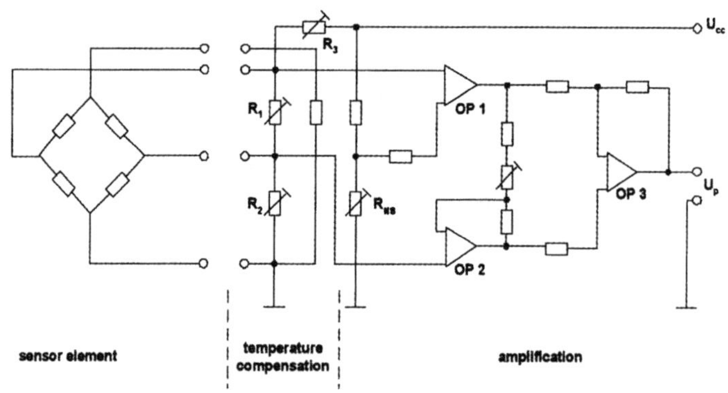

Fig. 6. Analogue signal processing unit for a piezoresistive pressure sensor [6].

4 Signal Transfer

Which kind of sensor should be used depends among others on the necessary features for signal transmission and on the kind of signal processing applied. The major signal forms can be described as follows:

- Amplitude analogue.
- Frequency analogue (modulation of frequency or pulse width).
- Direct digital.

The main characteristics of these signal forms are listed in table 1.

For direct digital sensor data transmission via bus there exists a multitude of so called "standards". A recent count showed about 43 buses proposed by different car makers and suppliers [11]. According to [12] the following standards have the potential for truly global distribution:

- Cheap subsystem interconnect up to 20kb/s (UART-type, i.e. LIN).
- Fault tolerant medium speed used for interconnecting modules (CAN).

- By-wire chassis interconnect for safe systems (FlexRay).
- Multimedia interconnects (MOST, IEEE 1394).
- Airbag (safety) specialty connection.
- Short range RF-links.

feature	amplitude analogue	frequency analogue	direct digital
possible static accuracy	restricted	unlimited	unlimited
dynamic characteristic	in general very good	limited by conversion speed	limited by net data transfer rate
noise immunity on signal transmission	small	good at FM transmission	good; option for digital error correction
possible arithmetic operations	limited or expensive	division/integration easily feasible	any operation with arithmetic processor
connection to digital control units	by analogue to digital conversion	e.g. by frequency counter	direct at suitable code and signal level
error correction	improvement only by analogue averaging	on frequency counting natural redundancy	with error-correcting codes
galvanic signal separation	very expensive	easy	easy

Tab. 1. Comparison of the major signal forms [6].

A discussion of these major standards and their compelling features can be found in [12]. For typical sensor signals LIN and CAN will be the favourites. In comparison to LIN the CAN-bus is faster, thus able to manage more signals. However the demand for system resources is higher, demanding a good crystal based oscillator at each node. Nevertheless fo sensor or actuator signals the CAN-bus remains a good fit as the CAN is in total well introduced.

A rather novel development, but with high potential to reduce wiring harness is the so called DC BUS. The DC BUS high-speed multiplex technology transfers digital information inside the car using a single wire which integrates the functionality of 14V or 42V power net and communication. The use of a single wire for battery power and for communication allows to reduce the amount of wiring and will also enable to add optional modules after assembly of the car [7].

Up to now mainly copper cables have been used for the transfer of safety relevant information within cars. On the other hand optical networks with a ring topology for media oriented data transfer like D2B and MOST (Media Oriented Systems Transport) are already in production since 1997. The byteflight system, which is ready to launch will be the first network with an optical physical layer for passive safety, comfort and body control [8]. Also telemetric communication is under investigation. The main obstacle for the introduction of telemetric data transfer are EMC related safety issues. Hence, this technology is up to now mainly discussed for the so called intelligent tire [9].

5 Outlook/Trends

In general smart sensors are likely to be used increasingly, because of their ability to compensate for errors and changes in measurement conditions. Their digital output renders them also capable of communicating directly with a bus, which opens up possibilities of system integration and the sharing of sensor data between subsystems.

This trend is assisted by a general trend towards a decentralised E/E architecture with distributed intelligence within almost all sensors and actuators. Blocking point for this trend up to now is the cost of distributed intelligence, especially if the potential to reduce wiring harness is not utilised. Regarding this issue, the market entry of the DC BUS as well as the optical networks will probably accelerate the trend towards smart sensors. With increasing complexity of the sensors, also integrated diagnostics will be a requirement of the future.

As a longterm trend, the additional optical and telemetric communication channels will also affect the data processing inside sensors. For using these channels, additional conversion stages have to be included in the sensors. Especially telemetric signal transfer seems attractive as it enables to create self sustaining sensor systems without any cables [10] which have a high flexibility for mounting locations.

References

[1] S. Bolz, G. Lugert "Aufbau moderner Steuergeräte" in Elektronik im Kraftfahrzeug, Walliser u.a., expert Verlag Renningen(2000).

[2] Sensortechnik, H.R. Tränkler, E. Obermeier (Hrsg.), Springer Berlin (1998).

[3] P. Cockshott "Automotive Sensors" in Sensors A Comprehensive Survey 8, edited by W. Göpel, J. Hesse, J.N. Zemel, VCH Weinheim (1995).

[4] "Electronic System Architecture", Automotive Engineering, April 1993, p. 15-17.

[5] S. Bolz "A Novel Interface for Linear Oxygen Sensors" in ATTCE 2001 Proceedings (2001).

[6] H.R. Tränkler "Signal Processing" in Sensors A Comprehensive Survey 1, edited by W. Göpel, J. Hesse, J.N. Zemel, VCH Weinheim (1989).

[7] Y. Maryanka "14V/42V Power Line Communication for Automotive" in Advanced Microsystems for Automotive Applications 2001, edited by S. Krüger, W. Gessner, Springer Berlin (2001).

[8] J. Wittl, K. Panzer, H. Hurt, E. Baur "Transceivers for Optical Networks in Automotive Applications" in Advanced Microsystems for Automotive Applications 2001, edited by S. Krüger, W. Gessner, Springer Berlin (2001).

[9] M. Löhndorf "Wireless Tire Sensors Based on Amorphous Magneto-Elastic Materials" in Advanced Microsystems for Automotive Applications 2001, edited by S. Krüger, W. Gessner, Springer Berlin (2001).

[10] G. Hettich, V. Vieweger, J. Mrowka, G. Naumann "Self Supporting Power Supply for Vehicle Sensors" in Electronic Systems for Vehicles, VDI Verlag Düsseldorf (2001).

[11] C.A. Lupini "Multiplex Bus Progression" SAE-technical paper 2001-1-0060, SAE2001 World Congress, Detroit Michigan.

[12] G.H. Teepe "Sensors in the Next Generation Automotive Networks" in Advanced Microsystems for Automotive Applications 2002, edited by S. Krüger, W. Gessner, Springer Berlin (2002).

T. Riepl, S. Bolz, G. Lugert
Siemens VDO Automotive AG
SV P ED T IC
Siemensstraße12
93055 Regensburg
Germany
thomas.riepl@siemens.com

Keywords: signal processing, signal transfer, smart sensors, monolithic integration, normalisation, compensation, calibration

Packaging of Micro Devices for Automotive Applications - Techniques and Examples

E. Jung, V. Grosser, K.-F. Becker, M. Koch, Fraunhofer IZM

Abstract

Micro systems and micro devices play an important role in automotive applications. Today, sensors and actuators allow to increase safety, comfort and economy of modern cars. In order to facilitate the use of such fragile devices, packaging of these structures is on major issue. Due to the cost and yield issues associated with this process step, significant evolution on top of the microelectronic packaging techniques is required.

1 Introduction: Role of MEMS in the Automotive Field

Micro Electro Mechanical Systems (MEMS) have gained a crucial role in the automotive industry over the last decade. Initially for sensing various environmental, motor and movement situations, today actuators become more part of the automotive as well. For sensing, early fields of application included pressure sensors and acceleration sensors. Today, a large number of safety and comfort features rely on sensors [1].

In the field of comfort and safety, speech recognition plays an important role (man machine interface, hands free communication, early defect recognition detecting characteristic sound patterns of defects, ...).

While hybrid electronics paved the way for the electronic integration into the automotive, MEMS/hybrids will play a more and more a dominant role in the future. Packaging innovations here are crucial to fulfil the stringent requirements for the automotive market like:

- High reliability of function.
- High availability of function.
- Ease of manufacturing.
- Low cost devices.

Aside, miniaturisation is a part of the development goals not to be neglected, as it is expected that future electronics will comprise a significant part of the automotive value. To integrate a high degree of electronic functionality into the already densely packaged electronic compartment of the car requires miniaturization at its best.

Microsystems are expected to increase their value within the automotive from 2002 figures of ~1 billion Euro to 2.5 billion Euro within 2006 at an astounding 37% annual growth rate, mostly cutting away market share from existing, conventional solutions, however opening up as well new fields of application [2]. It is also expected, that in a car from 2002 to 2010 the part of electronics will increase from today 22% to 35% [3]. Table 1 shows several fields of application for MEMS devices.

Pressure sensors	tyre pressure, injection pressure, exhaust gas pressure, manifold inlet, combustion pressure...
Acoustic sensors	Early defect recognition, speech recognition
Acceleration sensors	airbag ignition, smart suspension, ...
Gyroscopes	roll over detection, ESP,...
Optical sensors	rain drop detection, seat occupancy, auto-dimming mirror, ...
Flow meters	air mass flow
Shear sensors	oil quality
Temperatur sensors	air inlet temperature, exhaust gas temperature
Electric and magnetic field sensors	oil quality, child seat detection, ABS, oil sump level

Tab. 1. Fields of application for sensors using MEMS technology (table not exhaustive).

Not at last, integral solutions with sensor/acutator systems will have a significant role in the future of automotive MEMS, e.g.

- Auto cruise control (radar distance sensor coupled to a speed control/brake).
- Uncooled IR Sensors for night drive collision warning coupled to brake.

2 Role of MEMS Packaging

MEMS packaging fulfils a crucial role within the function of the MEMS device. Besides the electrical interconnection, the package provides mechanical protection, media separation or coupling (e.g. pressure), signal conditioning, etc. It must enable the resulting package to be manufactured and assembled at low cost. Also, high reliability requirements even under harsh conditions (e.g. oil sink with extreme temperature differences and aggressive ambient) must be fulfilled in order to be accepted in the automotive market.

Today, packaging relies either on using well established package types like TO, butterfly or ceramic packages, or – applicable only for very high volume products – dedicated package developments. Wafer level packages, as being developed in the microelectronic industry, are currently finding their niche and potential wide spread use also for the packaging of MEMS devices. This can incur substantial cost savings in the future, as seen in the microelectronics industry as well.

3 Sensor Packaging Options

Wafer Level Preparation of Sensitive Devices

Basis for the large majority of moving MEMS devices like accelerometers and gyros is the process of structure protection while still on wafer level. Therefore, a structured "cap" wafer is utilized (typically consisting of pyrex glass or silicon) to cover the delicate structures. This capping wafer is aligned to the device wafer and bonded to this via one of the following technologies (order according to share of use):

- Anodic bonding.
- Glass frit bonding.
- Adhesive bonding.
- Silicon direct bonding.
- Soldering.

Fig. 1. Micromechanical airbag sensor as system in package configuration (sensor die and readout circuitry connected in a standard package); courtesy Robert Bosch GmbH [4].

Each of the technologies has their distinctive advantages and disadvantages, so the choice depends on the product and product´s requirements in operation. The so protected device can be handled similar to a conventional IC and can e.g. be mounted in a standard package en-capsulated by transfer molding (figure 1).

Alternative approaches allow to realized the packaging and the interconnect structure to the outer world in a wafer level fashion, as described e.g. in [5]. Here an gyroscope was realized in a bulk micromachining technology, adding wafers to a stack and bringing the terminals via hermetic feedthrough contacts to the face of the MEMS device (figure 2). To these terminals, e.g. BGA balls for board assembly can be attached.

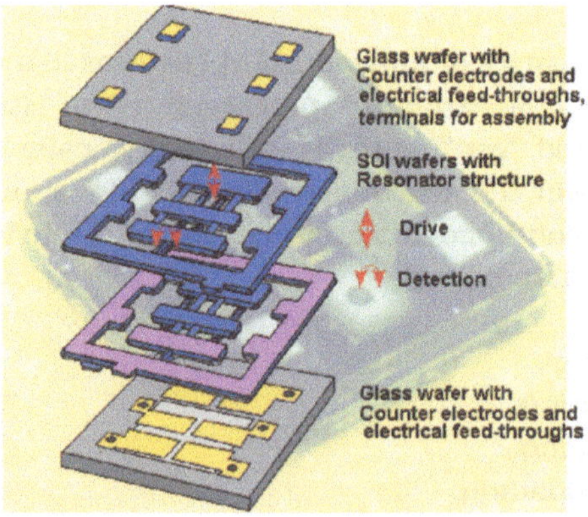

Fig. 2. Functional layering with SMD ready contacts.

Wafer level packaging including the terminals is expected to be one of the major technology trends in the near future of MEMS packaging [6].

Single Chip Preparation

The vast majority of devices is packaged using well established technologies known from the microelectronics industry since several decades. With movable MEMS devices, the first step is capping while still on wafer level. This process is part of the manufacturing process in the MEMS fab. With less sensitive devices like field detectors or thick film sensors, such rigorous protection is not necessary. After fabrication, the wafers are diced using standard dicing equipment. Then, they are singulated and packaged in individual housings like:

- Metal.
- Ceramic.
- Plastic.

packages.

Plastic packaging as used in the microelectronic industry has several draw backs like:

- Stress induced due to the electronic molding compound (EMC) process influencing sensor characteristics.
- Difficult media access (e.g. pressure of sampling gas).
- The number of devices to be packaged is several orders of magnitude smaller than those of microelectronic devices of the same size.

Therefore, pre-molded substrates have found widespread use in the manufacturing of MEMS packages. Here, the devices are mounted similar to package as within the ceramic package. Lid attach using injection molded caps provide mechanical protection while allow media access to the device. Instead of lid attach, for several applications like pressure sensors a low modulus gel is used to protect the wirebonds and hydrostatically transfer the external pressure to the device

Even laminar substrates using respective capping techniques can be an optimum solution. Infineon is currently developing [7] an acoustic sensor (microphone) packaging concept that evolves from a premolded package (figure 3) with a less favourable area and volume ratio of package to device to a much smaller laminate solution at lower cost and identical performance (figure 4) and lends itself to comfort applications in the automotive field like speech recognition systems.

Fig. 3. Pre-molded package with a silicon microphone.

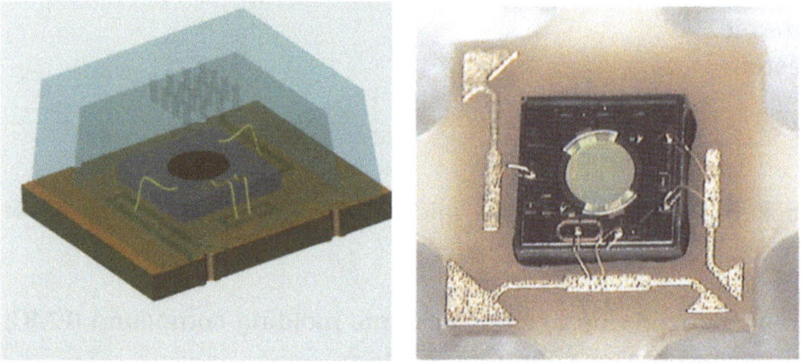

Fig. 4. Small laminate package of the like microphone (developed by Infineon and partners in the INFON project.).

For modular applications using MEMS, a recently commercialised technology "MATCH-X" lends itself to package MEMS devices into ceramic or laminate housing while allowing electrical buses, pressure, optical and fluidic access into the package (see figure 5 and figure 6) [8].

Sealing techniques used with ceramic substrates are soldering, glass frit bonding and adhesive bonding (non-hermetic).

With the laminate, the lid is non hermetically mounted to the frame using e.g. solder contacts or adhesive bonds. Especially with the laminate package, significant savings and competitive cost structure to discrete system assembly are expected.

Fig. 5. MATCH-X modular MEMS packaging using LTCC ceramic.

Further on, concepts for molding such stackable packages with media access holes to the sensitive structures have been proposed, allowing even lower cost while maintaining the benefits (figure 7 shows one of several concepts realized in rapid prototyping technique)

The evolution seen here for modular system configuration may be advantageous especially for the automotive sector, as platform concepts are more and more accepted in this field requiring modular means of adding functionality.

Application specific packages (i.e. packages designed for one specific device and purpose [9]), as they are known in e.g. the medical or optoelectronics industry are not simply applicable in do not lend themselves easily to the automotive sector, as the volumes involved and the price tag obtained with those ASIPs is typically not compatible with the automotive requirements.

Fig. 6. MATCH-X modular MEMS package using organic laminate.

Fig. 7. Stack pack modular package used for microelectronic and microsystem applications.

Outlook – Functional Packages

Conventionally, packaging of the devices is a separate process step that providinges a structure whichthat can be handled during the subsequent manufacturing sequence and provides the functionality required. However, for high volume products and for optimised performance, an entire (sub-) system may be designed from the beginning to provide highest cost efficiency, highest function density and optimum performance. Such devices are "functional packages" whichthat not only encase the devices´s function but add their own functionality to maximize the overall performance. So far, functional packages find their applications in the consumer industry (e.g. ink jet cartridges) and medical industry (blood sensors), e.g. in throw-away products. However, with "platform concepts" "re-use of technology" over the various automotive models, the automotive industry can take advantage of the concepts developed for these ultra-low cost products.

4 Acknowledgements

The authors would like to thank the partners in the Fraunhofer´s MEMS Packaging Programe for their contribution. The granted use of images of the Infineon´s silicon microphone and packaging concept developed together with IZM and partners in the INFON project (funded by the BMBF under the contract N° 16SV1273) is gratefully acknowledged as well as the stack pack contribution from the Micro Mechatronics Center, Oberpfaffenhofen.

References

[1] Krüger et al.,"New Challenges For Microsystems-Technology In Automotive Applications Applications", MSTnews 01/01

[2] Hoffmann, „Marktentwicklung für mikromechanische elektronische Sensoren", ZVEI Kompetenztreffen „Innovationsmotor Automobil", Jan., 17th 2003, Frankfurt

[3] Schleuter, „Innovationen durch Mikroelektronik und Mikrosystemtechnik im Automobil", ZVEI Kompetenztreffen „Innovationsmotor Automobil", Jan., 17th 2003, Frankfurt

[4] Trah, "Mikrosystemtechnik – eine Basistechnologie für das Auto der Zukunft", ZVEI Kompetenztreffen „Innovationsmotor Automobil", Jan., 17th 2003, Frankfurt

[5] Wiemer et al., "Application of high and low wafer bonding processes for bulk.micromachined components", Proc. of the 5th Int. Symp. on Semicond. Wafer Bonding: Science, Technology and Applications V; 1999, Honolulu, Hawaii

[6] see IMAPS Advanced Technology Workshop on MEMS Packaging, Denver, 6.-8. Sept. 2002

[7] Dehé et al,. Silicon Micromachined Microphone Chip at Siemens, 137th Meeting of the Acoustical Society of America and 2nd Convention of the European Acoustics Association, Berlin, 1999

[8] Hr. Reichl, V. Grosser, "Overview and development trends in the field of MEMS packaging", IEEE MEMS 2001, Interlaken, 2001, pp. 1

[9] http://www.infineon.com/products/chipcds/portfol/biometr/introduction.htm

Erik Jung, Volker Grosser, Karl-Friedrich Becker, Matthias Koch
Fraunhofer IZM
Gustav-Meyer-Allee 25
13355 Berlin
Germany
erju@izm.fhg.de

Keywords: packaging, yield, speech recognition, cruise control, sensor packaging, MATCH-X

Micro-Mechatronics in Automotive Applications

F. Ansorge, J. Wolter, C. Rebholz, H. Reichl, Fraunhofer IZM

Abstract

Micro-Mechatronics & System Integration is a multidisciplinary field, which offers low cost system solutions based on the principle of homogenising system components and consequent elimination of at least one material, component or packaging level from the system. These system approaches show, compared to the existing solutions, a higher functionality, more intelligence and better reliability performance.

The paper presents system solutions and manufacturing technology for mechatronic systems, developed at Fraunhofer IZM Micro-Mechatronic Center. Special emphasis is given to novel rapid prototyping processes enabling a faster system development as well as a more consequent integration of electronics in mechanics. The mechatronic packages will be discussed in detail. Especially cost, reliability performance and according "Design for Reliability" show the potential of micro-mechatronic solutions in automotive and industrial applications.

1 Basic Conditions for the Development of Micro-Mechatronics

The development of mechatronics micro-mechatronics an MEMS-Packages is a vital key for the future of electronic integration. Ever more complex applications require the processing of ever increasing data volumes. A corresponding increase in flexibility and functionality of packages is one of the challenges developing micro-mechatronics.

Synergistic co-operation between the individual disciplines concerned is also an indispensable condition. To yield added functionality within one package the effort of specialists from various fields of technology as physics, chemistry, electrical engineering, microelectronics, software design, biology, medicine etc. are needed [1, 2]. In figure 1 a mechatronic design procedure is shown. Simultaneously with the mechanical and electrical a thermo-mechanical simulation is performed at certain points, to ensure a lifetime estimation at critical points like "sandwich soldered semiconductors". Designs must be compatible for a proper realisation of a thermo mechanical concept a possible exchange format is IDF 2.0. The package developed fits to an optimum place within a macroscopic system of mechanical parts, which are especially designed to work with electronics as intelligent sub-system. A corresponding STL-Package will support the development process by an important physical hardware-check. It is of same importance to generate a software map of all requirements and functions, already at the planning and development stage.

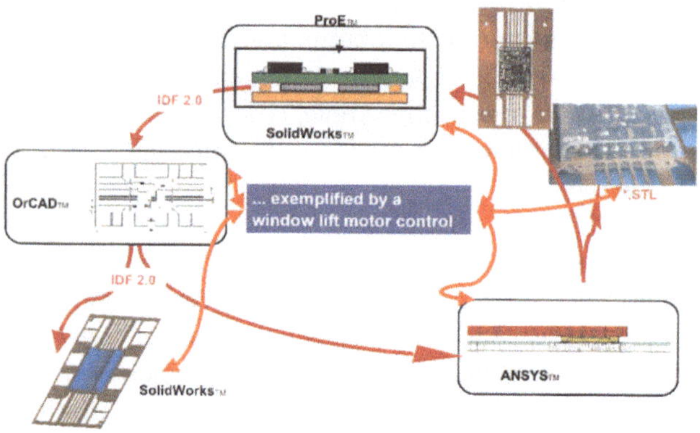

Fig. 1. Mechatronic design procedure.

1.1 Packaging Concepts and their Aspects on Polymer Materials for the Use in Automotive Applications

Focusing on micro-mechatronic applications there are additional demands to encapsulants commonly used for packaging. The encapsulants need a wide range of temperatures, a high resistance against harsh environment and the integration of moving and sensing elements without losing package functionality. Additionally for the use of optical components like micro-mirror devices clear compounds are crucial [7].

Typical materials used for microelectronics encapsulation are epoxy resins, where the chemical basis is a multifunctional epoxy oligomer based on novolac resins. These materials do have Tg's beyond 200°C and so they have the potential for short term/high temperature application. The evaluation of encapsulants for optimized longterm stability is one of the topics the micro-mechatronics center is focusing on. Clear compounds as potting or molding material are also already available, but due to the lack of fillerparticel there the CTE is much higher than for standard materials. To overcome this a more sophisticated selection of package outline and arrangement of the components is necessary and was developed at IZM-MMZ specialists.

Further potential for micro-mechatronics are in the use of advanced thermoplastic materials. In the past they have been used rather for electronics housing than for the direct encapsulation of microelectronics. The use of thermoplastic materials allows the integration of further mechanical functionality (plug housing, clip on mounting, ...) simultaneously with microelectronics encapsulation, but these materials do not cross-link and thus do have a high potential for recycling.

Research of Fraunhofer IZM-Micro-Mechatronic Center is performed in the fields of direct encapsulation of microelectronics using thermoplastic polymers, thermoplastic circuit boards a.k.a MID-devices for both, Flip Chip and SMD components. Combination of different types of materials, tailored for the application, is a key for highly reliable modules.

These applications are also capable for cross-linking the package geometry with mechanical functions. Thermoplastic materials are suited for the creation of scoop-proof connectors or clip-on-connections. In order to preserve a multifunctional package, both technologies would have to be joined in the forming of two- or multi-component-packages.

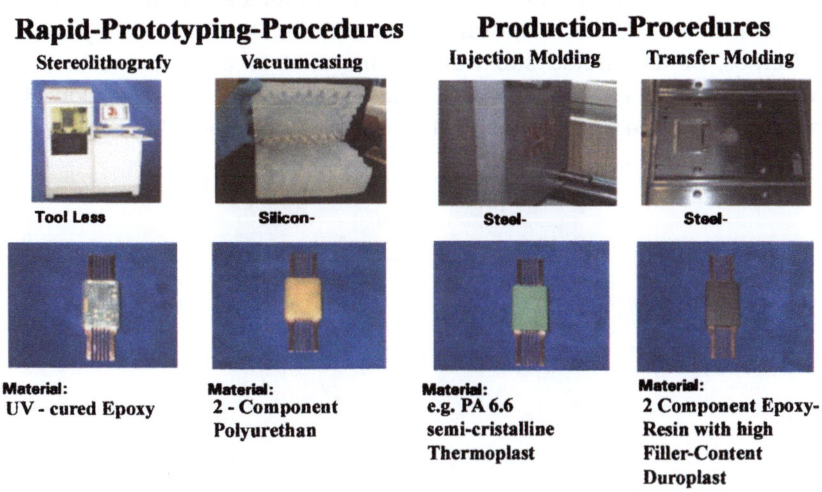

Fig. 2. Survey of different important encapsulation methods for mechatronic- and MEMS Packaging.

Figure 2 gives a survey of the addressed different encapsulation processes. The basis of this technology is a leadframe component, which is assembled on a printed circuit board. The first prototype is an encapsulation by stereo lithography. This is a tool less and thus a very fast process. The achieved modules have full mechanical an electrical performance but moderate thermal properties. The process is explained in detail in the following chapter. Vacuum casting is using a silicon tool from a master-form. These prototypes have full functionality. Production procedures like injection or transfer molding are available for highest reliability demands.

The use of thermoplastic hotmelt materials for micro-mechatronic packaging allows a decrease of processing temperature, the use of cost effective low temperature materials [8].

1.2 Functional Rapid Prototyping

Rapid prototyping, e.g. CNC milling/drilling systems and stereo lithography, round up the design flow, as explained in chapter 1.1 functional rapid prototyping is an important technology to get real prototypes for testing issues.

The prototypes are built up using geometric CAD-data and a slicing software to transform the 3D parts in 2D slices. Conventional technologies use a photopolymer to build up real models by a laser beam. Advantages of this technology are very detail-faithful prototypes and tool-less manufacturing. High requirements of micro-mechatronic demand for a direct integration ICs, Leadframes, or assembled substrates. Compared to other processes, a new developed method enables the direct embedding and encapsulation of bare semiconductors (see figure 3) as well as complete sub modules.

Fig. 3. Direct encapsulation of bare die using stereo lithography.

Figure 4 is giving a sequence of an encapsulation of a leadframe including the electronic by stereo lithography within 3h. The achieved package can be used directly as an functional prototype.

Fig. 4. Tool less encapsulation by stereo lithography.

2 Mechatronic Solutions for the Automobile

2.1 Modular Construction Systems for Mechatronics

In the course of a project sponsored by the Bavarian Government, the Fraunhofer Institute for Reliability and Microintegration (IZM) developed a modular construction system for sub-systems based on flex circuit integration. Spatially stackable single components include sensor elements, signal processing, actors and bus technology. The components are mounted taking into account the application they are supposed to serve. They produced standard components and can be mounted by soldering, leading to a minimized number of different interconnection technologies.

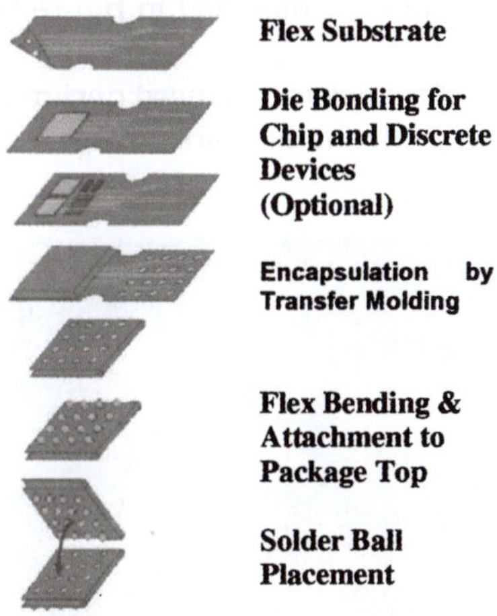

Fig. 5. StackPac™ production process.

The standardized dimensions in different size categories enable the user to combine the system components, which are part of the modular construction programe, in a most flexible way by defining his own functionality parameters [3-5]. Connection to the actorics can be effected by a direct integration of the components.

Based on molding-on-flex-technology the StackPac™ offer its advantages. This technological concept allows the insertion of individual ICs by contacting through either wire bonding or flip chip technology and integration of SMD-components. Multichip modules can be created using different contacting technologies.

Fig. 6. CavityPack: Molding on flex approach to sensor packaging.

Molding the package widens the range of possible package designs from simple patterns to additional functional geometries which are compatible with the modular construction system as depicted in figure 5.

The same top package of a stack can be used designing chip size packages. The necessary combination with the sensorics can be created by by means of a CavityPack (figure 6).

Fig. 7. Round shaped StackPac`.

The systems have already been used successfully in engineering applications. StackPac™'s achieve excellent lifetimes in reliability investigations, in particular for high-temperature applications, using the mentioned novel encapsulants.

This allows a reduction of development and qualification costs by using standard modules, which makes the use of advanced technology in confined spaces economically viable. In figure 7 a StackPac™ application for a round shaped fit is shown.

2.2 Mechatronic Combination of Power&Logic Electronics and Actuators

In the course of a project jointly run by BMW AG, BROSE, MOTOROLA and the Fraunhofer-IZM, an H-bridge circuit with control logic, bus connection and sensor facility has been developed. In figure 8 the principal process flow is depicted. First solder performs are soldered to a leadframe. The substrate with the soldered components is placed on the leadframe connecting the bus structures as well as power semiconductors with the leadframe. Thermal management functions are fulfilled perfectly by the heat sink comprising the inserted copper lead frame and the engine which has been mounted right at the bottom side of the package. The high degree of integration resulted in a reduction of spatial requirements by about 50%.

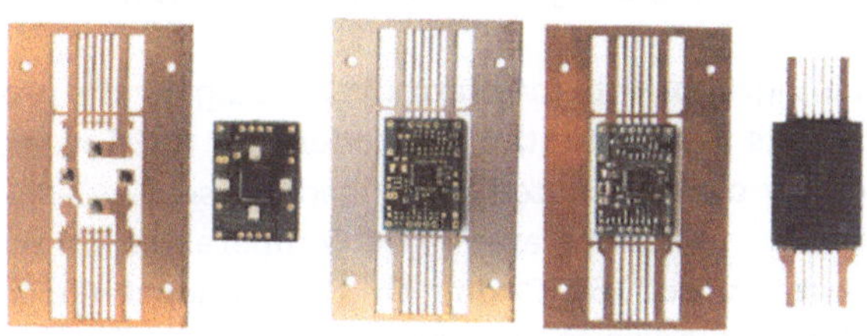

Fig. 8. Gemipak process flow.

2.3 Micro-Mirrors Used for Front Light Systems

New application for micro mirror devices will be the front light system for automobiles. In this field already crucial improvements of illumination were achieved by the development of "Xenon"-lamps. In contrast to traditionally automotive-illumination, these lamps have an electronic regulation whereby the changeover between low beam light and high beam is no longer realized by switching between two filaments. Thus there is additional potential for an electronic control. A brand-new functionality is controlling the lamp reflector while driving along curves by the view-direction of the driver or by speech. Additional projection of signs or illumination on the roadside would be conceivable [6, 7, 9].

In order to do this the reflector mirror from the front headlights can be replaced by a micro-mirror system (see figure 9). Due to the activation of single segments the driver will be able to project signs onto the road by the front light system (e.g. arrows, road information, etc.).

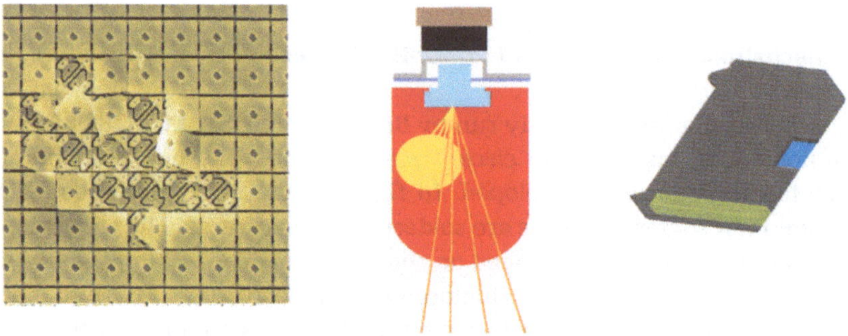

Fig. 9. a) SEM picture of an array of micro-mirror, which are partially disassembled to show structures underneath the mirrors (left); b) Integration of a mechatronic control module close to the reflector (middle); c) Design study of the control module (right).

The functional integration into mechatronic housing for the production of good value mass products is an essential component for the development of a new interactive security concept »intelligent headlights«. Key aspect is the harsh environment at the assembly area. The temperature of the micro mirror device can reach up to 150°C due to the under-the-hood environment. Therefore cooling strategies for the mirror especially for the mirror control seems to be very important. Another requirement of the system is a good stability against the vibrations of the automobile. First tests already showed that the DMD is very insensitive due to the compact design. In figure 9 a design study of a possible packaging strategy is depicted. Functional adaptations to robotic applications like navigation help for robotics etc. are easily possible.

A first prototype of the intelligent reflector was assembled and is depicted in figure 10. The necessary electronic is objective for further miniaturization and depicted on the right hand side of the picture.

Fig. 10. First prototype of the intelligent reflector.

3 Integration of Micro-Mechatronic in Mechanical Engineering Areas

»Smart Shaft« is a synonym for the integration of micro-mechatronic in mechanical engineering parts. This process demands the assembly of sensor systems and electronics directly into the mechanical engineering part. It offers therefore the possibility of in situ measuring of important mechanical datas as bending stress, torsion stress and vibration stress directly at certain interesting areas, as depicted in figure 11.

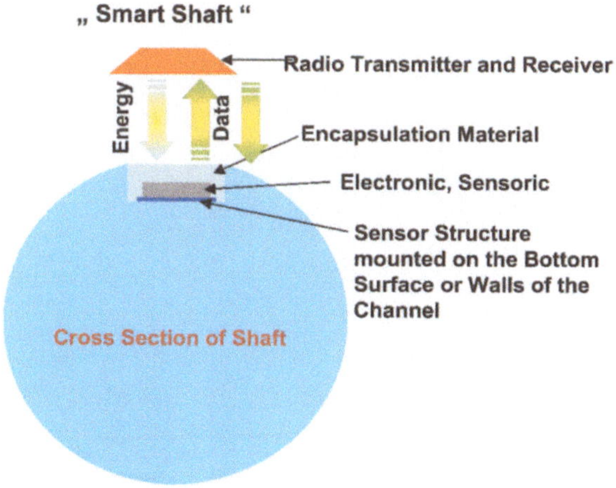

Fig. 11. Section of a smart shaft.

One main task of the encapsulation efforts for the electronics and sensors is the protection against mechanical, thermal and chemical stress. Due to a Mechanical engineering parts mostly work in harsh environment, therefore the requirements on the encapsulation materials are very high, that means the encapsulation materials need to have high temperature stability (up to 150 °C) and a good chemical resistance against solvents, oil and grease.

4 Outlook

According to intensive research studies, a large part of what is today still a vision of the future we also gave you a glimpse of that in our paper will be converted into reality by micro- machine technology already in the period between 2002 and 2007: the integration of sensorics, actoric functions and controllers will make it possible. Special emphasis will be given in future to a 3-dimensional assembly technology. Harsh engine environmental and overall vehicle systems complexity coupled with requirements for low costs, size, and weight plus high reliability make the automotive sector.

5 Acknowledgement

Parts of this work are funded by the "Bayerisches Kompetenznetzwerk für Mechatronik" part of the "High Tech Offensive Zukunft Bayern".

References

[1] Control and Configuration Aspects of Mechatronics, CCAM Proceedings; Verlag ISLE, Ilmenau, 1997

[2] F. Ansorge, K.-F. Becker; Micro-Mechatronics – Applications, Presentation at Microelectronics Workshop at Institute of Industrial Science, University of Tokyo, Oct. 1998, Tokyo, Japan

[3] F. Ansorge, K.-F. Becker, G. Azdasht, R. Ehrlich, R. Aschenbrenner, H.Reichl: Recent Progress in Encapsulation Technology and Reliability Investigation of Mechatronic, CSP and BGA Packages using Flip Chip Interconnection Technology, Proc. APCON 97, Sunnyvale, Ca., USA

[4] Fraunhofer Magazin 4.1998, Fraunhofer Gesellschaft, München

[5] F. Ansorge; Packaging Roadmap – Advanced Packaging, Future Demands, BGA's; Advanced Packaging Tutorial, SMT-ASIC 98, Nuremberg, Germany

[6] Michael A. Mignardi; "From ICs to DMDs", TI Technical Journal, July-September 1998, pp. 56-63

[7] Long-Sun Huang; "MEMS Packaging for Micro Mirror Switches", Abstract, University of California

[8] Jeff Faris, Thomas Kocian; "DMD Package – Evolution and strategy"; TI Technical Journal, July-September 1998, pp-87-94

[9] Fraunhofer Institut Mikroelektronische Schaltungen und Systeme; "Lichtmodulatoren mit mikromechanischen Spiegelarrays" 2001

Dr. Frank Ansorge, J. Wolter, C. Rebholz, H. Reichl
Fraunhofer Institute Reliability and Microintegration; Micro-Mechatronic Center
Argelsrieder Feld 6
82234 Wessling
Germany
ansorge@mmz.izm.fhg.de

Keywords: mechatronic, micro-mechatronic, MEMS, sensors, packaging

A Novel High Aspect Ratio Technology for MEMS Fabrication Using Standard Silicon Wafers

C. Lohmann, A. Bertz, TU Chemnitz
M. Küchler, T. Gessner, Fraunhofer IZM

Abstract

A novel CMOS–compatible technology for the fabrication of MEMS based on standard single crystal silicon wafers is available at the Center of Microtechnologies [1]. High Aspect Ratio Microstructures (HARMs) are manufactured using a three mask level technology and dry processing throughout. The released micromechanical components consist of monocrystalline silicon without additional thin films after processing. As a result of the novel process flow the structures are surrounded by air gaps and fixed by special anchors. These Air gap Insulated Microstructures (AIM) were fabricated and tested with respect to mechanical stability, temperature dependence and electrical behavior. The low production costs and high device performance of the fabricated sensors and actuators demonstrate the ability of the technology for high volume production.

1 Introduction

The search for the best compromise between low fabrication costs and high device performance is a driving force for the continuing technology development within the MEMS field. Low cost and high volume technologies for the fabrication of MEMS are essential to accommodate the increasing demand for automotive applications. Advanced bulk micromachining competes with surface micromachining even for integration with microelectronics [2, 3]. Other solutions are using high aspect ratio structures in combination with special release/isolation processes [4, 5] or on the base of SOI-Wafers [6].

2 The Basic Technology of Air Gap Insulated Microstructures

The mechanical components of the microstructures fabricated by using the AIM technology consist of single crystal silicon exclusively because of the excellent mechanical properties and its common use. Thin films covering silicon MEMS structures are avoided in order to minimize device bending [7]. Standard silicon wafers are used because they are less expensive compared to SOI substrates (by a factor of about 10). However, an advanced isolation technique is required. In reference [5] the anchors are isolated first by a release processing and trench fill followed by a second release process for the mechanical components.

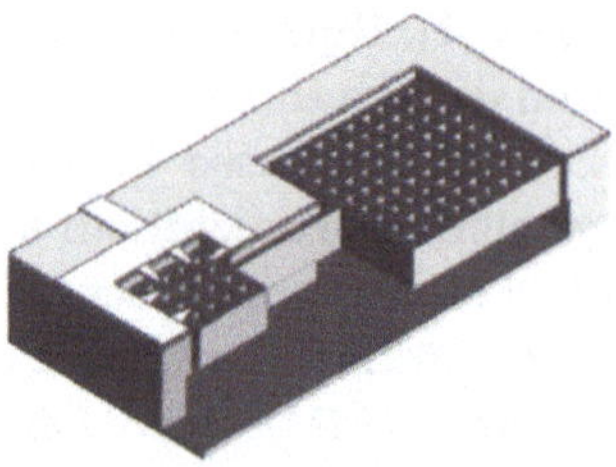

Fig. 1. Principle drawing of an AIM structure.

Our technology releases and isolates both in parallel. In contrast to SOI technologies special benefit is expected for the fabrication of a small number of pieces because the device height can easily be varied by the etched trench depth. Thus, mechanical properties can quickly be adapted to special applications maintaining the high performance of crystalline silicon devices. Figure 1 is a principle drawing of an AIM structure with a seismic mass, a spring and an anchor fixed by interconnection beams. Figure 2 is a photograph of a test structure and figure 3 is a SEM image of a test structure.

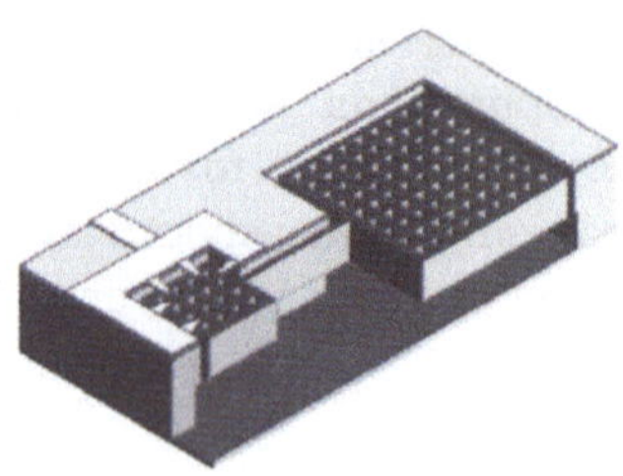

Fig. 2. Photograph of an AIM test structure (left).

Fig. 3. SEM image a of test structure (right).

3 Structure and Properties of the Interconnection Beam

The mechanical and electrical contact between the seismic mass and the bulk material, represented by the conducting layer, is realized by interconnection beams. Therefore, the electrical and mechanical properties of the beams are designed for both tasks. The interconnection beams consist of a SiO_2- and PE-SiN layer to ensure good mechanical stability, an aluminum layer for the electrical conductivity and PE-SiO_2 for passivation during processing and also to enhance the mechanical performance of a AIM structure (figure 7). This material stack provides tensile stress, which acts as a force on the released structure. For compressive stress an unsteady mechanical system would be obtained and tilting of the seismic mass is possible (figure 4). Tensile stress produces forces with opposite directions compared to the compressive stress. The mechanical system is in a steady state and the position of the seismic mass is fixed. The temperature dependence and long term stability of the tensile stress is proved and shown in figure 5 and figure 6. The measurements confirm an operation area up to 150°C and verifies a good long term behavior of AIMs.

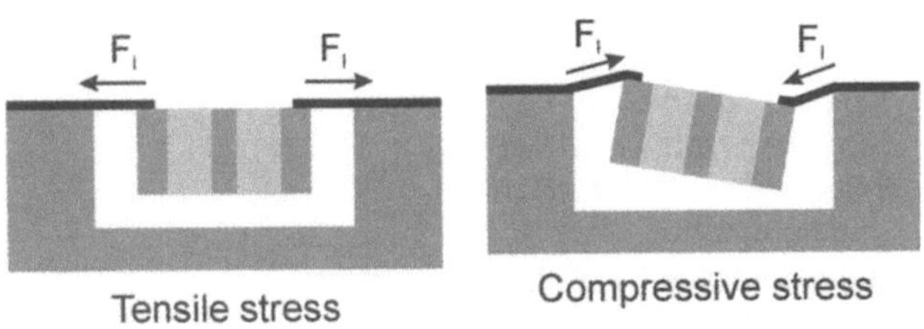

Fig. 4. Tilting of a seismic mass depending on tensile stress.

The mechanical stability and reliability of an AIM can be increased by an increasing number of interconnection beams. Tests confirm that the maximum load of an AIM is multiplied by the number of interconnection beams.

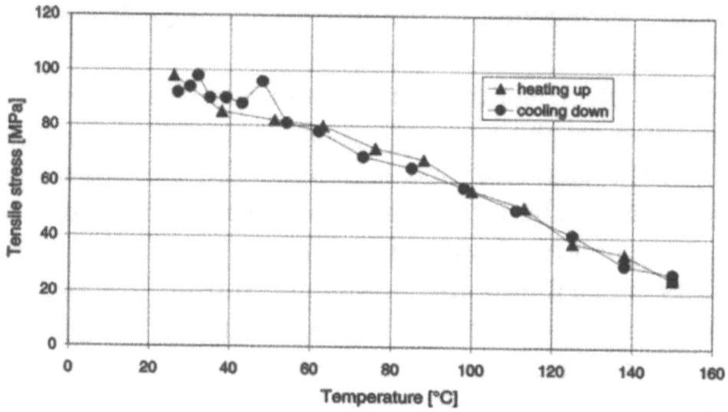

Fig. 5. Temperature dependence of the tensile stress.

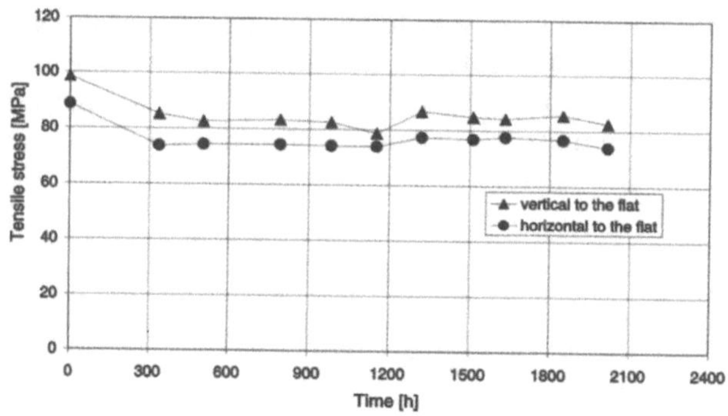

Fig. 6. Long term behavior of the tensile stress.

4 Process Flow

Starting with a standard silicon wafer, a material stack consisting of thermal SiO_2 and PE-SiN is deposited. The deposition process is followed by the first lithography and etch step. These steps define and open the contact holes between the metal layer and the doped bulk silicon. The following metal layer consists of a 40nm Ti/TiN barrier and 500nm aluminum, deposited by a PVD process step. The metal layer is defined by the second lithographic step. A self aligned etch process (the metal layer is used as mask for this process) for the opening of the silicon surface follows. The third lithography is used for the deep silicon etching. PE-SiO_2 is deposited and used as mask material for the deep silicon etch. The PE-SiO_2 also passivates the sidewalls of the material stack. This is especially important for the thin interconnection beams, because the following fluorine based dry etch process also attacks the PE-SiN layer depending on the selectivity, which may destroy the beams. The PE-SiO_2 is removed after processing. The deep silicon etch, passivation of the silicon sidewalls and release etch step are performed by an in situ process. A time multiplex deep etching (TMDE) process, called Bosch process [8], is used for the deep silicon etch. Aspect ratios up to 1:20 can be realized. In contrast to other technologies [4], a CF-Polymer is used for the passivation instead of SiO_2. The use of CF-Polymer enables the fabrication of structures free of any additional layer or coating because the CF-Polymer can be removed by an oxygen plasma process easily. Finally, a highly parallel isotropic silicon etch process is necessary for the removal of the silicon under the interconnection beams. Figure 3 is a SEM picture of released AIM structures.

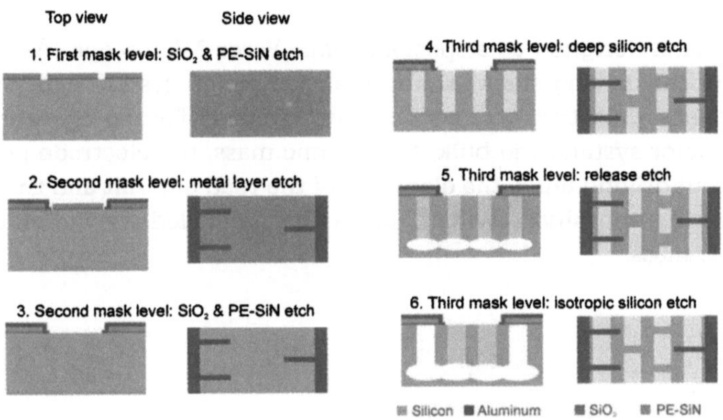

Fig. 7. Basic process flow for the fabrication of AIM.

5 Applications

Sensors, actuators as well as several test structures were designed and fabricated for testing and evaluation of the new technology.

A low-g sensor, based on a spring-mass system using capacitive position detecting, was fabricated and tested with respect to the temperature dependence. Figure 8 is a picture of a part of the sensor. Three different electrical areas and potentials, isolated against each other, are required for the operation of the sensor, the bulk potential, the seismic mass potential and the potential of the opposite electrode. Figure 10 shows the temperature dependence of the AIM sensor output. No significant change was detected. This result demonstrates one of the advantages of the new technology.

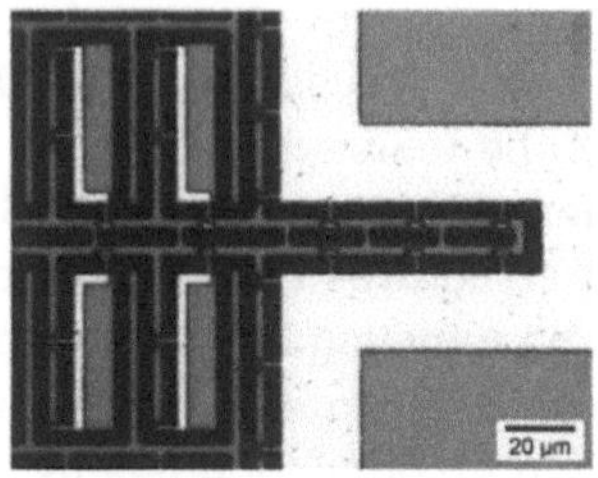

Fig. 8. Detailed view of a low-g sensor realized as AIM (left).

Fig. 9. Detailed view of an AFM actuator realized as AIM (right).

The actuator is designed as a special Atomic Force Microscope (AFM) system integrating the driving stage, a movement detection system and a sharp tip [9]. Six separate electrical connections are required for the operation of this sensor/actuator system: the bulk, the seismic mass, the electrode potential of two different driving units, the electrode of the read out unit and the potential for the maximum position switch. Figure 9 is a detailed view of an anchor of the seismic mass.

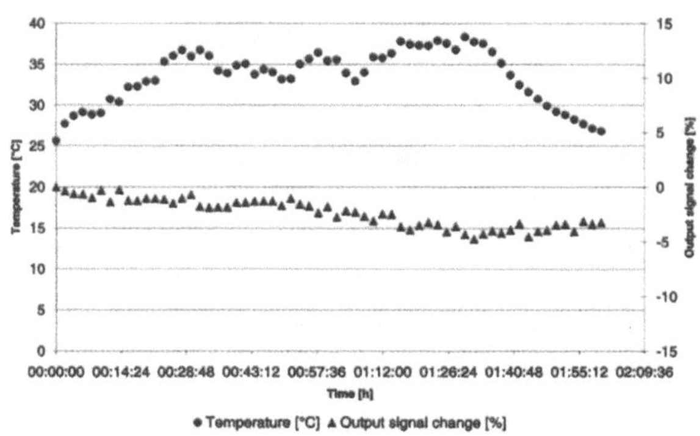

Fig. 10. Temperature dependence of the low-g sensor output.

6 Conclusion

The presented MEMS fabrication technology for high aspect ratio structures is based on standard silicon wafers and features a novel suspension and isolation technique. Dry and low temperature processing throughout are further features of this approach. The fabricated devices are characterized by low parasitic capacitance, high mechanical stability and improved temperature behaviour compared to silicon structures surrounded by thin films. Interconnect crossing and post-CMOS integration should easily be possible. The technology is designed for the fabrication of high volume and low costs products and therefore especially interesting for automotive application like acceleration, slope or yaw rate sensors.

Acknowledgement

The authors would like to thank the staff of the Center of Microtechnologies for assistance in processing the samples and devices.

This work was funded by the German Research Foundation (DFG) within the Collaborative Research Centre SFB 379 „Micromechanical Sensor- and Actuator-Arrays".

References

[1] A. Bertz, M. Küchler, R. Knöfler, T. Gessner, "A novel high aspect ratio technology for MEMS fabrication using standard silicon wafers", Sensors and Actuators A 97-98 (2002) 691-701.

[2] H. Takao, Y. Matsumoto, M. Ishida, "An integrated three-axis accelerometer using CMOS compatible stress sensitive differential amplifiers", IEEE Trans. Electron Devices, vol. 46 no.1 (1999) pp.109-116.

[3] H. Xie, G.K. Fedder, "A CMOS z-axis capacitive accelerometer with comb-finger sensing", Proc. IEEE: 'MEMS 2000', Miyazaki, Japan, 23-27 January, 2000.

[4] Z.L. Zhang, N.C. MacDonald, "An RIE process for submicron silicon electromechanical structures", J. Micromech. Microeng., 2 (1992) 31-38.

[5] U. Sridhar et al., "Single crystal silicon microstructures using trench isolation", Tech. Digest, Transducers '99, Sendai, Japan, 7-10 June, 1999, pp 258-261.

[6] M.A. Lemkin, T.N. Juneau, W. A. Clark, T.A. Roessig, T.J. Brosnihan, "A low-noise digital accelerometer using integrated SOI-MEMS technology", Tech. Digest, Transducers '99, Sendai, Japan, 7-10 June, 1999, pp.1294-1297.

[7] K. Griesbach, A. Bertz, W. Faust, R. Dudek, M. Kuechler, T. Gessner, "Investigation and Modification of the mechanical stress of single crystal silicon actuators based on surface micromachining", Proc. of 6th Int. Conf. on New Actuators, Bremen, Germany, 1998, pp. 82-85.

[8] A. A. Ayón, C. C. Lin, R. A. Braff, R. Bayt, H. H. Sawin, M. A. Schmidt, "Characterization of a time multiplexed inductively coupled plasma etcher", J. Electrochem. Soc. 146 (1999) pp. 339

[9] Roman Knöfler, Klaus Wolf, Enrico Ahl, Ulf Kotarsky, Wolfgang Manthey, Matthias Küchler, Andreas Bertz, Thomas Gessner, "Micromechanical Scanning Device for Large Vertical Range with High Resolution", Proceedings of the 7th International Conference on New Actuators, Bremen, Germany, 2000, pp. 83 - 86

Christian Lohmann, A. Bertz
TU Chemnitz -Zentrum für Mikrotechnolgien – ZfM
Reichenhainer Str. 70
09126 Chemnitz
Germany
Christian.Lohmann@zfm.tu-chemnitz.de

M. Küchler, Thomas Gessner
Fraunhofer Institut Zuverlässigkeit und Mikrointegration – IZM
Reichenhainer Str. 88
09126 Chemnitz
Germany

Keywords: micromachining, high aspect ratio microstructures

Interconnect Systems in Automotive Sensors at Elevated Temperatures

K. Gottfried, T. Gessner, Fraunhofer IZM
Ch. Kaufmann, T. Gessner, TU Chemnitz, Zentrum für Mikrotechnologien

Abstract

One limiting parameter for the application of MEMS in cars at places with elevated temperatures is the interconnect and wiring system of these devices. Currently used materials and technological concepts often are not sufficient for temperatures far above 125°C. This article attempts at first, to reflect main problems of device metallization at high temperatures. Finally, two application specific interconnect and wiring concepts are briefly introduced, which are able to work at temperatures up to 400°C and up to 500°C, respectively.

1 Introduction

Elevated temperatures are one of the main characteristics of car specific processes. Consequently, Micro Electro Mechanical Systems (MEMS), which are used to control i.e. the combustion process, the anti-lock brakes, and similar things must work reliable and precise under such temperatures. Due to that reason, the choice of appropriate materials for each part of a micro device becomes one of the crucial points. Beside the functional suitability of a material, its thermal and thermodynamic stability relating the other materials must be given.

A first topic is the semiconductor material itself. The small band gap of Silicon is limiting its electrical function to temperatures below 300°C [1]. Therefore, in many cases alternative substrates like Silicon On Insulator (SOI) or wide band gap semiconductors like Silicon Carbide (SiC), Gallium Nitride (GaN), and Gallium Arsenide (GaAs) have to be used. Thereby, SOI will play the important role in the future due to its full compatibility to the silicon technology. Furthermore, SOI has a low price compared to the wide band gap materials. Not at last, SOI is available in wafer sizes up to 300mm. However, wide band gap semiconductors will find their opportunity in particular cases.

The second problem area of temperature charged MEMS are the interconnect system, the connections to the next wiring level as well as the whole packaging concept. Presently, these items seem to be the most limiting factors. One main reason is the huge number of possible interactions within and between the several materials of the different functional parts at elevated temperatures.

2 Interactions in Temperature Charged Interconnect and Wiring Systems

Within temperature charged interconnect and wiring systems interactions like diffusion, solid state reactions, crack formation and propagation as well as film delamination have to be expected. Moreover, oxidation and corrosion problems are arising. Figure 1 summarizes the possible interactions. Diffusion processes and solid state reactions mainly take place at interfaces of different materials. The formation of cracks as well as the delamination of films are related to thermo-mechanical stress. Reasons for stress are a mismatch in the TCE values as well as changes in the crystalline structure of the materials itself. However, these problems are partly solvable by the choice of appropriate materials and/or introduction of supporting materials like diffusion or reaction barriers. Further serious problems are the oxidation and corrosion of the metal layers. The use of a hermetic package is mostly not possible for MEMS due to their desired function. Moreover, guaranteed hermetic packages for elevated temperatures are a problem in general.

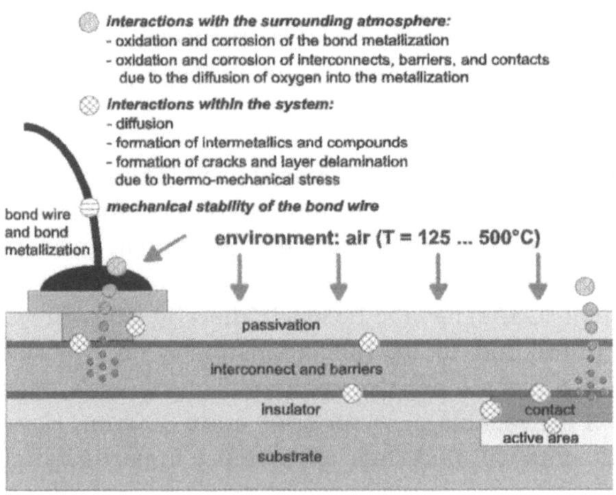

Fig. 1. Possible interactions in temperature charged interconnect systems.

Therefore, technological solutions on the chip level are required, which ensure the necessary stability against corrosion and oxidation. Basically, the whole interconnect system has to be encapsulated by passivation layers from silicon oxide (SiO_2) and/or silicon nitride (Si_3N_4). The limiting factor in this case is thermo-mechanical stress. Most of the passivation materials show compressive stress. This leads to tensile stress in the underlying conductors that enhances the formation of voids. To overcome these problems, an optimisation of the deposition technologies and of the film thickness is necessary. Finally, a passivation could not solve all corrosion related problems. The connections to the next wiring level require an opening of the passivation at certain points and the deposition of a bond metallization. This metallization either must not react with oxygen or the reaction must be self limiting. Furthermore, a diffusion of oxygen through the bond metallization has to be prevented.

3 Interconnect System for a SiC-Pressure Sensor

The sensor was designed to monitor the pressure during the combustion cycle within the cylinder of a car engine [2]. It has been chosen a piezo-resistive work principle because it is widely used for that purpose. A silicon carbide layer (3C-SiC), epitaxially grown on SOI-substrates [3], has been used for the fabrication of the piezo-resistors. Figure 2 shows the schematic structure of the sensor. The necessary interconnect system comprises ohmic contacts to the SiC-substrate, interconnects, and the bond metallization. Moreover, a wiring technology is needed, to realise the connection to the next wiring level.

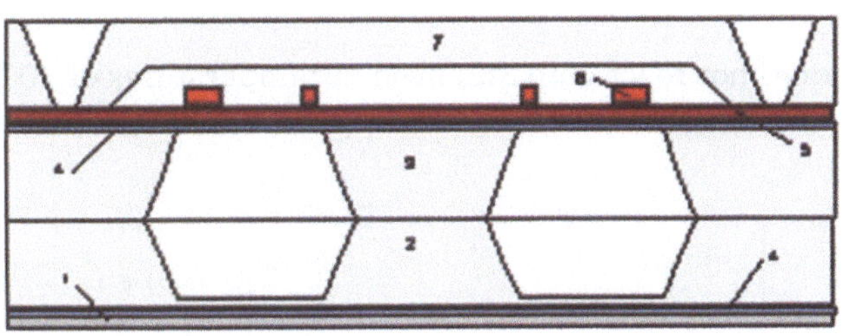

Fig. 2. SiC-Pressure sensor element. 1. SiC protection membrane, 2. Lower membrane SOI-Wafer, 3. Upper membrane SOI-Wafer, 4. Oxide layer, 5. SiC membrane, 6. Piezo-resistors, 7. Cap wafer.

For the ohmic contacts to 3C-SiC WSi_2 and $MoSi_2$ have been chosen. The reasons therefore are the high thermal stability of these silicides regarding SiC [4, 5] and promising results concerning the reachable contact resistivity [6, 7]. Both silicides have been used as interconnect, too. For the connection to the next wiring level an aluminium bond metallization has been used together with Al heavy wire bonding. The use of heavy wires is based on a simple reason. Commonly used bond wires with diameters far below 100µm are alloys of aluminium with up to 2% weight content of silicon. That is necessary to ensure the mechanical stability of the wire. At elevated temperatures the silicon starts to diffuse. The result is drastic loss in mechanical stability [8]. Heavy wires do not need that silicon content, because of the large diameter their own stability is sufficient.

Aluminium as bond metallization offers an another advantage. It is well known, that Al in connection with oxygen forms an extremely dense oxide (Al_2O_3). Only a few nanometers of this oxide are necessary, to prevent a further diffusion of oxygen through the oxide film itself. This leads to a self limiting oxidation process and ensures the protection of the underlying silicide against oxidation and corrosion.

3.1 Thermal Stability – Analytical Verification

The thermal stability of the fabricated contact and wiring systems was examined at first using AES depth profile analysis. In the left picture of figure 3 the AES depth profile of the interface SiC-WSi_2 after the silicide and contact formation is shown. The right picture shows the AES depth profile of the layer stack Al-WSi_2-SiO_2 after a thermal storage of 150 hours at 400℃ in air. For both layer stacks no material intermixing or interface reactions could be found. All layers are well separated, the interfaces are sharp and small. Therefore it can be assumed:

I WSi_2 does not react with SiC, even at temperatures of 1000℃.
II The layer stack Al-WSi_2-SiO_2 is usable for applications up to 400℃.

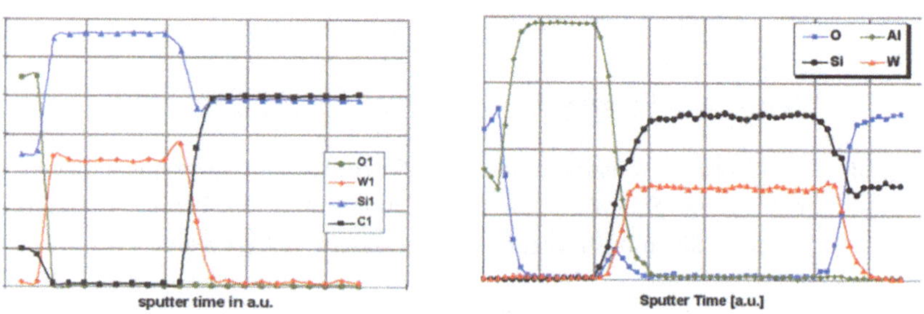

Fig. 3. Left: AES depth profile interface SiC-WSi_2 after the contact anneal (20min / 1000℃ / H_2); Right: AES depth profile Al-WSi_2-SiO_2 after a storage of 150 hours at 400℃ in air.

The aluminium top layer protects the stack against corrosion by the formation of an aluminium oxide as visible in figure 3. The thin oxidised layer on top of the WSi2 was already found after the contact annealing and is not due to the high temperature storage. The same results were obtained for $MoSi_2$ instead of WSi_2, but here not shown. The results for the interface SiC-Silicide are in accordance to other works [4, 5].

3.2 Mechanical Stability of the Wire Bonds

The mechanical stability of the bonds has been investigated by pull and shear tests. These tests were performed immediately after the bonding, after a thermal storage for 48 hours at 400°C, and after a thermal storage for 120 hours at 400°C. For each type of bonding and for each storage cycle 120 bonds have been prepared (see figure 4). Table 1 summarises the results. After the pull test for all annealed ball/wedge bonds a decrease of 25% in the wire strength was found compared to the none annealed. It is remarkable that this behaviour is independent from the time of the storage. It can be assumed this decrease is caused by a recrystallisation within the wire. The shear test shows no significant changes in the strength for the first and the second bond. In the case of the wedge/wedge bonds unfortunately, wrong parameters for the second bond have been chosen. So these bonds all failed during the pull test.

		Ball / Wedge			*Wedge / Wedge*		
		0 h	*46 h 400°C*	*120 h 400°C*	*0 h*	*46 h 400°C*	*120 h 400°C*
pull test	F [cN]	823	611	604	530	537	532
	S [cN]	66	48	44	150	37	32
Shear	F [cN]	708	813	850	-	-	-
ball	S [cN]	124	71	80	-	-	-
Shear	F [cN]	897	870	910	894	850	910
	S [cN]	145	122	110	81	100	120

Tab. 1. Results of the mechanical stability tests of the heavy wire bonds.

But the shear test for the first bond shows the same behaviour as above. So it can be assumed with proper bond parameters the strength will be the same as for the ball/wedge bonds. The parameter for the evaluation of these results is the wire tear strength. For the used wire this value is 600cN. The measured strength of the pull test must be corrected by a factor depending on bond geometry [9]. For these bonds 0.6 has to be applied. The corrected strength should be equal or larger then half the value of wire tear strength. This is well given for all bonds.

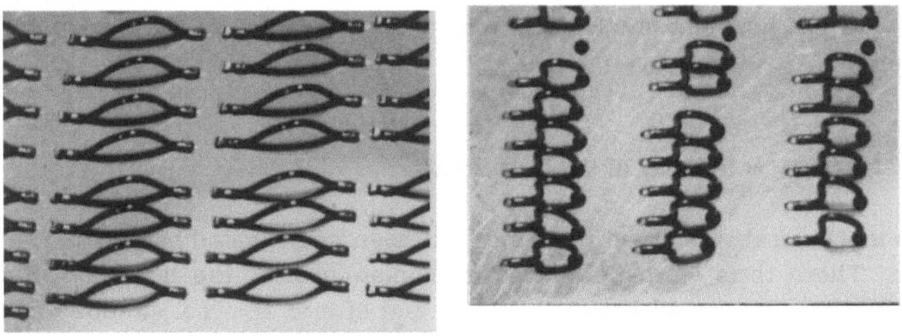

Fig. 4. Left: wedge-wedge Al-heavy wire bond samples; Right: Ball-wedge Al-heavy wire bond samples.

3.3 Characterisation of the Electrical Behaviour

The electrical behaviour of the metallization schemes was characterised using contact resistivity and overall resistance measurements. The contact resistivity as well as the overall resistance remained nearly unchanged after a storage for at least 1000 hours at 400°C in air. Table 2 summarises the results.

3.4 Summary

The refractory metal silicides WSi_2 and $MoSi_2$ together with an aluminium bond metallization are suitable material combinations to build up interconnect systems for MEMS using SiC as active semiconductor material. The thermal stability has been proven by analytical and electrical measurements for temperatures up to 400°C. Oxidation or corrosion effects have not been observed. Al heavy wire bonding is suitable for temperature stable inter chip connections in the same temperature range.

	contact resistivity		*over-all resistance*	
	Before	*After*	*Before*	*After*
Al-WSi_2	2,4 *10^{-4} Ωcm^2	1,1*10^{-4} Ωcm^2	19,8 Ω	19,7 Ω
Al-$MoSi_2$	5,2*10^{-5} Ωcm^2	1,7*10^{-5} Ωcm^2	14,3 Ω	14,3 Ω

Tab. 2. Results of the mechanical stability tests of the heavy wire bonds.

4 Metallization of a TiO_2-Gassensor

The sensor has been designed to analyse car exhaust fumes directly inside the exhaust [10-12]. For that purpose the sensor is placed near the exhaust end. That is important insofar, because the maximum surrounding temperature must not exceed 400°C, due to the work principle and the sensitive material. The data delivered by the sensor are resistance changes of a gas sensitive material, i.e. TiO_2. The absolute value of these changes depends on the applied temperature and the surrounding atmosphere. To distinguish between several gases like CO, CO_2, NO, NO_x, and THC, the sensor is arranged as an array of at least nine independent cells. Each cell is kept on a certain temperature between 400°C and 500°C. Because the sensitivity for each gas type depends strongly on the cell temperature, signal patterns will be obtained. With these patterns it is possible to calculate the concentration of the several gases from a mixture.

To fulfil its function, each cell needs heaters and electrodes. Furthermore, temperature sensing resistors (TSR's) are integrated in the heater level of each cell to ensure the needed accuracy in temperature controlling. Figure 5 shows a cross section of one sensor cell. The TSR's are not shown in this figure.

The metal of choice for the heaters and for the TSR's is platinum, because of its excellent thermal stability. Furthermore, platinum does not react with oxygen. Therefore, corrosion related problems have not to be expected. A disadvantage of platinum is its poor adhesion on dielectric layers like SiO_2. To overcome that, a titanium adhesion layer has been introduced. Several tests using standard tape test method [13] before and after a thermal storage for 8h at 500°C resulted in a sufficient adhesion. No film peeling has been observed.

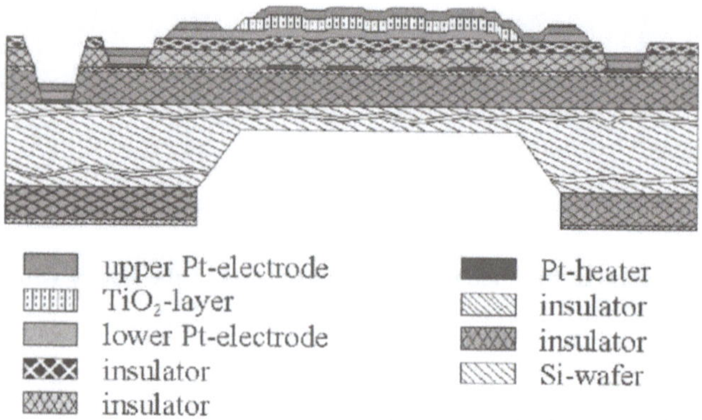

Fig. 5. Dross section of a sensor cell, the thickness of the heater layer is 300nm, the thickness of the electrode layers is 100nm, the titania layer has a thickness of 300nm, the TSR's are not shown in this picture.

Possible electrode materials are both platinum and gold. The experiments have been started with platinum. The reasons are given above. For the lower electrode a titanium adhesion layer has been used, too. The upper electrode was deposited directly on the titania. During measurements with these sensors in several gas measurement systems and on engine test beds no fail in the lower electrodes was found. However, in the upper electrode level sporadic failures (film peel off) were detected.

To save heating power a well insulation between the chip and the package is required. To ensure that, a self-supporting arrangement was chosen. This means the chip is fixed in the carrier only by the bond wires as shown in figure 6. The fixation of the chip during the wire bonding was realised using a re-workable glue. After bonding the glue can be removed using a simple temper step at 300°C for 20 minutes. As bond wire a heavy wire has been chosen, to achieve a sufficient mechanical stability.

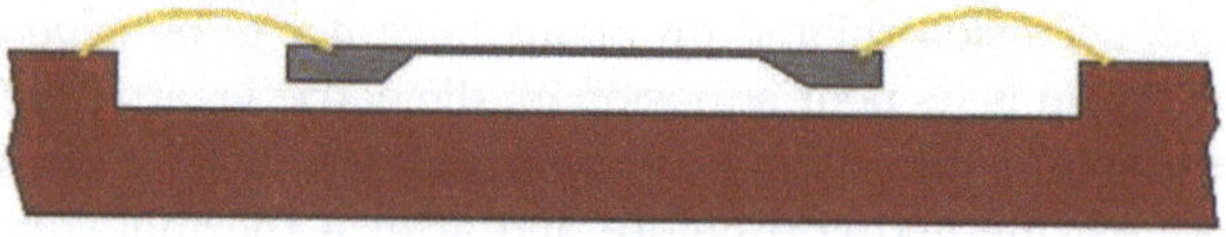

Fig. 6. Schematic of the self supporting arrangement.

Aluminium seems not be useful because it starts to form alloys with platinum at the given temperatures [14]. Alternative materials are platinum and gold. Both are not available as standard wires, but deliverable on demand. Wire bonding with platinum was possible, however a reproducibility could not be reached. In contrast to platinum wires, promising results were obtained using a 125μm gold wire. The mechanical stability was tested in the same manner as described above. In first investigations only wedge/wedge bonds were considered. Table 3 summarises the results.

	0 h	*24 h 400°C*	*70 h 400°C*	*24 h 600°C*	*70 h 600°C*
pull test F [cN]	248	171	175	114	140
S [cN]	28	17	14	55	41

Tab. 3. Results of the stability tests of the heavy wire bonds with 125μm Au-wire.

The remaining strength is sufficient in all cases. However, for the samples stored at 600°C a complete strip off of the bond metallization has been found, as shown in right picture of figure 7. The reason for this loss in adhesion are interactions between the gold wire and the platinum film. Gold diffuses into the platinum and starts likely to build up an intermetallic. In the left picture of figure 7 the diffusion areas around the wire are well visible. Consequently, the material system gold wire / platinum metallization is not suitable to build up long term stable and reliable devices for temperatures in the range of 600°C. Therefore, experiments are under way to use a gold bond metallization and introduce a diffusion barrier between this metallization and the platinum.

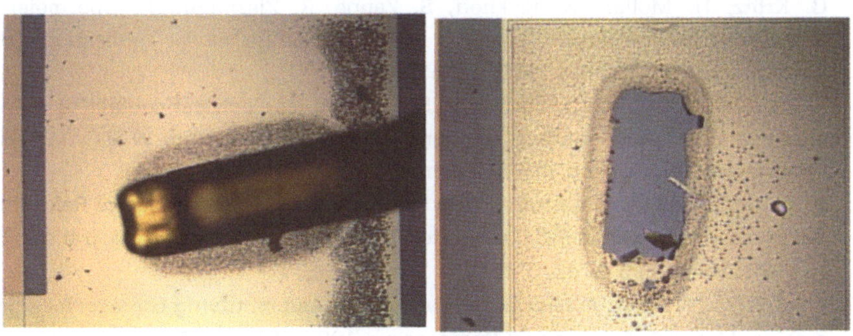

Fig. 7. Left: Diffusion areas around the gold wire; Right: Complete film strip off after the pull test on a 600 °C tempered bond.

Summary

Platinum is a favourable candidate to realise a metallization for an exhaust gas sensor working at temperatures up to 500 °C. The advantage of platinum is its inert character concerning oxygen. This leads to an excellent corrosion stability. Using gold wires with diameter of 125µm a mechanical stable, self supporting arrangement of the gas sensor chip could been realised. However, gold heavy wire bonding directly on a platinum metallization is not suitable for long term temperature stable applications. Additionally work has to be done to overcome this problem.

However, for sensors fabricated in this manner the ability to quantify gases in a temperature range up to 400 °C (by sensor working temperatures between 400 °C and 500 °C) could be proofed during several measurements on test systems as well as in a real exhaust atmosphere on engine test beds. Results of these measurements are published [10-12].

References

[1] H. Morkoc et al., Large-bad-gap SiC, III-V nitride, and II-VI ZnSe-based semiconductor device technologies, J. Appl. Phys. {76} (3), 1994, pp. 1363-1399

[2] J. von Berg, Piezoresitiver Brennraumdrucksensor auf der Basis neuer Substratmaterialien für den Serieneinsatz im Automotor, Dissertation TU Berlin, UFO Atelier für Gestaltung und Verlag GbR, Allensbach, ISBN 3.930803-88-7, 2000

[3] G. Krötz, H. Möller, M. Eickhoff, S. Zappe, R. Ziermann, E. Obermeier, J. Stoemenos, Hetroepitaxial Growth of 3C-SiC on SOI for sensor applications, Mat. Science and Eng. B61-62, 1999, pp. 516-521

[4] F. Goesemann, R. Schmid-Fetzer: Stability of W as electrical contact to 6H-SiC, phase relations and interface reactions in the ternary system W-Si-C, Materials Science and Engineering, 1995, Volume B34, pp. 224 - 231

[5] A. Costa, Silva, M. J. Kaufman: Phase Relations in the Mo-Si-C System Relevant to the Processing of MoSi2-SiC Composites, Metallurgical and Materials Transactions A, 1994, Volume 25 A, pp. 5 - 15

[6] L. M. Porter, F. Davis: A critical review of ohmic and rectifying contacts for silicon carbide, Materials Science and Engineering, 1995, Volume B34, pp. 83 - 105

[7] N. Lundberg, M. Östling: CoSi2 ohmic contacts to n-type 6H-SiC, Solid-State Electronics, 1995, Volume 38 No.12, pp. 2023 - 2028

[8] J.T. Benoit, R.R. Grzybowski, D.B. Kervin, Evaluation of aluminium wire bonds for high temperature (200 °C) electronic packaging, Third international high temperature electronics conference, 1996, Trans. vol. 1, pp. III-17 - III-23

[9] F. Rudolf, K.-D. Lang, G. Jarney, Test methods for wire bonds (Part I: Mechanical strength)

[10] T. Gessner, K. Gottfried, R. Hoffmann, C. Kaufmann, U. Weiss E. Charetdinov, P. Hauptmann, R. Lucklum, B. Zimmermann, U. Dietel, G. Springer, M. Vogel, Metal oxide gas sensor for high temperature application; Microsystem Technologies Vol. 6, No. 5, August 2000

[11] K. Gottfried, M. Vogel, R. Hoffmann, C. Kaufmann, St. Günther, F. Dieckmann, Gas Sensor for High Temperature Application, mstnews, No. 4/01, September 2001, pp. 10-12

[12] R. Hoffmann, K. Gottfried, M. Vogel, F. Dieckmann, A. Streit, U. Dietel, Vollstrommessung mit Titanoxidgassensoren, MTZ – Motortechnische Zeitschrift, 5/2002, pp. 400-409

[13] Standard Methods for Measuring adhesion by Tape Test, American National Standard ASTM 3359-7

[14] M. Hansen, Constitution of Binary Alloys - Second Edition, McGraw-Hill Book Company, New York, Toronto, London, 1958

Knut Gottfried, Thomas Gessner
Fraunhofer Institut Zuverlässigkeit und Mikrointegration - IZM
Reichenhainer Str. 88
09126 Chemnitz
Germany
Knut.Gottfried@che.izm.fhg.de

Christian Kaufmann, Thomas Gessner
TU Chemnitz
Zentrum für Mikrotechnologien – ZfM
Reichenhainer Str. 70
09126 Chemnitz
Germany

Keywords: interconnect, high temperature, sensor

The 6th Framework Program: Microsystem and Micro@Nano-Technology in IST

T. Reibe, D. Beernaert, European Commission

Abstract

The 6th Framework Program covering the 2003-2006 time interval, is defined and ready to be implemented. Seven thematic areas have been chosen: a limited number but well representative of those sectors of excellence where Europe intends to maintain and grow its leaderships. IST is one of the seven thematic areas. In IST, micro and nano-systems and Micro@NanoTechnology include topics where the priorities are concentrated on research integration and networking, multi-disciplinarity and applied driven research, increase in functionality, more functions, extreme miniaturization, low cost. The research activities will be driven towards building the knowledge society for all, setting up Interactive, Intelligent & Emotional Environment (IIEE) and implementing the vision of the Ambient Intelligent landscape. The research will be organised around major integrated projects (IP), networks of excellence (NoE) and traditional projects. The RTD work will be complemented by initiatives aiming to easy transfer the outcome of the results into innovation.

1 Introduction

The 6th multi-annual Framework Program (6thFP) for Community research, development and demonstration activities is at the terminal and ready to become operative. The FP aims are those set out in the art. 163 of the Treaty. Namely, strengthening the scientific and technological base of Community industry and encouraging it to become more competitive at international level while promoting all the research activities deemed necessary. To achieve these goals in a very efficient way and to contribute to the creation of ERA, the 6th FP is structured around 3 major issues:

- Focussing and integrating the Community research.
- Structuring the European Research Area.
- Strengthening the foundations of the European Research Area.

The activities under these issues will contribute to the integration of research efforts and activities on an European scale as well contributing in structuring of the various dimensions of the European Research Area. A shift from "project thinking" towards "initiative thinking" and the use of new instruments, as integrated projects and networks of excellence, represent those further elements needed to better deal with the attainment of the FP strategic objectives. Co-ordination of activities will be also ensured. Furthermore, the overall efforts, in the 6th FP structure and pursuant the ERA strategy, are directed to establish a close synergy with several other initiatives in the field (from national, to Eureka, EIB, etc). All this means that, with respect to the past, the 6th FP is certainly characterized by a wider scope, ambition and more pronounced future oriented vision.

2 The IST Vision

For the achievement of the 6th FP objectives, the Commission has proposed a set of interventions covering a limited number of strategic objectives. Among them, the IST, thematic area 2, and "Nanotechnologies, intelligent materials new production processes", thematic area 3.

The IST overall objectives, pursuant the Lisbon European Council and the objectives of the e-Europe initiative, are intended to stimulate the development in Europe of both technologies and applications at the heart of the creation of the information society in order to increase the competitiveness of European industry and allow European citizens in all Union regions to fully benefit from the development of the knowledge based society. Concentration on the future generation of IST will bring IST applications and services to everyone and enable the development of the next generation of technologies to be more user friendly.

The focus for the first technological priority is represented by 4 major topics:

- Research on technologies addressing the key security challenges posed by "all-digital "world.
- Ambient intelligence landscape offering access to the information society for all.
- Electronic and mobile commerce as well as technologies for secure transactions and infrastructure.
- Large scale distributed systems and platforms which includes the global resource information database (GRID) based systems.

For the second technological priority, the miniaturisation, low-cost component, performance, integrating extended functionality, reducing the power consumption, bringing multidisciplinary applied research into microsystems, nano-electronics and displays represent the key factors to foster the components and microsystems integration into a wide range of intelligent products and networked applications.

It is important to highlight that most of the R&D work on micro&nano-science and technology is carried out in the priority area 2 and 3, while the results deriving from the activities in these two areas can be also exploited on the thematic area 1 (genomics), 4 (space), 5 (food and quality) and 6 (sustainable development), which represent the other thematic priority areas where the FP6 is structured.

3 The Ambient Intelligence Vision

Looking at the research supported under the priority 2 (IST) in FP6 it is worthwhile to mention that this is oriented towards building the knowledge society for all and the realization of the vision of ambient intelligence. This latter concept refers to an environment that is sensitive, adaptive and responsive to the presence of people. It emphases on greater user-friendliness, more efficient services support, and support for human interactions. In this vision, people will be surrounded by intelligent and intuitive interfaces, supported by computing and networking technology which is everywhere, embedded in everydays objects such as furniture, home walls, clothes, vehicles roads and smart materials, embedded in any objects around us and in the environment recognizing and responding to the presence of individuals in an invisible way. An ambient intelligence so conceived will take advantage from all those technologies which major aim is to scale down the overall dimensions of the developed/functionalized materials and devices, as well as to add more functionality to smart micro & nano systems, while reducing their cost and the power consumption. All this new technological scenario will integrate sensing, actuation, electronics, smartness, emotion into final products. To this end and to fulfill the ambient intelligence requirements, the availability of multidisciplinary expertise appears to be mandatory.

4 Microsystems and Micro@NanoTechnologies

Microsystems and Micro@NanoTechnology (M@NT) for sensors, actuators and for integration in micro and macro applications is a high-tech very fast evolving area, pervasive in all industrial sectors. The microsystems, and M@NT are introduced to have an increased functionality, intelligence, more functions, portability or networking capability. They represent an essential element to build the "ambient intelligent" landscape.

Mainly in the past decade, Europe has proved to have built excellent research competence as well as the required system knowledge to capture a good part of this market. In order to further gain positions in the global microsystem market, Europe must improve the capability to convert the results of the research into final applications, in a much shorter time, in a reliable way and looking at more visionary applications. By the nature of the field itself, Europe need to invest in multidisciplinary co-operative research. Complementary to national funding, often oriented to device technology or RTD fundamental, and in order to transfer results into practice earlier, more applied co-operative research and faster take-up of innovation need to be stimulated. Today a more critical mass is necessary in several application domains. Institutes need to be stimulated to better focus their activities on multidisciplinary applied research and to co-operate covering different application fields with more common goals, offering challenging results and intellectual property to industry. At the same time, the Industry is expected to cooperate more intensively with institutes to transfer results faster and to provide a reliable guidance. Industry should also work together to offer more global solutions to their customers and to bring more innovative products on the market. As the field is still relatively "immature" and moving fast, a flexible approach and the stimulation of entrepreneurship is found essential. Training and education in multi-disciplinary sectors of competence need to be encouraged as well, to co-operate in a more efficient and synergic way and "produce" the skills necessary.

Nowadays, those sectors taking most advantage of the outcome of activities in microsystems are the medical, communication, automotive and, in general, the biochemical market. The developments are driven by the use of the silicon as basic material to work on, even if recently the use of advanced polymer technology, functional and improved material performance have further improved and generated new devices and systems. With a further miniaturisation of the devices and taking advantage of the work done and of the achievements reached in other sectors (see f.i. the microelectronic sector), the microsystem is entering slowly but surely into the nano regime (in our vision, what is below the 50nm range). A mixed approach top-down and bottom-up, generating small devices/structures, the availability of nano structured materials micro and nano electronics and molecular technology, will bring to the possibility to explore the nano domains, providing the capability of developing nanosystems, by introducing the nano inside the "macro", which will allow to explore new physical effects for switching, processing, storage and much more. A roadmap for bringing microsystems into the nano-regime is forecasted as well as the possibility, in a future not too far away, to have microsystems with improved functionality due to the incorporation of nanotechnology or to have Micro@NanoTechnology incorporated in nearly everything.

5 Microsystems and Micro@NanoTechnologies in the 6th FP - IST: Position, Expression of Interest and Instruments

In the 6th Framework, the Commission will offer a diversified approach to cooperate, focus and bring result into practice. In particular integrated projects (IP), clustering projects around a roadmap towards a (single) application objective within a broader application vision/application domain, will be promoted to stimulate co-operative multidisciplinary applied research. The main objective of an IP will be the application addressed, its social or market potential, but even more important its capability to drive a set of microsystems, microsystems' technology, microsystem functions and underlying competencies that may also have a wider applicability in other application fields. Transfer of results in several other applications fields will be organized via trials within the IP, which particularly will benefit SME participation. Networks of Excellence comprising of a few institutes or centers of competence co-operating at organizational level within the application visions envisaged will be set-up to promote further co-operation. These have the possibility to expand later to larger NoE to increase the critical mass. More speculative and exploratory projects, and more generic activities to the total field of microsystem will be addressed using specific targeted research projects which may thematically be clustered. Two calls for proposals, with fixed deadlines, are planned in IST for the time interval 2003 and 2004, which may be extended by a 3rd call in December 2003 timeframe.

The definition of the calls can be found on our CORDIS website under http://www.cordis.lu/ist. For "Micro and nano systems" we launched a call for expressions of interest (EoI) in 2002.

The automotive sector has gathered 12 (11 IPs and 1 NoEs) EOIs. Safety in the car, more functionality and smartness added by MEMS, MOEMS and 3D modules and devices are the main issues covered by the proposals. Four are based on the development of MST devices having magnetic sensors as traditional core sensing/transducting material. Most EoIs are structured more as thematic projects rather than fully integrated IP. These should combine their activities and target more visionary applications. This analysis also takes into account those EOIs submitted to transport (application area within IST) and to IST (1.1.2.) in general with significant activities on MST.

MAGMAS	Magnetic Micro-Actuators & Systems	Automotiv, medicine, Telecom., IT
INTEGMAGSENS	Micro and Nano Magnetic materials in Integrated Sensors and Devices	Automotiv, medicine, Telecom., IT
HITEMP	Technologies for Advanced Devices in High Temperature Ambient	Automotive
N/a	Magnetic Sensor Microsystems for Industrial and Automotive Applications	Automotive, Consumer
ARS-LIM	Advances in Reliability and Sustainability through Life Cycle Information Modules	Automotive, medical biotechnology industrial equipment and aeronautics
SPLAT	Sensor Platform Design for Future Automotive Applications	Automotive
ICARIS	In Car Road Conditions Information System by opto-electronic	Automotive (safety)
MST Automotive	Safe, Clean and Economic Automobiles using Microsystems technologies as an enabling Technology	Automotive environment for clean, safe cars
ADVMOEMS	Advanced Micro Opto Electro Mechanical devices and processes	Automotive, aerospace, medical and communications
EU-ESS-Transport	Enhanced safety and security for transportation in an enlarged Europe	Automotive (transportation of person and goods)
NewMagSens	New generation of magnetic sensors for biology, RF Identification and other applications	Automotive, data storage, aeronautics, magnetic imaging, watch industry
Ecubes	3-D-integrated micro/nano modules for easily adapted applications	Automotive and personal communications

Tab. 1. EoI's in the automotive field.

Technology driven proposals with no specific applied field driving or where the application field is only very generically mentioned.

- The reliability issue is addressed by 8 EOIs (6 NOEs and 2 IP) which enforce the importance of the topic.

- 6 EOIs are addressing the MEMS/MOEMS/NEMS design, analysis, technology, characterisation, standard definition issues. They are technology driven EOI proposals.
- Developments on ink-jet printing technologies is addressed by 5 IPs. The applications range from inkjet printers to medical microdosing systems and electronics printing on textile.
- The remaining 24 EOIs (mostly IPs) are aiming to cover different aspects of material, technologies and techniques related to MST. These EoIs are often limited in scope of activities and could benefit from integration in a more structured NoE and/or in focused IP.

Table 1 describes the EoI expressed the automotive field as the main target area. All non-confidential EOIs are listed in the document 'EOIs related to Microsystems and Micro@Nano Technologies' on http://www.cordis.lu/ist/ka4/supermic/index.htm. A complete set of all EOIs can be found on http://eoi.cordis.lu//search_form.cfm.

6 Conclusions

The 6th FP is at the terminal and the first call had been launched. From now ahead, the implementation ERA objectives will constitute the driving force to move forward the cooperation and integration of the scientific and industrial Communities and, in the near future (by 2010), this challenging experience will trigger a modification into our daily life as well as in the way Academia, Industries will cooperate among them to achieve common strategic objectives in a more efficient and productive manner. With priority 2, the objective is to enlarge and install the "ambient intelligence" landscape to bring the knowledge society in the hands of all people. In this context, the microsystem field and the exploration of the opportunities offered by the Micro@NanoTechnology world will play a crucial role in pursuing and achieving the overall mission. In fact, this will open opportunities to integrate sensing, actuating and intelligence in nearly everything, everywhere and will enable application of nano-technology never thought before.

Thomas Reibe, Dirk Beernaert
European Commission
DG-INFSO
Micro and Nano-Systems and Displays
049 Brussels
Belgium

Keywords: 6th framework program, integrated projects (IP), networks of excellence (NoE), european research area, ambient intelligence vision, Micro@NanoTechnology, intellectual property

Functions and Applications

Safety

Advanced Solution for Pressure Based Side Airbag Systems

M. Kandler, Infineon Technologies AG

Abstract

Side airbags in the front and rear side doors are getting more and more standard in modern cars. Up to now mainly acceleration sensors are utilised to detect a side crash. In recent years a new concept for side crash detection has been used by some car manufacturers: The detection of the pressure pulse which is generated within the side door during the side crash. Compared to acceleration based systems the pressure detection method has some advantages. The firing decision can be made very fast and reliable. Furthermore the differentiation between misuse cases, i.e. non fire crashes, and fire crashes is very good. Infineon produces pressure sensor ICs for this application since 1998.

1 Introduction

In the present paper specialised pressure sensor IC's for side crash detection are described in detail. A partitioning for a side airbag system consisting of the door module, the transmission line and the airbag ECU is presented.

The core of the pressure sensor IC's are surface micromachined capacitive sensor cells, monolithically integrated with an A/D converter and digital signal conditioning circuitry. The pressure sensor IC's offer a digital interface which is compatible with standard µCs. Furthermore self diagnosis features are implemented which are essential for safety relevant applications.

Two types of pressure sensors are available. A basic type providing an absolute pressure signal via SPI and an advanced type providing a Manchester encoded pre-processed relative pressure signal. The advanced type allows to reduce the number of devices and hence the construction of cost effective door modules. Both types have a small SMD package which is proven for automotive requirements.

2 Technology and Signal Processing

The automotive safety market demands cost effective and highly specialised semiconductor components complying with high quality standards. These requirements can only be met by using a standard VLSI process for high volume production.

The fabrication process for the presently available pressure sensor IC's is a 0,8µm BICMOS process running on standard equipment for high process stability.

The basic element of the sensor IC is the pressure sensor cell. In comparison to bulk micromachining, where a pressure sensing membrane is obtained by anisotropic etching of the wafer backside, the surface micromachined sensor cell is fabricated by using standard deposition and etching steps.

Figure 1 shows a principle cross section of a sensor cell consisting of an elastic membrane of conductive polycrystalline silicon (poysilicon), a sealed cavity and a conductive electrode on the substrate surface (well).

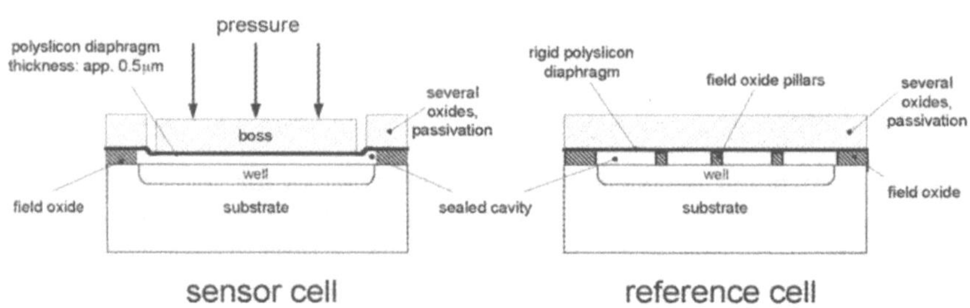

Fig. 1. Cross-section of the surface micromachined capacitive sensor and reference cells (not in scale).

The intermediate cavity is formed by a particular sacrificial layer etching technique. In this process step the previously deposited field oxide is locally removed through small holes in the polysilicon membrane. The resulting cavity is sealed by following oxide deposition steps. To achieve the desired pressure sensitivity in a final procedure a groove is etched into the oxide layers down to the polysilicon. This results in a quadratic boss in the centre of the membrane. The typical size of a cell is 70µm x 70µm.

Almost identical reference cells are also implemented. They incorporate small pillars of field-oxide which prevent the membranes from bending. Additionally the groove etching is omitted. From hence the reference cells are not sensitive to pressure but have a quite similar capacity and temperature characteristic like the pressure dependent cells. Therefore the reference cells are well suited for compensation purposes, e.g. in a bridge circuit (see below).

3 Functional Principle

The conductive polysilicon membrane and the conductive substrate surface are forming the plates of a capacitor. If the outside pressure is increased the elastic polysilicon membrane of the sensor cell is deflected towards the substrate. This causes an increase of the capacitance in the order of several fF/bar.

In order to achieve higher sensitivity and higher SNR, arrays of several cells (e.g. 14 cells) are used. Apart from this a particular bridge configuration is implemented, whereby two sensor arrays and two reference arrays yield the four elements of a capacitive bridge circuit.

The bridge signal is converted into a digital signal by a switched-capacitor fully differential 2nd order ΣΔ-ADC (see block diagram figure 2). The output of the ADC is a 1 bit stream. The ratio of high bits and low bits in this stream is a measure for the pressure.

The ADC is followed by a digital averager (FIR filter) which increases the resolution of the bitstream from 1 to 16 bits, while reducing the sampling rate. The following digital low pass filter (IIR filter) gives the signal a cut-off frequency of 380Hz, which is optimal for the application, and removes remaining out of band frequency components.

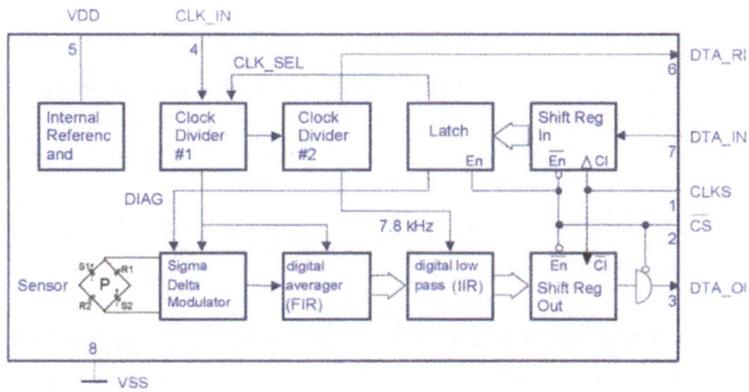

Fig. 2. Block diagram of the basic pressure sensor for side crash detection (KP100).

Via the data-in pin the diagnosis modes can be selected allowing the testing of the sensor operativeness at regular intervals. In principle two diagnosis modes are available. One mode checks the sensor cells for mechanical or electrical defects by comparing the sensor arrays against each other. Another mode checks the electrical and the digital circuitry.

A block diagram of the signal path of the advanced pressure sensor is shown in figure 3. In comparison to the basic type the absolute pressure signal p is additionally processed.

1) A measure for the ambient pressure is derived from the pressure signal p. This is done by averaging the signal p with a time constant of several seconds, resulting in a slowly changing pressure signal p_0.
2) In an arithmetical block the operation $(p-p_0)/p_0$ is carried out. This relative signal r only becomes unequal zero for pressure pulses which are short in comparison to the p_0 time constant, e.g. for pressure bursts due to a side crash. Furthermore the signal r only depends on the intensity of the volume compression inside the door during the crash. It is independent of the ambient pressure and hence from altitude above sea-level or meteorological pressure changes.
3) In a last stage the relative signal r is converted into a Manchester encoded signal m. This is the output signal of the IC.

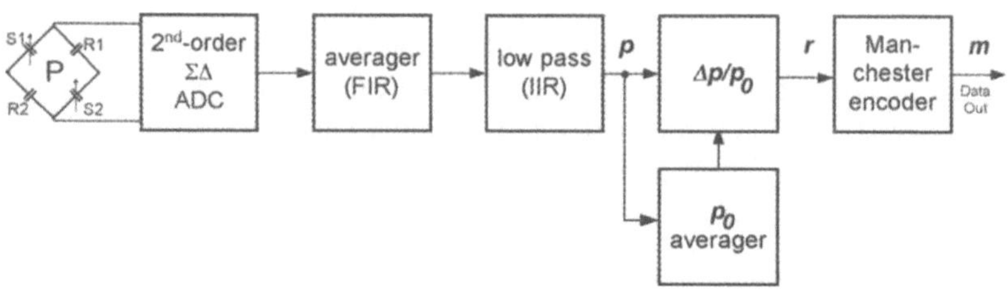

Fig. 3. Block diagram of the signal path of the advanced pressure sensor IC.

4 Data Transmission

To meet the requirements for a safe data transmission, the relative signal is encoded into a Manchester encoded signal where a high bit is represented as a rising edge and a low bit is represented as a falling edge of an electrical voltage or current signal. Therefore even if a series of low bits is transmitted, the level of the electrical signal is not constant, but changes permanently. So the lack of single data bits can be detected in any case by the central airbag unit. This makes the Manchester encoded signal safer than a non return zero signal.

To make the transmission line less sensitive against electrical interference a current modulated signal is preferred. The current modulation can be achieved by a few discrete components connected to the voltage output of the pressure sensor IC. The transmission line has to be monitored with a high sampling rate in order to detect every signal change properly.

Therefore a special receiver IC is also available, which is capable of monitoring up to four transmission lines at the same time. When a Manchester telegram has been received the receiver IC activates an interrupt pin to tell the central airbag µC that a data word can be fetched from the SPI Interface of the receiver. This reduces the needed µC performance drastically. The receiver IC also protects the µC against high voltage on the transmission line.

Furthermore the transceiver IC is capable to send a synchronisation pulse to the sensor IC. This pulse triggers the sensor to send a data telegram to the ECU. A asynchronous data transmission is also feasible.

5 Side Airbag System

Figure 4 shows a block diagram of a side airbag system. The door module consists of the pressure sensor IC (advanced type) a voltage regulator IC and a few discrete devices like a transistor for the current modulation. A quartz is not necessary, since the sensor IC has a digitally trimmed on chip oscillator to meet the timing requirements of the Manchester output signal.

The airbag ECU consists of the receiver IC, a power supply IC and a µC. IC's for the firing of the airbags are also part of the airbag ECU.

The firing decision is made by a special algorithm implemented in the µC. This algorithm evaluates the pressure data received from the sensor IC. Height and slope of the pressure pulse are important criteria for the decision. In advanced systems not only the data of the pressure sensor are taken into account but also data from a central saving sensor or additional acceleration sensors located in the cars side at stiff structures like the pillars. The combination of acceleration and pressure sensors enhances the performance of a side airbag system significantly and allows a better differentiation of must fire and no fire cases.

Compatible acceleration modules can also be connected to the receiver IC to realise such combined systems.

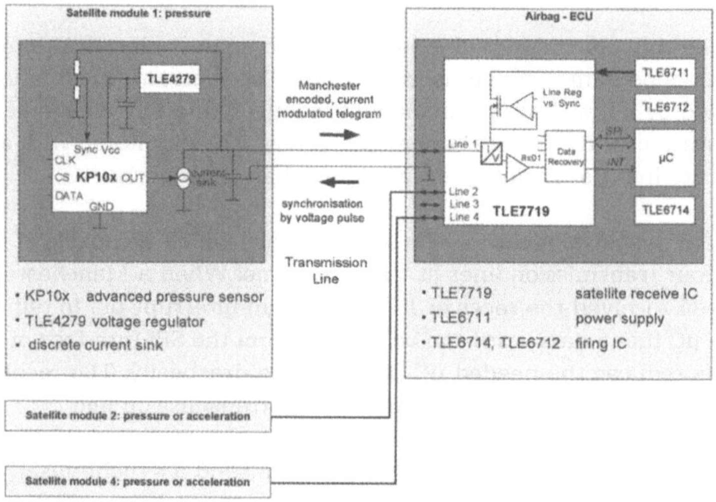

Fig. 4. Side airbag system.

6 Safety Features

For traceability reasons the advanced pressure sensor IC has a unique identification number. This number is transmitted in a start-up sequence after the chip is powered on. For safety reasons a diagnosis value is also transmitted to the ECU after power on. This diagnosis value indicates a malfunction of the sensor cells. Other self diagnosis functions permanently check parts of the electronic circuitry. If a failure is detected an error code is transmitted to the ECU.

7 Package

The sensor IC's are packaged in a special SMD package, which allows the construction of small modules. The chip surface must be accessible to the ambient pressure. Therefore the chip can not be covered with a stiff material but is protected by an inert gel. This gel is resistant against the chemicals in the automotive environment. It also prevents the condensation of liquids on the chip surface. The chips are shipped with a protection cap, which can be removed after chip soldering. Figure 5 shows the SMD package without protection cap.

Summary

For side impact airbag systems specialised semiconductor components are available. Particularly high integrated, advanced pressure sensing IC's with on chip signal processing, on chip trimmed oscillator, on chip ID-number and various self diagnosis features make it possible to construct cost effective, reliable and fast side airbag systems.

References

[1] Combining Acceleration and Dynamic-Pressure Sensing For Side-Impact Restraint Activation" G. Winkler, Th. Stierle, Th. Malbouef. Airbag 2002, 6th International Symposium, Karlsruhe Dec. 2002

[2] Advanced Components for Pressure Based Side Airbag Systems" M. Wycisk. Airbag 2002, 6th International Symposium, Karlsruhe Dec. 2002

[3] KP100 datasheet, Infineon technologies, December 2002

Michael Kandler
Infineon Technologies AG
Balanstrasse 73
81541 Munich
Germany
Michael.Kandler@infineon.com

Keywords: pressure sensing, side airbag system, VLSI process, BICMOS, safety

Prototypic Realisation of Millimetre Wave Radar Imaging

R. Schneider, J. Wenger, DaimlerChrysler AG

Abstract

In future novel driver assistance, comfort, and safety systems the needs for environmental sensing and traffic scene interpretation are increasing. The reliability of these systems will be mainly determined by the perception of the sensors and the capability of automatic data interpretation, and it will be a differentiating factor in the competition amongst automotive OEMs. Automotive radar is expected to be a key sensor technology for future active safety systems, especially due to its specific physical properties (direct acquisition of range and velocity). This contribution demonstrates the potential of radar sensors with high resolution in azimuth and range in traffic environment and addresses some aspects of their practical implementation. A prototype of an automotive instrumentation radar is presented, which allows on-line acquisition and real-time visualisation of radar images with an update rate of up to 10Hz. The imaging capabilities of high resolution radar in traffic environment will be pointed out with respect to new comfort and safety features. In addition practical realisation approaches will be discussed.

1 Current Status of Automotive Radar

In 1999 Mercedes-Benz has been the first to introduce a millimetre-wave radar for autonomous cruise control (ACC) and radar has proven to be an ideal sensor in traffic environment [1]. Meanwhile some car manufacturers like Jaguar, BMW, or Volkswagen in the new Phaeton offer ACC for highway operation as an option, others plan to do this in the near future. Further development activities of the suppliers are primarily focused on reducing costs and size of the sensor. Although announced on their roadmaps, no additional functionality beyond ACC, like stop-and-go support, obstacle detection, and collision avoidance seems to be in sight for mass products, as this would require clearly enhanced sensor capabilities. Due to the moderate resolution of the employed sensors, insufficient information about the environment is delivered in order to enable advanced driver assistance functions like reliable obstacle warning or emergency breaking.

A conventional ACC radar sensor delivers a target list with distance, azimuth angular position, relative speed, and reflectivity of significant objects, which typically are smaller than the spatial resolution of the radar. Hence a classification of these point-targets generally is not possible. Table 1 lists some characteristics of an ACC radar sensor (Distronic by Mercedes-Benz). In current ACC systems the handling of stationary targets is difficult, as they cannot be separated by differences in their Doppler-shift.

ACC Radar Sensor	
frequency	76 – 77 GHz
radar principle	pulse Doppler
number of separate beams	3
beam width (symmetrical)	3.4 deg
range resolution	5 m
Doppler resolution	1 km/h
system sensitivity	0 dBsm @ 150 m
cycle time	70 msec

Tab. 1. System parameters of a typical ACC radar sensor.

2 Specific Properties and Potential of Millimetre-Wave Radar Imaging

The use of a multi- or scanning-beam radar with high resolution in range and azimuth, which generates real-aperture radar images containing considerably more information about the ambient traffic situation, could overcome the missing capability of object classification and scenario interpretation in first generation ACC sensors. At the DaimlerChrysler Research Centre in Ulm the fundamental potential of high resolution mm-wave radar with respect to automotive applications has been investigated since several years [2-4]. One of the most important results was that vehicles appear as laterally and longitudinally extended objects in high resolution radar images in range versus azimuth. Caused by reflections from the road surface, the complete two-dimensional extension of bottom sides is perceived, allowing for the recognition of vehicles and distinction of different classes (cars, vans, trucks) [5]. Even obscured road users become visible, which provides very important information for the operation in stop-and-go traffic. Another benefit is that road environment like any kind of vegetation and man-made ground structures (slopes, guard rails, reflection posts, traffic signs, etc.) can clearly be detected, enabling recognition and prediction of the road course.

As the essential information about the environment is contained in a radar image with the spatial co-ordinates azimuth and range, no resolution in elevation and no large vertical antenna aperture is required. The corresponding line antenna would be well compatible with the vehicle front design and space requirements.

A crucial advantage of radar sensing is its direct acquisition of speed via Doppler processing, which could be taken as a third dimension in the radar images. Effective data compression is achieved by reducing the velocity information of an azimuth-range resolution cell just to a single value representing the dominant reflection object in this cell. The resulting information is described in two 2D image matrices of radar reflectivity and Doppler. For visualisation purposes both matrices can be combined into a single image using a pseudo-colour coding of Doppler and radar intensity into hue and saturation, for example.

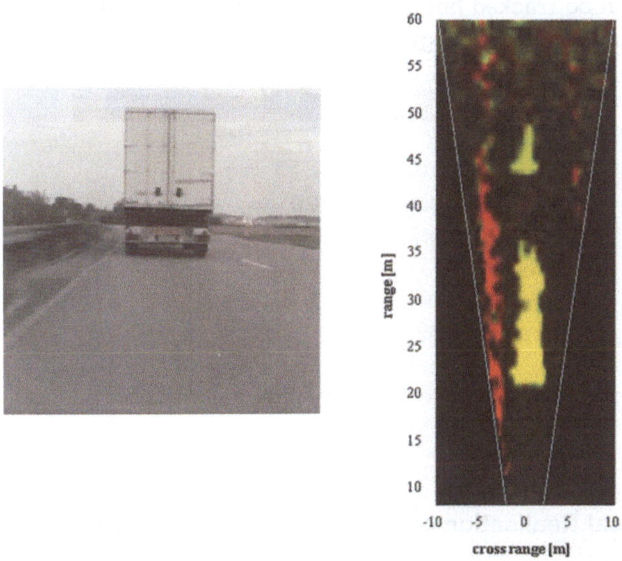

Fig. 1. Traffic scene and corresponding high resolution radar image in PPI representation.

Figure 1 shows an example of such a kind of radar image and exhibits the specific properties of high resolution radar sensing at millimetre waves in traffic environment. The colour coding of relative Doppler shift is: yellow for the same speed, red and green for closing and opening Doppler, respectively.

The spatial radar data acquired in polar co-ordinate grid has been transformed into a Cartesian PPI (planar position indicator) representation in range and cross-range, which delivers a geometrically correct bird's eye view of the scene. The camera for video documentation was located close to the antenna position 60cm above ground. Radar acquisition parameters as listed in table 2 have been set. Details about the radar set-up will be discussed in the following sections.

Figure 1 clearly depicts the left roadside with guard rail, which approaches while driving and appears as a red line. Although the right roadside is merely covered with grass, it can be recognised. The preceding truck with trailer appears yellow due to its zero relative Doppler shift, and as an extended area with true dimensions as a result from the road surface reflections. Even a car in front of the truck becomes visible by the wave duct effect. Its green colour indicates an increasing distance from the truck.

Here a very attractive feature becomes evident: In platoon traffic not only the next vehicle can be tracked but also at least another one ahead. Perceiving its manoeuvres could improve longitudinal vehicle control considerably.

High Resolution Radar	
centre frequency	76.5 GHz
field-of-view in azimuth	± 10 deg
azimuth beam width	1.0 deg
elevation beam width	5.0 deg
range resolution	1 m
velocity resolution	1 km/h
system sensitivity	-20 dBsm @ 150 m

Tab. 2. Acquisition parameters for high resolution radar data.

3 Prototypical Realisation of an Automotive Imaging Radar

Based on the fundamental findings about millimetre-wave imaging a prototype of a mobile instrumentation radar system for on-line radar image acquisition and real-time visualisation was specified and realised. Although the overall system design aims at high performance and versatility, its basic features have the perspective of mean to long term commercialisation. Installed in a test car the system serves for the detailed evaluation and demonstration of the capabilities of radar imaging, data acquisition, design, and test of processing algorithms, and as a basis for the implementation of new comfort or safety applications. Its operation parameters can be varied in wide ranges in order to achieve different image areas, resolution, and sensitivity.

3.1 System Design

Imaging of vehicle underbodies and detection of hidden objects via reflections from the road surface require higher sensitivity with respect to a conventional ACC radar sensor. Preliminary investigations resulted in the need of about 20dB additional sensitivity in order to enable these features up to 150m range, as those targets are characterised by radar cross section (RCS) values of down to -20dBsm. An increased target dynamic of 50dB has to be assumed. Together with the dynamic of 60dB due to propagation attenuation in 5-150m range an overall requirement for system dynamic results, with certainly cannot be handled by a FMCW system. Hence pulsed waveform has been chosen for the imaging radar system design because of its robustness against saturation.

The radar transceiver lay-out has been based on practical imaging parameters (see table 2). Range resolution of 1 m corresponds to 7 nsec pulse duration and to 150MHz bandwidth of a matched IF filter. The gain of an antenna with beam widths of 1deg · 5deg is assumed to be at least 38dBi, which contributes considerably to the additionally required sensitivity. Speed resolution in the Doppler processing depends on the number of range bins N and the pulse repetition frequency PRF. A practical choice is N = 512 and PRF = 70kHz, which provides sufficient unambiguous speed (50 km/h), unambiguous range (2km), and processing gain (27dB).

A minimum signal-to-noise ratio of 10dB should be present in the radar images in order to achieve acceptable detection and false alarm rates. The overall system losses of radome, front-end, and signal processing are estimated not to exceed 8dB. The transceiver design finally has to find a trade-off between transmit power and receiver noise figure F. Assuming a value of F = 8dB achievable with state-of-the-art balanced Schottky mixers, a transmit power of 19dBm (80mW) would be necessary. As this hardly can be realised with today's MMIC technology, a low-noise receive amplifier (LNA) with a noise figure of 4.5dB was provided to reduce the required transmit (peak) power to a practical value of about 35mW. The resulting mean transmit power is below 20μW.

The front-end design should demonstrate a long term potential with respect to producibility (cost aspect), which won't allow for parallel transceiver concepts. This means that the complete field-of-view (FOW) has to be scanned sequentially with a single antenna beam, provided that time-on-target requirements can be satisfied. The above-mentioned Doppler acquisition of 512 samples with a PRF of 70kHz delivers full integration gain for a minimum time-on-target of about 7msec. The required FOW is deduced from the demand of operation in urban area or narrow road curvature. Case studies have shown that a coverage of $\pm$15deg is fully sufficient, $\pm$10deg is acceptable with restrictions in urban (crossing) environment. Assuming an azimuth beam-width of 1deg, the maximum achievable image update rate would be $(7\text{msec} \cdot 30)^{-1} = 4.8\text{Hz}$ in the first and 7.1Hz in the second case. Compared to a video system, the update rate of an imaging radar is allowed to be clearly lower, as the velocity information is directly available and needs not to be derived from differentiation of subsequent images. Hence the mentioned update rates are assumed to be sufficient.

At the present time the antenna constitutes the technological key component in a real aperture imaging radar. The challenge is to combine fast beam scanning of a wide FOW at good electrical properties (efficiency, side-lobes) with the potential of low-cost production, compactness, and mechanical robustness. Phased array concepts fail due to their complexity, focal plane concepts have shortcomings in achieving large FOVs, beam switching leads to high insertion loss in the required switch cascade, and pure mechanical scan isn't practicable for the large aperture.

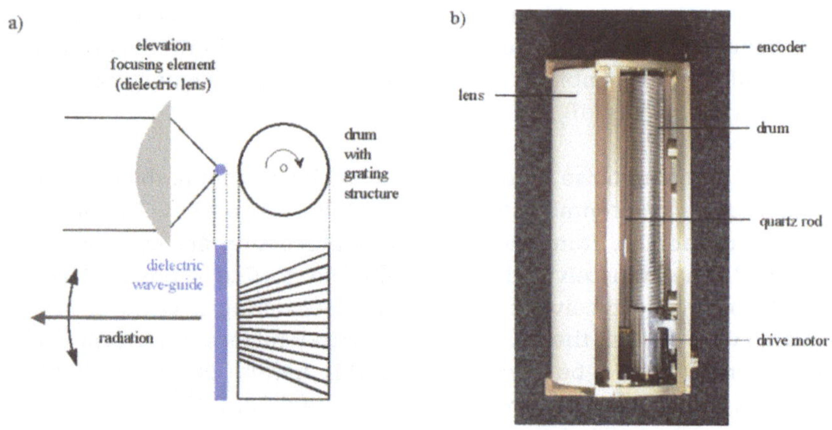

Fig. 2. Schematic diagram (a) and implementation (b) of azimuth scanning fan beam antenna.

A very attractive approach, which doesn't completely dispense with moving parts, is a tricky advancement of the image line antenna [6], which exploits the fact that a material inhomogeneity close to a dielectric wave-guide will cause radiation. An equidistant structure (grating) parallel to the wave-guide effects the radiation of a wave-front with a direction depending on the grating period. If now gratings with varying spacing are arranged on the circumference of a drum, sawtooth-wise scanning of the radiated wave-front is achieved by acceleration-free rotation of the drum (see figure 2a) [7]. For beam forming in the orthogonal (elevation) plane either a cylindrical parabolic reflector or dielectric lens can be used.

Figure 2b shows the prototypic implementation of such a 'spinning grating' antenna with lens (design and manufacturing by WaveBand Corp., USA), which was applied for the instrumentation radar set-up. The grating structure is realised by grooves with different pitch in a metallic cylinder resulting in a continuous beam scanning behaviour. A rigid quartz rod of about 1mm diameter forms the dielectric wave-guide. A mode launcher allows the antenna to be connected via a standard WR-12 wave-guide flange. The drum driving is performed by a small DC gear motor (which could be integrated into the drum to increase compactness), an angular encoder serves for rotation speed control and exact reference for the beam pointing direction.

Practical experience with the antenna proves well applicability in harsh automotive environment. Some electrical and mechanical characteristics are compiled in table 3 (verified by measurements). The azimuth beam-width is somewhat larger and accordingly the gain lower than postulated in above system design, but for availability reasons this had to be accepted.

Antenna Characteristics	
frequency	76 – 77 GHz
field-of-view in azimuth	± 11.5 deg
azimuth beam width	1.5 deg
elevation beam width	4.2 deg
gain	≥ 35.5 dBi
side-lobe suppression	≤ - 15 dB
maximum scan rate	20 Hz
dimensions (width × height × depth)	24 cm × 6 cm × 10 cm

Tab. 3. Antenna characteristics.

3.2 Description of the Hardware and Vehicle Installation

The instrumentation radar is designed as a versatile test system with real-time processing and visualisation features, which requires an open and universal hardware structure. It consists of a base unit in a standard VME chassis and the remote front-end (transmitter and receiver modules plus antenna). Figure 3 depicts the overall block diagram. The central component is an embedded Pentium III PC with Windows NT performing system controlling, radar image composition and pre-processing, visualisation, and data handling. Its main periphery is formed by timing&control electronics generating all the radar timing and a number of direct control signals, an IF processor module for converting the IF signal from the receiver (Rx) module into base-band I/Q signals and digitisation, a digital signal processor (DSP) performing the Doppler FFTs and other numerically extensive pre-processing (details see below), and some external standard components.

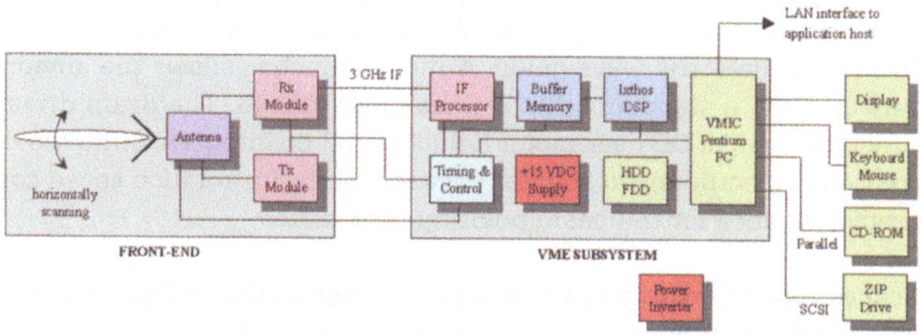

Fig. 3. Block diagram of the complete instrumentation radar system.

System operation is done via a graphical user interface (GUI) allowing the adjustment of pertinent radar parameters like transmit pulse width, PRF, IF filter bandwidth, range sampling, number of Doppler bins, and receiver gain.

A ferrite circulator for transmit and receive duplex is mounted together with a receive LNA directly at the antenna and connects to the Tx and Rx modules. The transmit signal is generated in a phase locked dielectric resonator oscillator (DRO) at 19.125GHz, pulse gated in a two-stage PIN-modulator, and multiplied to 76.5GHz in an active quadrupler with an output power of about 16dBm. The minimum pulse length is 3.5nsec. Key component in the Rx module is a biased, balanced fundamental mixer (measured noise figure F = 7.6dB). It converts to an IF signal of 3GHz, which is led to the base system on a high power level in order to minimise interference in the long transmission line.

The detailed design and implementation of the VME base system and the Tx/Rx modules as well as programming the control software of the VME components were performed by Innovative Technology, USA.

Figure 4 shows the integration of the system into a test car. The front-end components are mounted centred at the front bumper in a weather-proof housing (figure 4a). The antenna height is positioned 60cm above ground, which is a good compromise between exploiting multi-path effects by road surface reflections (hidden cars, vehicle underbodies) and roadside detection. Also with respect to shadowing effects of road crests [5] the antenna position shouldn't be to low. The radome is made of 5 mm polypropylene (according to the multiple half-wavelength matching condition) and showed less than 0.2dB two-way insertion loss.

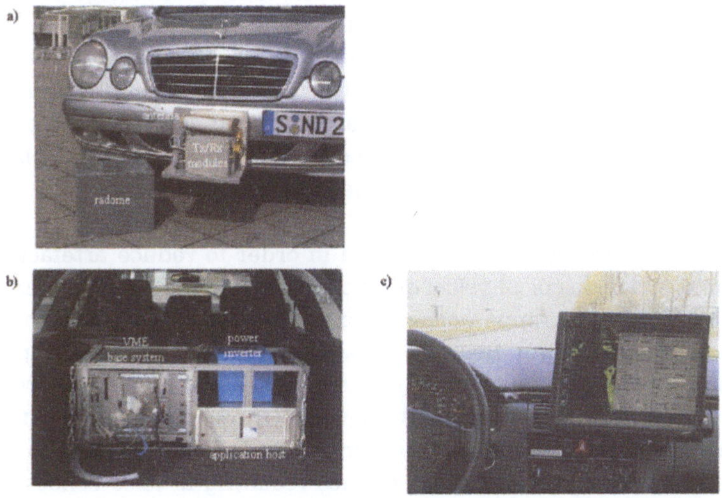

Fig. 4. Test car equipped with imaging radar: (a) front-end, (b) base system, (c) operation and visualisation display.

The base system is rack-mounted together with its AC 230V power supply in the trunk of the estate car (see figure 4b). An additional component placed there is a second PC provided for the intended implementation of safety or comfort applications. In the current status of the set-up it serves for synchronised video documentation and is connected to the radar control PC via Ethernet LAN.

For GUI operation of the system and the on-line radar image visualisation a display is attached at the dashboard (figure 4c). Note the junction on the left hand side recognisable in the instantaneous radar image.

3.3 Data Processing

The imaging radar data processing is performed in the following subsequent stages: radar data pre-processing, composition of radar images, image pre-processing, and conditioning for visualisation. Most of the pre-processing is performed in the DSP card, primarily computation of the Doppler FFTs for each resolution cell and maximum search in the Doppler frequency spectra. The resulting range profiles of reflectivity and associated Doppler values have to be assembled into the actual 2D radar images. As radar acquisition is not synchronised with the antenna pointing direction, this is done based on the encoder values and interpolating onto a fix azimuth grid. The antenna scan speed is controlled in a feed-back loop including the complete processing chain, so that the azimuth increment between subsequent range profile measurements is exactly the 3dB beam width (Nyquist condition). With the currently used antenna this results in 16 azimuth directions per scan. Range is typically sampled from 5m to 150m at 1m resolution, resulting in raw radar images of $16 \cdot 145$ pixels. Range dependent amplitude correction is performed by R^p with an exponent $p \sim 1$ as a good compromise between point target detectability and noise limitation at larger ranges.

Some image pre-processing is performed in order to reduce artefacts caused by system deficiencies or physical effects. Cross-coupling in the transceiver or antenna leads to near range echoes at zero Doppler. They are blanked by a kind of adaptive MTI filter. Because of the relatively low side-lobe suppression of the antenna cross-range blooming of strongly reflecting objects occurs. It is reduced by appropriate lowering the level of those pixels having the same range and Doppler as this object. Another physical effect occurring in the presence of strong reflections is double bounce resulting in false echoes at twice range and twice Doppler in the images. They can be eliminated by pixel-wise checking for this condition.

In order to achieve a proper mapping of intensity values into indices suited for an image colour coding, low and high level clipping has to be performed on the logarithmic values. A heuristically optimised choice is: low level clipping at 10dB above noise floor (this is the above specified SNR) and establishing a dynamic head room of 20dB. The noise floor can be computed directly from the measured data by averaging the intensity of target-free resolution cells, which allows a brightness control of the radar images independent of any system parameters.

For visualisation conditioning 2D interpolation of the radar data has to be performed, which favourably is done directly in the desired co-ordinate system, either Cartesian (PPI plot) or vanishing point projection with direct comparability to optical vision. Due to the moderate FOV, trigonometric approximations can be applied in the transform formulas, which minimise the computational effort. Final step is combining intensity and Doppler matrices into a single one for pseudo-colour coding of a specific colour-map.

All of the described data processing could be implemented on the base-system PC controller without significant reduction of real-time capability. With the above mentioned radar and imaging parameters an update rate of 3Hz results.

By reduced range and Doppler processing settings it can be increased to 10Hz. In addition to visualisation the acquired data can be recorded either in MATLAB binary file or TIF image format for off-line processing. For future on-line data evaluation in the application host the image data can be transferred via the LAN interface.

3.4 Imaging Performance

The sensitivity of the mobile instrumentation radar could be verified as predicted in the system design. Hidden vehicles proved to be detectable in all relevant traffic situations. Roadsides can typically be recognised up to a range of 70m, provided they consist of grass or gravel. Road curvature prediction based on real clothoid parameter modelling and tracking in consecutive images will work clearly farther. In many cases the prediction is eased by the presence of structures parallel to the road (trees, shrubs, fences, guard rails, poles, walls, slopes, etc.).

Radar echoes from the underbodies reliably appear in two-dimensional representation with correct dimensions and alignment in the radar images and encourage the development of corresponding classification algorithms for traffic scene interpretation.

Spurious objects are to be found regularly in the case of guard rails, walls, metallic fences, or kerbstones parallel to the road. Due to detour propagation vehicles appear virtually mirrored on the other side of these reflecting areas.

The forward reflectivity even of low kerbstones is very high, explicable by their dihedral appearance. But here an exploitable effect becomes evident: Although nearly no backscatter can be observed from kerbstones, occurrence of multipath reflections of preceding vehicles allows localising the roadside in the middle between direct and virtual image. This expands the capability of roadside recognition.

4 Applications

In the previous sections some novel applications for an automotive high resolution radar with imaging capabilities have already been addressed. Currently it is presumed that the following functionalities can be achieved or at least supported by such an environmental sensor:

- Improvement of ACC function (higher speed, wider range, country road use).
- Obstacle warning (including stationary objects - for example rear end of traffic jam).
- Emergency breaking.
- Collision avoidance.
- Stop-and-go traffic support.
- Detection of pedestrians.
- Identification of parking space.
- Curve warning.
- Visibility enhancement.

Some of these are not stand-alone applications, but have to be seen in combination with others. Parking space identification would be an interesting enhancement for guided or automatic parking systems, which currently are under development. The sensing capability of high resolution radar has been investigated by measurements in according situations and is illustrated in figure 5. A free parking space can be identified clearly earlier than by visual inspection and a rough size determination is possible (the free space in the analysed scene was 5m long).

With respect to safety applications pedestrian detection also can't be an autonomous feature, as the spatial resolution of the imaging radar won't allow classification of persons. But localisation of potentially relevant objects would be a very important input for sensor fusion approaches. Radar can identify regions of interest (ROIs), in which detailed analysis based on video image processing , e. g., is performed [8]. This provides also the information redundancy, which is required in safety systems, as radar and video sensing is largely complementary.

A specific contribution of radar can be the detection of pedestrians hidden by vehicles, for example at a bus stop. Here again the wave duct effect between road surface and vehicle underbody becomes beneficial. This could help to recognise situations of persons abruptly entering the roadway (playing children) at an early stage. Radar measurements have been performed indicating that persons can be detected reliably in front of a vehicle, provided the spacing is wider than the range resolution (1m in the current system realisation). Figure 6 shows an example with a person in front of a minivan.

An interesting application based solely on radar information could be a kind of "curve assistant", which warns the driver in the danger of overshooting a curve. Exploiting the capability of continuous roadside detection even in the absence of any border structures, the road course and its curvature can be determined and related to the actual speed, which is directly available from the Doppler data. Figure 7 gives an impression of the roadside detection performance in the situation of a narrow country road. Maximum acquisition range has been 100m. In this example a vanishing point projection has been applied to the radar image.

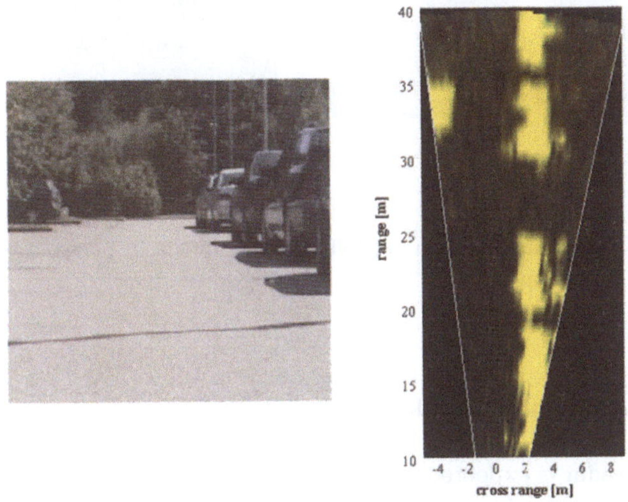

Fig. 5. Capability of free parking space identification (radar image in PPI representation).

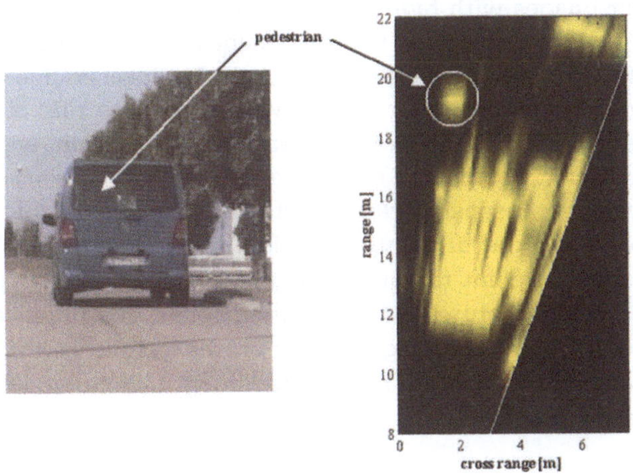

Fig. 6. Detection of a pedestrian hidden by a parked vehicle (radar image in PPI representation).

Fig. 7. Roadside detection capability (radar image in perspective representation).

5 Conclusion and Outlook

Automotive radar is expected to be a key technology improving driving safety in future, especially due to its inherent advantages (weather independence, direct acquisition of range and velocity). In this contribution the potential of radar sensors with high resolution in azimuth and range has been pointed out. They deliver images with high information content and considerably smaller amount of data in comparison to video images. A prototype of an automotive instrumentation radar has been realised, which allows on-line acquisition and real-time visualisation of radar images with an update rate of up to 10Hz. It could be shown that by moderate increasing of resolution, angular coverage, and sensitivity with respect to existing automotive radar sensors a new quality of environmental sensing can be established. It enables a fundamentally new approach of radar data interpretation: image processing based scenario analysis instead of point target tracking. This opens up new perspectives of comfort and safety features in future automobiles. The practical implementation of a prototypic mobile imaging radar proves technical feasibility at system design parameters, which are not contradicting to commercialisation aspects like complexity, size, and cost.

Future activities will be hardware- and software-related as well. Mass production realisation aspects of the front-end (antenna and millimetre-wave electronics) have to be clarified. Autonomous implementation of the above mentioned comfort or safety features requires the development of radar image processing algorithms (based on the adaptation of established methods like edge detection, segmentation, clustering, tracking, object classification) and scene interpretation according to the particular application. An important aspect will also be fusion of radar and vision data, for which an optimised concept has to be worked out.

References

[1] J. Wenger, R. Schneider, M. Stotz, B. Danner, W. Lauer, R. Nieuwenhuizen, R. Adomat: Forward Looking Radar Activities at DaimlerChrysler; Workshop on Automotive Millimetre-Wave Radar Systems (WS 3), EuMC 2000, Paris, France, Oct. 3, 2000, pp. I,1-I,8.

[2] R. Schneider, G. Wanielik, J. Wenger: Millimeterwave imaging of traffic scenarios; Intelligent Vehicles '96, Tokyo, Japan, Sept. 18-20, 1996, pp. 327-332.

[3] R. Schneider, J. Wenger, G. Wanielik, H. Neef: Millimeter-wave images of traffic scenes and their automatic interpretation; Conf. Proc. Microwaves and RF, London, UK, Sept. 30 - Oct. 2, 1997, pp. 239-244.

[4] R. Schneider, J. Wenger: System Aspects for Future Automotiv„e Radar; Proc. of the IEEE MTT-S Microwave Symposium, 1999, pp. 293-296.

[5] R. Schneider, D. Didascalou, W. Wiesbeck: Impact of road surfaces on millimeter wave propagation; IEEE Trans. Veh. Tech. Vol. 49, No.4, July 2000, pp. 1314-1320.

[6] K. Solbach, I. Wolff: Dielectric Image Line Groove Antennas for Millimeter Waves, Part II: Experimental Verification; IEEE Trans. AP, vol. AP-33, No. 7, July 1985, pp. 697-706.

[7] V. Manasson, L. Sadovnik, R. Mino: MMW Scanning Antenna; IEEE AES Systems Magazine, October 1996, pp. 29-33.

[8] B. Heisele, N. Neef, W. Ritter, R. Schneider, G. Wanielik: Object Detection in Traffic Scenes by a Colour Video and Radar Data Fusion Approach; First Australian Data Fusion Symposium ADF-96, Adelaide, Australia, November 1996, pp. 48-52.

Robert Schneider, Josef Wenger
DaimlerChrysler AG
Research & Technology, REM/CR
Wilhelm-Runge-Str. 11
89081 Ulm
Germany
robert.r.schneider@daimlerchrysler.com

Keywords: automotive radar, environmental sensing, driver assistance, autonomous cruise control, obstacle warning, collision avoidance.

New Generation of Inertial Sensor Cluster for ESP- and Future Vehicle Stabilizing Systems in Automotive Applications

R. Willig, M. Mörbe, Robert Bosch GmbH

Abstract

In 1995 Robert Bosch GmbH (RB) started the mass production of the first VDC-System (Vehicle Dynamics Control system) for vehicles, today called ESP (Electronic Stability Program). This ESP-System went beyond ABS and Traction Control Systems and offered consumers unsurpassed driving confidence and safety. The key part of this system was a first generation Yaw Rate Sensor DRS 50/100, based on a metal vibrating cylinder. The second generation DRS MM1, introduced in 1998, based on silicon micromachining and included an integrated linear acceleration sensor element. For new additional functions of ESP and of future high dynamic and high performance vehicle stabilizing systems, like Hill Hold Control (HHC) or Steer by Wire (SbW) BOSCH develops the third generation, a flexible and cost-effective Inertial Sensor Cluster with a modular concept for hard- and software, called DRS MM 3.x. Design, basic functions, modular concept, safety features and system requirements of the new Inertial Sensor Cluster DRS MM 3.x are presented.

1 Introduction

In 1995 Robert Bosch GmbH (RB) started the mass production of the first VDC-System (Vehicle Dynamics Control system) for vehicles, today called ESP (Electronic Stability Program). The key part of this system was a first generation Yaw Rate Sensor DRS 50/100, based on a metal vibrating cylinder with piezo-electric transducers (figure 1) [12].

The second generation DRS MM1, introduced in 1998, based on a combination of silicon bulk and surface micromachining with a electromagnetic drive and capacitive detection and included an integrated linear acceleration sensor element for measuring the lateral acceleration of the vehicle. (figure 1) [13].

ESP is a safety system for road vehicles which controls the dynamic vehicle motion in emergency situations. Vehicle handling at the physical limit of adhesion between the tires and the road is extremely difficult. In such situations the driver may be supported by controlling the longitudinal and lateral forces on the tires. The ESP- system of Bosch does that by controlled braking of individual wheels which makes the vehicle motion approach the nominal motion intended by the driver. It uses signals to determine the driver's intent, like steering wheel angle, brake pressure, engine torque and other signals to derive the actual motion of the vehicle, like angular velocity of the car around it's vertical axis (Yaw Rate) and lateral acceleration. The support of the driver is not limited to coasting conditions. Also during full braking, partial braking, engine drag, free rolling and acceleration of the vehicle the system supports the driver in all safety critical situations, within the physical limits.

For new additional functions of ESP and of future high dynamic and high performance vehicle stabilizing systems or comfort systems, like Hill Hold Control (HHC), Navigation (Navi, Travel Pilot), Adaptive Cruise Control (ACC), Four Wheel Drive (4w), Roll over Mitigation (ROM), Electronic Active Steering (EAS), Roll Over Sensing (ROSE), Active Suspension Control (ASC) and Steer by Wire (SbW) BOSCH develops the third generation, a flexible and cost-effective Inertial Sensor Cluster called DRS MM 3.x.

Fig. 1. Yaw Rate Sensors, DRS 50/100 (1995) and DRS MM 1.x (1998).

These new system-functions require additional inertial measuring data of the dynamic behaviour of the vehicle like angular acceleration, longitudinal acceleration, tilt angle, acceleration of the z-axis, angular velocity of the x-axis and redundancy of the inertial data in the case of high safety relevant systems.

The demands for this new sensor cluster are higher robustness, more compactness, easier applicability, higher accuracy and better signal to noise ratio (SNR), insensitivity against external interference, high signal refresh rate, extendability for cross-system applications, higher system availibility and reliability and an integral safety concept. This leads to a modular concept for hard- and software of the sensor cluster.

Design, basic functions, modular concept, safety and monitoring concept and system requirements of the new Inertial sensor cluster DRS MM 3.x are presented.

2 Basic Function and Design of the Inertial Sensor Elements and Readout Electronics

2.1 Angular Velocity Sensor Measuring Element and Readout Electronic Module

The new designed sensor measuring element for measuring angular velocity or rate of rotation belongs to the type of the well known CVGs, Coriolis Vibratory Gyroscope [1-19].

It works like an inverse tuning fork configuration with two linear in-plane orthogonal oscillation modes, a so called primary or drive mode and a secondary or detection mode, where the electrostatically excited motion (v) of the primary mode (Xdrive) is a linear antiphase oscillation of two coupled spring-mass structures. The secondary in-plane mode (Ydetection) of the two coupled spring-mass structures is excited due to Coriolis inertial forces produced as a result of applying a rate of rotation (W) perpendicular to the plane. The oscillation of the drive mode is provided by electrostatic forces at comb drive electrodes as well as the sensing of the motion of this mode while the detecting of the Coriolis acceleration (ac) in the secondary mode is accomplished capacitively via interdigitated electrodes. (figure 2)

Coriolis acceleration: $-a_c = 2\,\Omega\,v$

with
a_c : Coriolis acceleration (m/s^2)
Ω : Angular velocity (yaw rate) (s^{-1})
v : Vibration velocity of drive mode (m/s).

To fullfil various demands the sensor measuring element is designed as two spring-mass structures, which are mechanically coupled. Both modes are oscillating in plane at one natural resonant frequency of the structures, nominal 15 kHz with constant amplitude. Silicon surface micromachining is used for a more compact size and cost reduction according to the improved Bosch foundry process (figure 2).

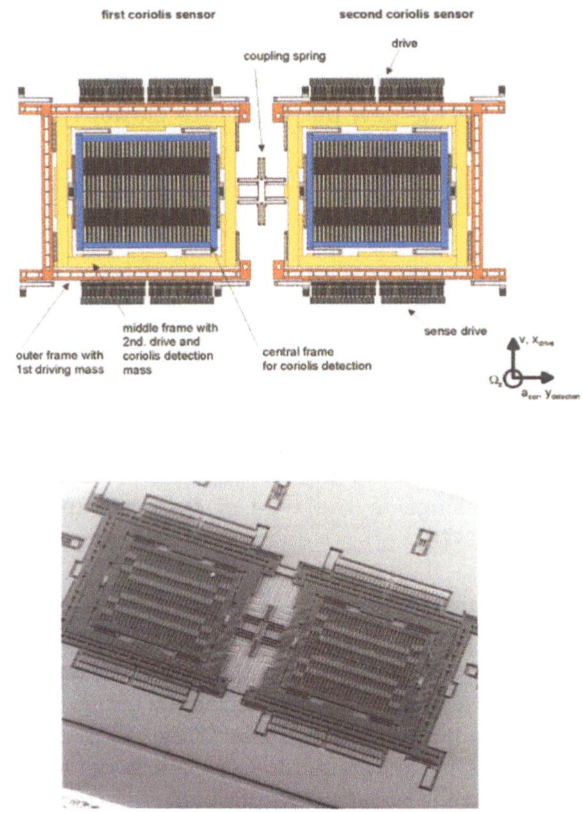

Fig. 2. Sensor measuring element, design (top) and SEM photo (bottom).

To be insensitive against external mechanical interferences and overloads in the vehicle like linear and angular vibrations a high natural resonant frequency of the micromachined structure is used, which is out of the range of power spectrum of the interference in the vehicles. To increase resolution, signal to noise ratio, linearity and bandwidth a high Q-factor, a double resonant oscillation (only one frequency of drive and detection mode) and a closed loop principle are neccessary and are implemented into the sensor readout electronic module (ASIC). A so-called multibit force rebalance control loop including an electromechanical $\Delta\Sigma$-modulator (high sampling rate and quantisation, noise shaping) will electrostatically drive the Coriolis force to zero. For a high Q-factor residual dynamic gas damping must be minimized due to an oscillation at low internal pressure. Mass and stiffness imperfections during manufacturing processes require an electrostatic stiffness tuning for the double resonant oscillation, a mechanical decoupling of the drive and detection mode of both spring-mass structures to suppress mechanical crosstalks and parasitic effects and an electronic out-of-phase (quadrature) rejection.

The new sensor readout electronic module (ASIC) contains a drive and detection mode loop with the following functional blocks: analog frontends, digital backend, programmable read only memory (PROM) and serial Peripheral Interface (SPI, figure 3).

The analog frontends are for fully differential C/V converting and common mode controlling with SC-technique (switch capacitor technique) and mainly for early digitizing the signals with a differential multibit flash-A/D converter. The drive loop consists of an analog frontend with the ADC, a digital bandpass filter, an AGC (automatic gain control), a lowpass filtering, a PI controlling and a DAC, a charge pump, a PLL (phase locked loop) with 8 and 40MHz clocks and a start up circuit. The functions of the drive loop guarantee a very fast open control loop for starting up the oscillation of the micro-mechanical sensor measuring element in x-direction and then the controlling of the oscillation at one resonant frequency with constant amplitude.

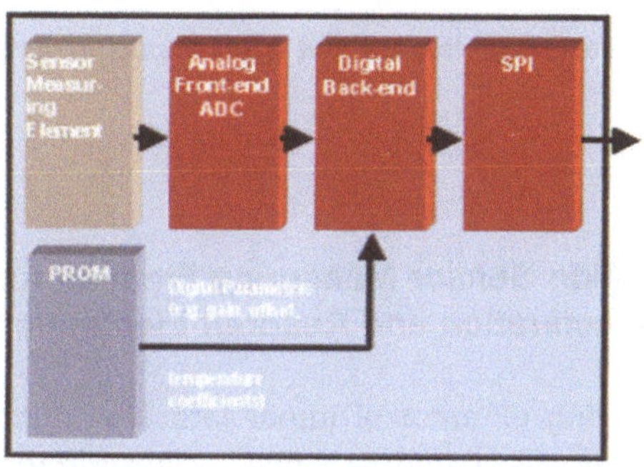

Fig. 3. Functional blocks of the sensor readout electronic module.

In order to measure the Coriolis force the electronics of the detection mode loop and the micro-mechanical sensor measuring element are combined to form the electromechanical $\Delta\Sigma$-modulator. The detection loop contains the same analog frontends as in the drive loop and a forced feed back loop. A synchronous demodulation and filtering, a frequency adjustment done by electrostatic stiffness tuning of the detection mode, quadrature control loop, output filtering, compensation of parameter shifts with temperature and calibration of offset and scale factor, a 40 bit wide PROM to program the coefficients for adjustments, compensations and calibrations, and a temperature sensor belonging to the digital backend. A SPI (Serial Peripheral Interface = synchronous serial data link with 2 MBaud) ensure the communication and data transfer between the ASIC and the sensor internal microcontroller with 8 or 40 bit frames. Self-testing during power on and self- monitoring on ASIC level are realized for 5V supply voltage, bond wires, range overflow of control parameters and a digital backend test.

The sensor measuring element and the ASIC are cased in a plastic surface mount package. This packaged device is fully calibrated, temperature compensated and tested (figure 4).

Fig. 4. Sensor measuring element and ASIC in plastic surface mount package.

2.2 Linear Acceleration Sensor Measuring Element for Lateral and Longitudinal Acceleration and Readout Electronic Module

For the sensor measuring element of linear lateral and longitudinal acceleration an in plane spring-mass structure is used according to the improved Bosch silicon surface microfabrication technology (figure 5).

The deflection of the spring-mass structure in the sensitive axis due to external acceleration forces is detected with a differential capacitive comb structure. The spring-mass structure is designed for a high mechanical g-range (70g) and together with the new sensor readout electronic module (ASIC) a variable low g measuring range (2g) with low noise (10mg rms) and high accuracy is possible. In this way external high mechanical interferences cause no clipping of the mechanical structure or of the signal processing.

The ASIC contains the same analog frontend like the ASIC of the sensor measuring element for angular velocity, with a fully differential charge to voltage conversion and a correlated double sampling (CDS) to achieve high dc stability. The fully digital signal processing includes a mechanical open loop electrical closed loop DS-modulation, filtering, offset adjust with temperature compensation, output filter and finally a scale factor calibration. A temperature sensor, 8 MHz oscillator and a 24 bit PROM are realized just as self-testing and -monitoring and data transfer via SPI.

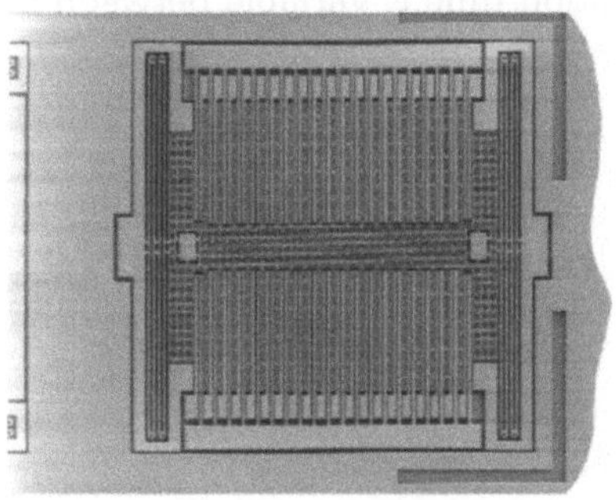

Fig. 5. SEM photo of the sensor measuring element.

The packaging type is the same as the package of the angular velocity sensor element. Also a two channel version is developed with two separate spring-mass structures on one chip with sensitive axes 90 degree to each other and one sensor readout electronic module with two output signals.

3 Design, Modular Concept and Safety Feature of Inertial Sensor Cluster DRS-MM 3.x

The packaged sensor measuring elements and ASICs, see figure 4, now called sensor elements are mounted on a PCB together with a microcontroller, EEPROM, voltage regulator, watch dog, quartz, etc.

The PCB is fixed in a plastic Sensor Cluster housing which is sealed by a protective cover. A four pin connector is used for external power supply (12V) of the Sensor Cluster and for data transfer via CAN. The Sensor Cluster can be installed in passenger compartment or trunk and the housing can be fixed with two holes for bolts (figure 6).

Microcontroller, sensor elements, etc. are supplied with controlled and monitored 5V power supplies. The internal data transfer and communication between the various and extendable sensor elements and the microcontroller are done by SPI, a synchronous bidirectional serial data link with 2 Mbaud, with four wires. Various data and status informations, like the measured inertial signals, control loop parameters and temperature signals, are transmitted via SPI to the microcontroller, checked by the microcontroller and transformed in a CAN-matrix and transmitted via CAN to the System-ECU after a request. The content of the CAN-identifier-matrix is also extendable and the synchronous CAN communication time is variable between 5 and 20ms (figure 7).

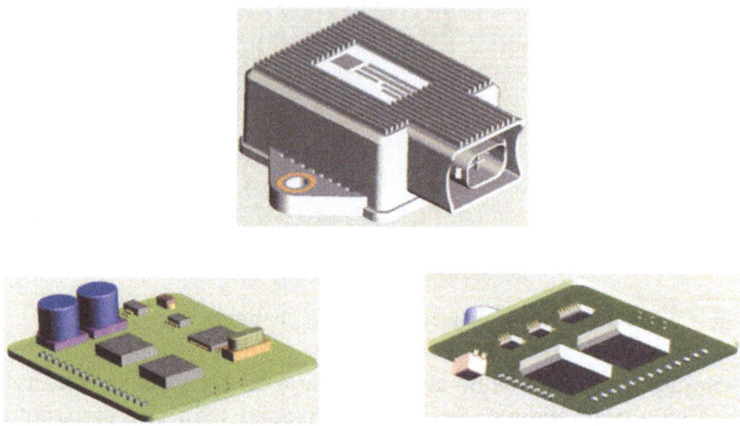

Fig. 6. Housing of the Sensor Cluster DRS MM 3.x and PCB with components.

Depending on system requirements various sensor elements can be mounted accordingly to the PCB, all SPI participants are addressed and the software (SW) of the microcontroller can check and handle all these different configurations. Due to this a modularity in HW and SW is cost-effectively possible and offers many different extendable versions of the Sensor Cluster DRS-MM 3.x with distinctive features in different inertial measuring data.

The Sensor Cluster internal safety and monitoring concept bases on

- Sensor element internal self-testing and -monitoring of control parameters at the drive mode (AGC, PLL), the detection mode, the digital backend and SPI- communication on ASIC level (sensor element and SPI failures) during power-on and normal mode.

- Furthermore a sensor element initialisation and signal monitoring on microcontroller level (sensor element and SPI failures) with sensor element ordinary inertial signals (Yaw rate or Lateral acceleration), additional signals and control loop parameters and temperature signals.

Microcontroller itself is testing and monitoring (µC and CAN failures) periodically CPU, CAN, SPI, Oscillator, ADC, watchdog, undervoltage and overvoltage (12V and 5V), RAM, ROM and EEPROM. The Sensor Cluster internal various status informations of the results of internal testing and monitoring are transmitted via CAN together with each inertial measuring data and the internal failure detection time is less 25ms.

An additional step for high safety and availibility of the system is a high CAN communcation time and a signal monitoring, filtering and plausibility checks on System-ECU level and a redundancy of sensor elements in the sensor cluster DRS-MM 3.x.

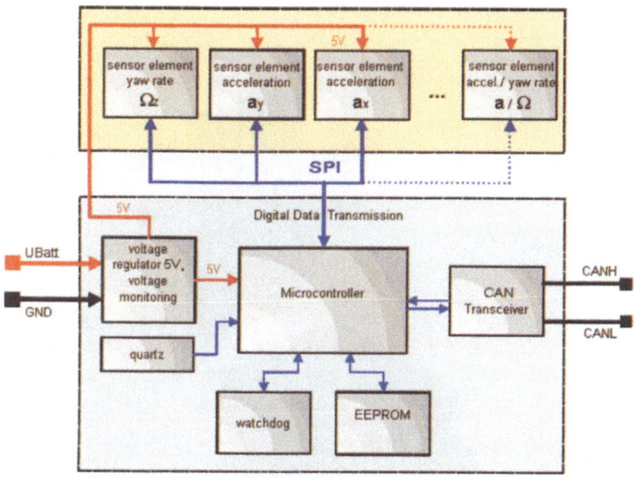

Fig. 7. Schematic circuit diagram of Sensor Cluster DRS MM 3.x.

4 System Requirements and Aspects

Different system applications with new additional functions require the simultaneous measurement of inertial signals, rates of rotation and linear accelerations, of the dynamical behaviour of the vehicle about the three orthogonal axes with different measuring ranges, bandwidth or accuracy.

Comfort systems like car navigation (Travel Pilot) and vehicle guidance (ACC), where Yaw Rate and linear longitudinal and lateral acceleration with a small signal bandwidth and high accuracy is required, or passive safety systems like restraint systems (Airbag) and Roll over sensing (ROSE) with Roll and Yaw Rate and low and high g-acceleration in all three axes or active safety systems like ESP (incl. ABS and TCS), HHC, ROM, EHB, ASC and EAS or future high safety relevant, high dynamic and high performance stabilizing systems like four wheel steering and steer by wire (SbW) with Yaw, Roll Rate and angular acceleration and low g-acceleration in redundancy and with high signal bandwidth, high accuracy and very low noise, needs often the same inertial signals but with different measuring ranges, signal quality, safety and dynamic demands.

The Inertial Sensor Cluster DRS MM 3.x is fit for these requirements of a growing diversity. The modular concept of HW and SW has furnished information for different systems and different equipment rate for vehicle platforms or models and is extendable in sensor measuring element equipment rate to enable add-on functionality and adaptation to new system architecture (figure 8).

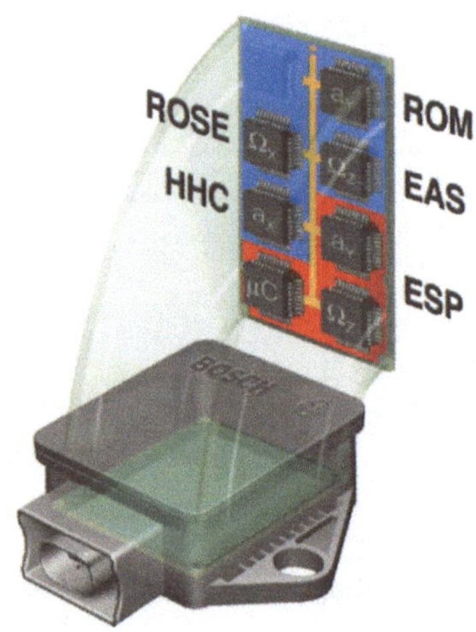

Fig. 8. Modular Concept of Sensor Cluster DRS MM 3.x.

5 Future Trends

The first step in the direction of a kinematic sensor platform for sensing the complete vehicle dynamics is presented with the Inertial Sensor Cluster DRS-MM 3.x. Future complex cross-system applications will require an intelligent sensor platform (ISP), which is sensing all inertial values of the three main axes internally, receiving external sensor signals like wheel speeds and forces, steering angle, engine torque, braking pressure, tire pressure, actuator states and informations of the driver assistance systems (radar, video, GPS) and then calculating or estimating the dynamic values like side slip angle, speed over ground in longitudinal and lateral direction, yaw rate, road inclination, road uphill gradient, etc. and controller values. These applications will need high data volume, high update rates and safe data transfer, why time triggered data networks like TT-CAN, TTP or FlexRay will be required [20]. For applications using only ESP without additional system functions the sensor elements of Yaw Rate and Acceleration can be integrated into the Electronic Control Unit of the ESP Hydraulic Unit.

6 Summary

For new additional functions of ESP and for future high dynamic and high performance vehicle stabilizing systems, like Hill Hold Control (HHC) or Steer by Wire (SbW) BOSCH develops the third generation, a flexible and cost-effective Inertial Sensor Cluster called DRS MM 3.x. These new system-functions require additional inertial measuring data about various axes of the dynamic behaviour of the vehicle.

The requirements for this new Sensor Cluster are higher robustness, more compactness, easier applicability, higher accuracy and better SNR, insensitivity against external interference, high signal refresh rate, extendability for cross-system applications, higher system availibility and reliability and an integral safety concept. This leads to a modular concept for hard- and software of the Sensor Cluster.

For all inertial sensor measuring elements silicon surface micromachining is used. To be insensitive against external mechanical interferences and overloads in the vehicle like linear and angular vibrations the micromachined structures are carefully designed. To increase resolution, signal to noise ratio, linearity and to suppress mechanical crosstalks and parasitic effects a fully digital signal processing with SC-, CDS-, and $\Delta\Sigma$-technique in closed or open loop control is implemented into ASICs.

The communication between the sensor elements and the internal microcontroller takes place via a SPI-interface. Internal logical links, self testing and self monitoring, CAN-interface between the Sensor Cluster and the System-ECUs and the possibility of internal redundancy for high safety-relevant, high dynamic and performance systems meet the future safety relevant requirements. Design, basic functions, modular concept, safety and monitoring concept and system requirements of the new Inertial Sensor Cluster DRS MM 3.x are presented.

7 Acknowledgments

The authors would like to thank all their colleagues of Corporate Research and Advanced Engineering and of Automotive Equipment, Division Automotive Electronics and Division Chassis Systems for the design of the complete sensor system, design and layout of the sensor measuring elements and sensor readout electronic modules.

References

[1] M.W.Putty, K.Najfi, "A Micromachined Vibrating Ring Gyroscope", Solid-State Sensor and Actuator Workshop, June 13-16,1994.

[2] J. D. Johnson, S. Z. Zarabadi, D. R. Sparks, "Surface Micromachined Angular Rate Sensor", SAE Technical Paper Series, 950538.

[3] J. Bernstein S.Cho, A. T. King, A. Kourepenis, P. Maciel, M. Weinberg, "A Micromachined Comb-Drive Tuning Fork Rate Gyroscope",0-7803-0957-2/93, 1993 IEEE.

[4] K. Funk, A.Schilp, M. Offenberg, „Surface-micromachining of Resonant Silicon Structures", Transducers `95, 519-News, page 50.

[5] M. Hashimoto, C. Cabuz, K. Minami, M. Esashi, "Silicon Resonant Angular Rate Sensor Using Electromagnetic Excitation and Capacitive Detection", Technical Digest of the 12th Sensor Symposium, 1994.

[6] K. Tanaka, Y. Mochida, M. Sugimoto, K. Moriya, T. Hasegawa, K. Atsuchi, K. Ohwada, " A Micromachined Vibrating Gyroscope", Sensors and Actuators A 50 (1995).

[7] Y. Cho, B. M. Kwak, A. P. Pisano, R. Howe, "Slide film damping in laterally driven microstructures", Sensors and Actuators A, 40 (1994).

[8] M. Offenberg, F.Lärmer, B.Elsner, H.Münzel, W. Riethmüller, "Novel Process for a Monolithic Integrated Accelerometer", Transducers 95, 148 - C4.

[9] K. H.-L. Chau, S.R. Lewis, Y. Zhao, R. T. Howe, S. F. Bart, R. G. Marcheselli, "An integrated Force-Balanced Capacitive Accelerometer for Low-G Applications", Transducers 95, 149 - C4.

[10] M. Offenberg, B. Elsner, F. Lärmer, Electrochem. Soc. Fall-Meeting 1994, Ext. Abstr. No 671.

[11] M. Offenberg, H. Münzel, D. Schubert, " Acceleration Sensor in Surface Micromachining for Airbag Applications with High Signal/Noise Ratio, SAE Technical Paper, 960758.

[12] A. Reppich, R. Willig, " Yaw Rate Sensor for Vehicle Dynamics Control Systems", SAE Technical Paper 950537 (1995)

[13] M. Lutz, W. Golderer, J. Gerstenmeier, J. Marek, B. Maihöfer, D. Schubert, " A Precision Yaw Rate Sensor in Silicon Micromachining ", SAE Technical Paper 980267 (1998)

[14] A.W. Leissa: Vibration of shells, NASA SP-288 (1973)

[15] G.B. Warburton: Vibration of thin cylindrical shells,Journal Mechanical Engineering Science, Vol 7 No 4 (1965)

[16] C.H.J. Fox and D.J.W. Hardie: Harmonic response of rotating cylindrical shells, Journal of Sound and Vibration, 101 (4) (1985)

[17] J.S. Burdess: The dynamics of a thin piezoelectric cylinder gyroscope, Proc. I. Mech. E.,Vol. 200 No c4 (1986)

[18] P.W. Loveday, " Analysis and Compensation of Imperfection Effects in Piezoelectric Vibratory Gyroscopes ", Diss. Virginia Poytechnic Institute and State University, Blacksburg, Virginia, 1999.

[19] J. A. Geen, "A Path to Low Cost Gyroscopy ", Solid-State Sensor and Actuator Workshop,June 8-11, 1998 .

[20] D. Sparks et al., "Multi-sensor modules with data bus communication capability", SAE paper 1999-01-1277, 1999.

Rainer Willig, Matthias Mörbe
Robert Bosch GmbH
Automotive Equipment, Division Chassis Systems, Product Group Sensors
P.O, Box 300240,
70442 Stuttgart
Germany
rainer.willig@de.bosch.com;
matthias.moerbe@de.bosch.com

Keywords: inertial sensor cluster, yaw rate sensor, acceleration sensor, silicon surface micromachining, modular concept

A New Micromachined Gyroscope with Digital Output

M. Lang, M. Großmann, O. Prütz, M. Koster, J. Franz, G. Wucher, T. Mayer, A. Gangei, E. Steiger, Robert Bosch GmbH

Abstract

In 2001 volume production of the first MEMS-based angular rate sensor for roll-over sensing and navigation control started [3, 8]. In this paper the next generation of a micromachined gyroscope is presented. The new sensor again uses an oscillating disc which is deflected by a coriolis force in case of an angular rate applied. This deflection is measured capacitively by electrodes underneath the disc. The resonance frequency has been increased from 1.5kHz to 5kHz compared with the previous sensor generation. The sensor is packaged in an PLCC44 together with a fully digital signal processing unit. In order to facilitate a broad application the sensor is equipped with an digital output (SPI) as well as with an additional analog output.

1 Introduction

The sensor technology in automotive applications has undergone a tremendous development in the recent years. Each car is equipped with several sensors for different purposes: pressure- and airflow-sensors for motor-management, acceleration sensors for airbag control, angular rate sensors for vehicle dynamic control as well as for roll-over detection and navigation control. Together with new concepts and applications not mentioned here, a car without sensors could not be imagined anymore [1-4, 7]. And the added value due to sensors and electronics in a car still increases year by year.

In spring 2001, Bosch started the series production of the worldwide first angular rate sensor SMG040 in surface micromachining technology for Rollover sSnsing (RoSe) and navigation control. Since then more than half a million cars from different manufacturers have been equipped with this safety feature and the production volume is increasing rapidly. Future versions of rollover sensing (RoSe II) combine roll-over signals from a ECU with signals delivered from the system for vehicle dynamic control in order to discover dangerous driving conditions as early as possible. This leads to an earlier reaction for a better occupant safety. RoSe II makes higher demands on different parameters of an angular rate sensor, e.g. noise, temperature stability, mechanical robustness. The next generation of our angular-rate sensor has been developed to fulfil the requirements of RoSe II and other advanced applications. The results are presented in this paper.

2 Micromechanical Sensor

2.1 Functional Principle

As the current sensor SMG040, the new sensor consists of an oscillating disc, which is driven by comb drives in an in-plane movement [3]. Upon applying an angular rate across the in-plane axis of the sensor it reacts with a tilt movement due to the conservation law for angular momentum (see figure 1). This tilt movement can be detected by underlying electrodes on the substrate which form a differential capacitor with the oscillating rotor. Due to the tilt movement, the distance to the substrate on one side of the disc increases whereas it decreases on the other side. Therefore a differential signal can be detected by evaluating the capacitances (figure 2).

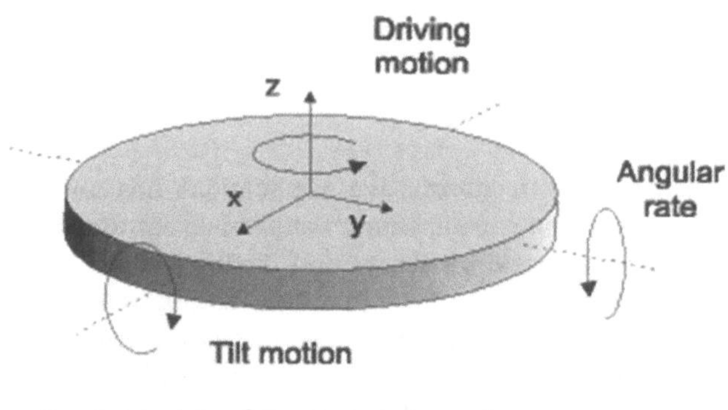

Fig. 1. Functional principle of the sensor element.

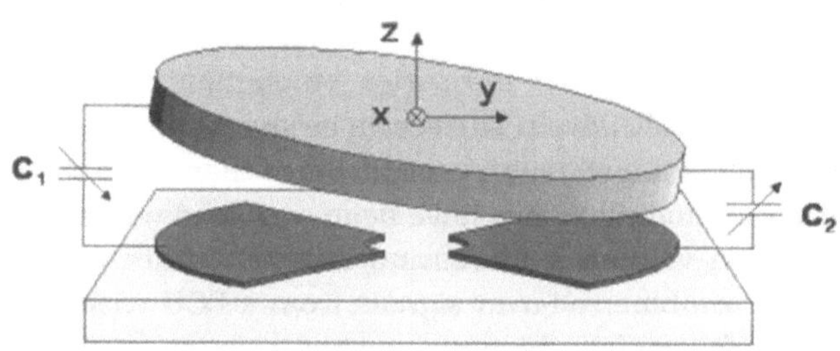

Fig. 2. Tilt movement and change of capacitance.

2.2 Theory

As mentioned before the sensor uses the conservation law for angular momentums, which can be described by the following differential equations (where J_i denotes the moment of inertia in each direction, Ω_Y the applied angular rate, a the tilt angle of the detection, β the tilt angle of the perpendicular direction and g the in-plane oscillation angle):

$$(J_x + J_y - J_z) \cdot \Omega_y \cdot \dot{\gamma} + (J_z - J_y) \cdot \dot{\beta} \cdot \dot{\gamma} + J_x \cdot \ddot{\alpha} + d_x \cdot \dot{\alpha} + k_x \cdot \alpha = 0$$

$$(J_x - J_z) \cdot \dot{\alpha} \cdot \dot{\gamma} + J_y \cdot \ddot{\beta} + d_y \cdot \dot{\beta} + k_y \cdot \beta = 0$$

The simplified solution for the tilt angle (detection) is:

$$\alpha = \frac{2}{\omega_z} \frac{\gamma_0 \Omega_y}{\sqrt{\left(\frac{\omega_x^2}{\omega_z^2} - 1\right)^2 + \frac{\omega_x^2}{Q_x^2 \omega_z^2}}}$$

where ω_z is the in plane resonance frequency and ω_x the out-of-plane resonance for the sensing mode. As can be seen, besides the angular rate, the tilt angle is directly proportional to the in-plane driving angle γ. Therefore the amplitude of this movement has to be kept constant in order to deliver constant sensitivities. This is achieved by an automatic gain control, which will be described below. The input signal for this AGC is measured by additional comb structures which are formed like the comb drives. These comb structures evaluate the change of the in-plane capacitance which is proportional to the in-plane driving angle. For further details on the theoretical consideration please refer to [5].

2.3 Technical Features and Technology

The sensor is fabricated in the Bosch surface micromachining technology [2, 6]. In order to protect the sensor itself and to encapsulate a vacuum, a cap is sealed on the sensor structure on wafer level. A pressure well below 3mbar is achieved in this process. This low pressure is needed for generating a sufficient oscillation amplitude of the disc.

In order to increase the mechanical robustness and facilitate application the in-plane resonance frequency has been increased up to 5kHz (SMG040: 1.5kHz).

As this increase reduces sensitivity (see eq. (3)) by a factor of 3.3, the relation of out-of-plane to in-plane resonance frequency (ω_x/ω_y) as well as the distance between rotor and substrate (capacitance) have been optimised to compensate this reduction. A SEM-picture of the sensor can be seen in figure 3.

Fig. 3. SEM-picture of the sensor element.

3 Electronic Circuit

For the electronic signal processing a completely new ASIC has been designed. It is fabricated in a High-Voltage CMOS process and is based on a fully digital concept. The signal processing unit consists of two separate control loops, one for the driving stage and one for the detection path. In the driving loop, the change in capacitance of the in plane comb structures is measured and used for an automatic gain control in order to keep the oscillating amplitude constant over temperature and lifetime.

The detection path in the front-end consists of the same structure: directly after a C/V-converter an ADC is placed for generating the digital signals. After demodulation of the high frequency carrier, the digital signal is amplified in different stages. At the final test, a variable amplifier can be use to calibrate the overall sensitivity. As the whole concept is based on a digital signal processing it is at least consequent to deliver the output data as digital words. This is realised with an 10 bit SPI interface. As an additional feature, an DAC has been added to provide an analog output, too. A block diagram of the circuit can be found in figure 4.

Besides the general advantages of a digital output – direct data processing through a microprocessor, precise data acquisition, less disturbances – several monitor functions have been implemented, which can be read out through the

SPI interface. The correct function of internal controllers as well as correct sensor performance is monitored constantly and can be checked via SPI. In an analog application with separate output, an additional pin is used to monitor the same functionality.

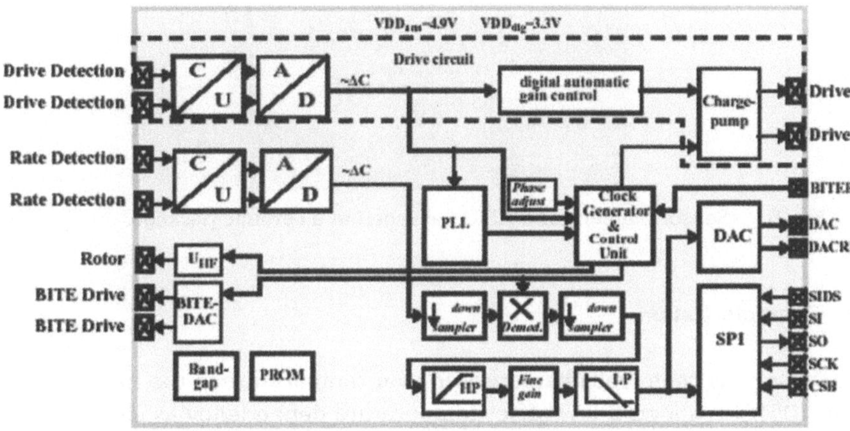

Fig. 4. Block diagram of the electronic circuit.

4 Sensor Module

4.1 Packaging

The micromechanical sensor element together with the signal processing unit is packaged in a two chip PLCC44. In order to reduce stress influences a low stress glue as well as a dedicated leadframe is used. The sensor and the ASIC are connected via chip-chip-bonds. This technique is well known from the former sensor and other micromechnical acceleration sensors fabricated by Bosch. Figure 5 shows a sensor module packaged in ceramics for test purposes.

Fig. 5. Sensor element and ASIC packaged in a ceramic package.

4.2 Characterization

The sensor has been well characterised over temperature in the range of -40°C up to 105°C. It shows a low offset temperature dependency as can be seen in figure 6, which represents a typical behaviour. The full scale signal is defined by ±480 LSB (2 LSB/°/s * ±240 °/s for RoSe and 6 LSB/°/s * ±80°/s). Because of its small offset, an offset calibration during final testing is not necessary.

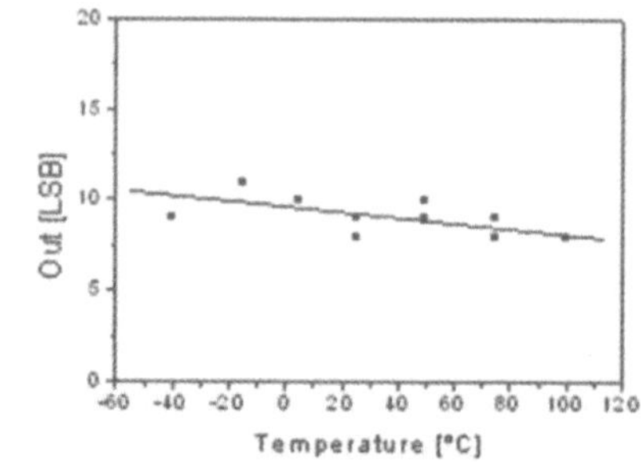

Fig. 6. Offset behaviour without offset cancellation.

In figure 7, there can be seen a typical temperature behaviour of the sensitivity.

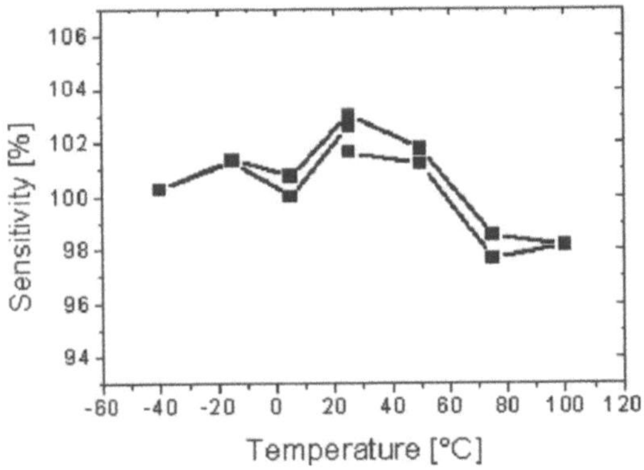

Fig. 7. Sensitivity behaviour.

4.3 Technical Data

Besides EMC capacitors no external components are necessary for the digital application. If the analog output is used, a passive R-C-filter is recommended for smoothing the output signal. As the former sensor, the new device is equipped with a comprehensive through-the-loop self-test functionality. This Built-In-Test can be triggered externally by the application. The BITE-signal causes the sensor to deflect out-of plane due to an applied electrostatic force and in return delivers a response through the whole signal path of the ASIC. Therefore the mobility of the sensor as well as the signal path is tested. The system has been designed so that the BITE can be triggered continuously. A use during the start-up phase is also possible.

The technical data of the sensor are listed below.

Parameter	*Value*
Power supply	V_{DDA} = 5V ± 0,25V, V_{DDD} = 3,3V ± 0,3V
Working temperature	-40°C .... + 105°C
Output	Digital (SPI, 10 bit), analog
Measurement range	± 240 °/s (ROSE), 80°/s (Navigation)
Sensitivity	2 LSB/°/s (ROSE), 6 LSB/°/s (Navigation)
Sensitivity tolerance (temperature, lifetime)	± 7 %
Noise	< 2 °/s rms (RoSe), < 0,5 °/s (Navi)
Self-test	60 % full-scale (RoSe)
Output-bandwidth	30 Hz (ROSE), 10 Hz (Navigation)
Slow offset cancellation	1/8 LSB/s
Start-up time	< 1s

Tab. 1. Technical data of the new sensor.

5 Application

Rotation about the driving axis of a vehicle (roll rate) is the main parameter used in rollover detection systems. Systems of the second generation like RoSe II include advanced functionality for instance the early detection of soiltrip-caused rollover or nearly-rollover events.

As much as the functionality is enhanced the requirements on an angular rate sensor applied for roll rate measurement increase. On the one hand resolution must be advanced in order to allow the detection of even small angular rates which occur in early soil trip stages. On the other hand robustness against interferences of mechanical an electrical type as well must increase for the same reason. Furthermore the demands on EMC of automotive products are rising continuously.

The new gyroscope is designed in a way to satisfy these manifold requirements:

- ▶ The oscillation frequency of the rotor is set on 5kHz. At the same time this frequency represents the lowest frequency with which the sensor may be interfered noticeably by mechanical vibrations. From the view of automotive application 5kHz is high enough to mount the MM2R on a standard PCB in a standard ECU without need for additional mechanical decoupling or damping.

- The complete sensor circuit is highly integrated in an CMOS-ASIC. Most functional blocks are realized as digital circuits. This enables high robustness against interference in the sense of EMC.
- The interface is realized as a standard SPI. All information transmitted to and from the sensor are digital. This means significantly increased data transmission robustness and monitoring depth compared to analog interfaces.
- Resolution of the angular rate signal is 10 bit.
- The sensor has an enhanced continuous monitoring functionality. 16 different on-chip failure modes are discriminated an can be read out via SPI. This allows a deep sensor monitoring and an adequate fail-safe application.
- Standard SMD-packaging (PLCC44) allows easy processing during PCB-mounting.

All in all the new sensor is designed to enable a robust and safe application of rollover detection systems of the next generation.

6 Conclusion

In this paper we presented a new generation of gyros in silicon surface micromachining. The sensor is designed for rollover detection and navigation control. It can be used with a digital interface with SPI protocol as well as with an analog output. It is equipped with several safety features to monitor a correct function permanently or in the start-up sequence. As the sensor is packaged in a standard PLCC44 housing it can be easily mounted on PCBs. Start of production of this new sensor will be end of 2003.

7 Acknowledgement

The authors would like to thank all the helping hands which support us in this new sensor, especially the technology team U. Bischof, T. Ritzi, R. Hausner, H. Höfer.

References

[1] M.Lutz, W. Golderer, J. Gerstenmeier, J. Marek, B.Maihöfer, S. Mahler, H. Münzel, U. Bischof: A Precision Yaw Rate Sensor in Silicon Surface Micromachining, Proceeding of Transducers '97.

[2] M. Offenberg, H. Münzel, D. Schubert, O. Schatz, F. Lärmer, E. Müller, B. Maihöfer, J. Marek: Acceleration Sensor in Surface Micromachining for Airbag Applications with High Signal/Noise Ratio, Robert Bosch GmbH, SAE960758, 1996.

[3] R. Schellin, A. Thomae; M. Lang, W. Bauer, J. Mohaupt, G. Bischopink, L. Tanten, H. Baumann, H. Emmerich, S. Pinter, J. Marek, G. Lorenz, R. Neul: A low cost Angular Rate Sensor for Automotive Applications in Surface Micromachining Technology, AMAA'99, pp.239-250

[4] J. Marek: Silicon Microsystems for Automotive Applications, Proceedings of Eurosensors XIII '99, pp. 1-2

[5] D. Teergarden, G. Lorenz, R. Neul: How to model and simulate microgyroscope systems, IEEE Spectrum, pp.66-75, July 1998.

[6] M. Offenberg, F. Lärmer, B. Elsner, H. Münzel, W. Riethmüller: Novel Process for a Monolithical Integrated Accelerometer, Transducers '95, Eurosensors IX, 148-C4, pp. 589-592.

[7] C. Song, M. Shinn: Commercial vision of silicon-based inertial sensors, Sensors and Actuators A66 (1998), pp. 231-236

[8] A. Thomae, R. Schellin, M. Lang, et al.: A Low Cost Angular Rate Sensor in Si-Surface Micromachining Technology for Automotive Application, SAE'99

M. Lang, M. Großmann, O. Prütz, M. Koster, J. Franz, G. Wucher, T. Mayer, A. Gangei
Robert Bosch GmbH
Department AE/SPP, /STZ, /EIS, /ERP
Tübinger Str. 123
72703 Reutlingen
Germany

E. Steiger
Herrenwiesenweg 24
71701 Schwieberdingen
Germany
Markus.Lang3@de.bosch.com

Keywords: gyroscope, angular rate sensor, digital output, rollover

Infrared Microbolometer Sensors and Their Application in Automotive Safety

JJ. Yon, E. Mottin, CEA
L. Biancardini, L. Letellier, CEA
JL. Tissot, ULIS

Abstract

Recently the emergence of a new generation of infrared sensors – the microbolometer technology – based on an infrared thermal detection mechanism which is particularly suited to operate at ambient temperature has opened the opportunity for achieving low cost infrared imaging systems for both military and commercial applications. In a first part, this paper gives an overview of this challenging technology highlighting the main characteristics of the sensors developed by LETI that are particularly relevant to automotive applications. A special highlight on recent results concerning the 160x120 focal plane array with a pixel pitch of 35µm is given. In a second part, the use of this technology in automotive safety field is illustrated through an application of detection of moving objects in front of a vehicle. The results shows that infrared sensors based on well-designed microbolometers represent a real middle-term alternative to usual video sensors.

1 Background

The automotive industry increasingly looks to microsystems to put intelligence into cars. Safety improvement is particularly concerned with this trend: acceleration sensors for airbags, tire pressure monitoring and collision avoidance radar system. However, despite all of the automotive safety breakthroughs of this last decade, drivers still face potential hazards during conditions of darkness or obscured visibility such as is present with fog, heavy rain or snow. A challenging concern for the next few years is to improve vehicle safety in such adverse conditions with the operation of front-hazard warning devices and reliable collision avoidance systems.

One of the major issues of such safety systems largely deals with the availability of adequate sensors that allow an early and reliable detection of road obstacles in front of the car. Infrared thermal imaging is particularly suited for this purpose as it provides an effective night-time viewing system that could tackle the inefficiency of the usual sensors and fulfils the night driving safety requirements. Indeed, thermal imaging systems detect the electromagnetic radiation emitted by any object at room temperature whatever its natural or artificial illumination. As a result, infrared sensor intrinsically offers large advantages in comparison to alternative sensors working in the visible spectrum or in the millimetric wavelength range such as radars do. This statement is clearly illustrated considering that a visible vision camera exhibits poor efficiency in bad weather conditions even if it is coupled with automobile headlights illumination. Moreover the range of the road that can be covered by headlights at night is much less than the eye can see during daylight. Unlike visible vision, Infrared vision enhances the range of visibility at night up to six times further than standard headlights. On the other hand, radar systems typically have poor resolution because of their long wavelength. Consequently radar gives limited information regarding the shape of the detected object in comparison to infrared imaging.

For various technological and financial reasons, infrared imaging has been primarily developed for military applications. Such systems were originally based on quantum devices that typically operate at liquid nitrogen temperature [1]. This low temperature requirement leads to high cost systems and has dramatically restricted the use of thermal imaging. But recently the emergence of a new generation of sensors – the microbolometer technology – based on an infrared thermal detection mechanism which is particularly suited to operate at ambient temperature has opened the opportunity for achieving low cost infrared imaging systems for both military and commercial applications [2].

In this context, CEA/LETI has been involved in amorphous silicon uncooled microbolometer development since 1992 [3]. This high performance infrared technology is now commercialized in mass production by the French company ULIS and it will rapidly meet the market ramping up demands like car safety applications. In order to prepare the next infrared launch into automotive industry, CEA/LETI is involved in two European projects that aim at improving automotive safety. In EURIMUS framework, a project named ICAR is under progress to develop a specific camera for an affordable Driver Vision Enhancement (DVE) systems [4]. Besides, the SAVE-U project, partially funded by the European Commission INFSO DG under IST program aims to develop an enhanced vulnerable road users (VRU) detection system based on several detectors: a 24 GHz radar network coupled with a vision part composed of both visible and infrared imaging sensors [5].

In a first part, this paper gives an overview of microbolometer technology highlighting the main characteristics of these sensors that are particularly relevant to automotive applications. Then the paper will focus on recent results obtained from a 160x120 microbolometer infrared focal plane array (IRFPA) with a pixel pitch of 35µm that has been specifically designed for automotive Driver Vision Enhancement in the scope of ICAR project. In a second part, the use of this technology in automotive safety field is illustrated through an application of detection of moving objects in front of a vehicle highlighting the potential of this technology for pedestrian detection in the context of the SAVE-U project.

2 Microbolometer development at CEA/LETI

2.1 Thermal Detector Structure

The schematic structure of an uncooled thermal detector is shown in figure 1. As a general rule, these detectors measure the temperature rise due to IR radiation absorption by a thermally insulated element. For this purpose, thermal detectors are mainly composed of an infrared absorber embedded in closed contact with a thermometer element. The thermometer element senses incoming IR induced temperature rise and converts it into an electric signal. The most common detection mechanism is the resistive bolometer whose resistance changes with temperature, but various other mechanisms can be used, such as pyroelectric effect [6, 7], thermoelectric junction [8], P-N junction conductivity [9] or thermal stress induced mechanical deflection [10].

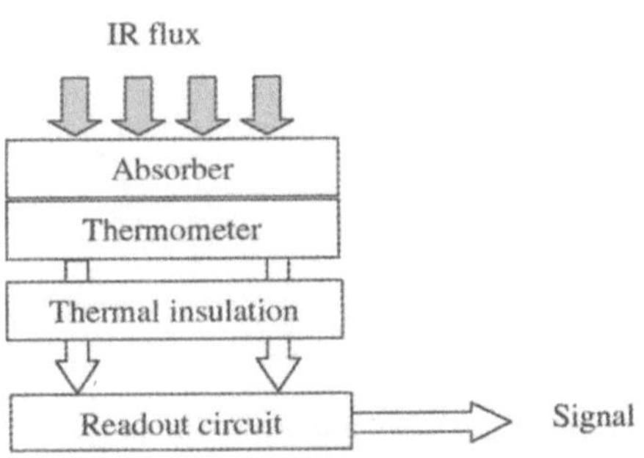

Fig. 1. Schematic structure of thermal detector.

Considering a two dimensional array of detectors, a readout integrated circuit (ROIC) is generally designed to measure the resistance of each bolometer and to format the results into a single data stream for video imaging purpose.

Finally, due to the strong correlation between thermal insulation and sensitivity, the high performance uncooled IR detector must be operated under vacuum – typically 10^{-2} Torr – in a specific package supplied with an infrared window.

2.2 CEA/LETI Technology

In the footsteps of MEMS devices, microbolometer sensors have taken benefits from the latest silicon technology advances. Unique surface micromachining techniques have been developed at CEA/LETI in order to produce above the full custom CMOS readout circuit, very thin membranes made from amorphous silicon that are very sensitive to infrared incoming radiations heating. Figure 2 shows schematically the structure of such a pixel whereas figure 3 describes its manufacturing process flow.

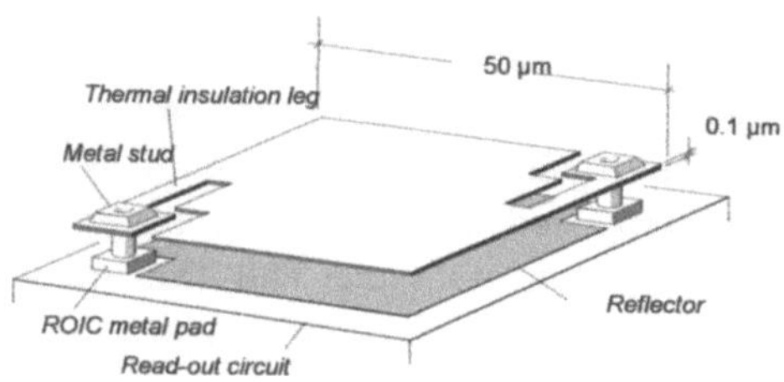

Fig. 2. Schematic of microbolometer pixel.

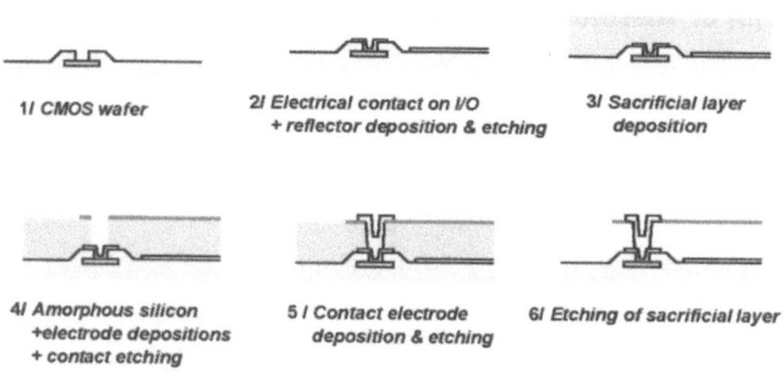

Fig. 3. Process flow of microbolometer technology.

In a first step a thin aluminium reflective layer is deposited and delineated directly on top of the ROIC. A 2.5µm thick polyimide sacrificial layer is then spun and cured. An amorphous silicon film 0.1 µm thick is deposited over the polyimide layer and covered by a metallic electrode obtained by reactive physical vapor deposition. Vias are opened by dry etching throughout the structure down to the ROIC pads, and metal deposition and etching achieves electrical continuity between the underlying substrate and active bolometric structures at the surface of polyimide. At this point electrode delineation is done by wet etching of the metallic film selectively over the amorphous silicon. The pixel contour is delineated and dry etched to the polyimide, and a final local polyimide etch over testing pads is carried out. At this stage the wafers are tested for standard automatic electrical functionality and acquisition of array parameters. Finally the microbridge arrays are created by polyimide removal in conventional resist etching equipment. Figure 4 shows scanning electron microscopy pictures of a pixel and the detail of the metallic stud that interconnect the microbolometer detector to the ROIC through the thermal insulation leg.

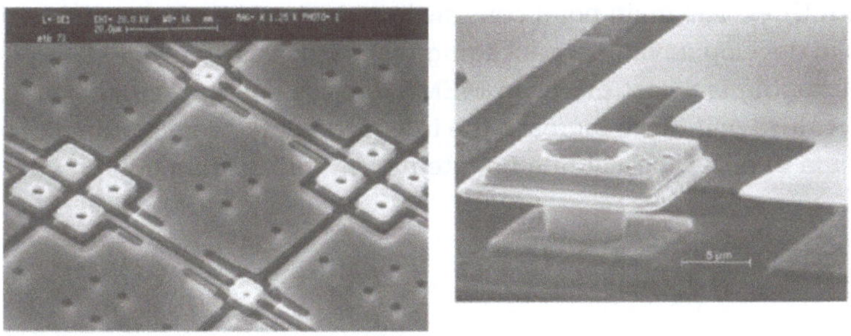

Fig. 4. SEM views of pixels and their electrical interconnection.

On this technological baseline, a first generation of bolometer technology focused on 45µm pitch was developed and transferred to ULIS in early 2000 [11]. ULIS is currently manufacturing and commercialising two different products (figure 5) based on a 320x240 focal plane array. These are both packaged under vacuum in a metallic package. The UL01 01 1 device is a general purpose imaging uncooled infrared array whereas UL01 02 1 E device, as it is supplied with an extra internal thermal shield, is more suited for radiometric applications.

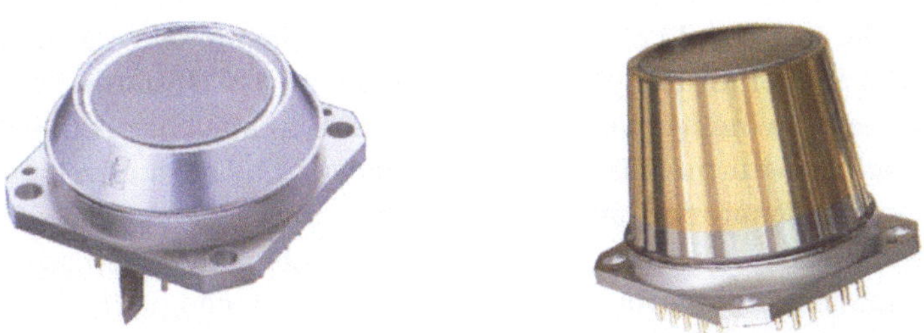

Fig. 5. ULIS uncooled staring arrays UL 01 01 1 (left), UL 01 02 1E (right).

2.3 Cost Reduction Studies

The requirements of automotive application like the Driver Vision Enhancement system is mainly constrained by objective cost of the overall system. CEA/LETI and ULIS technology is particularly designed to meet these requirements. In fact, one of the key point of CEA/LETI and ULIS microbolometer technology has been to elect a thermometer material made from amorphous silicon that features absolute compatibility with standard silicon processing. This basic option leads to a high yield monolithic arrangement fully compatible with commercially available CMOS silicon wafers. This feature intrinsically guaranties low cost attainment ideally suited for large market distribution. Nevertheless, to extend this low cost high volume approach even more some further developments are under progress at CEA/LETI in partnership with ULIS. The main point consists in reducing the pixel size. Another key point is to develop advanced packaging techniques as it is well stated that vacuum packaging is a cost driver in MEMS devices and particularly in uncooled IRFPA. A third point is to increase the integration of advanced functions on the focal plane in order to facilitate its integration into system equipments.

2.4 Pitch Reduction Studies

Cost reduction has multifold benefits from this pitch reduction approach. Obviously, owing to the increase of the number of dies per wafer, this will reduce the cost of the bolometer array itself. Besides, given a processing defects density, the manufacturing yield is directly linked with the focal plane array size. Furthermore, this size reduction will impact on both the dimension of the bolometer packaging and the form factor of the camera and consequently their cost. Finally, we can expect a dramatic drop of volume, weight and cost of the infrared optics as the diameters of the lens are directly linked with the size of the pixel for a given field of view and optical aperture.

To maintain a high level of performance despite the decrease of the pixel size, CEA/LETI in partnership with ULIS have engaged in deep technological developments for the last couple of years. These developments aims at increasing the thermal insulation of the pixel and at reducing the 1/f noise. In order to address this issue, an innovative second generation technological embodiment, totally compatible with the ULIS industrial process, has been developed. This so called second generation amorphous silicon microbolometer technology exhibits dramatically enhanced sensitivity and enables the decrease of pixel pitch to 35µm, keeping a level of performance entirely compatible with automotive night vision requirements.

This second generation technology is now completely matured and will be transferred from CEA/LETI research line to ULIS production line in 2003.

It is well stated that thermal insulation is the most critical parameter defining the performance of uncooled detectors. In order to improve this point, the microbolometer designer can rely on two different options.

A first option consists in increasing the length of thermal insulating legs that sustain the microbolometer above the CMOS substrate. The main drawback of this approach is the damaging decrease of the fill factor of the detector induced by the increase of the area devoted to the insulation legs implementation. A two level pixel arrangement has been proposed to tackle this issue but at the cost of a more complex and costly manufacturing process [12].

A second option followed at CEA/LETI is to achieve an enhanced thermal insulation by an advantageous reduction of the section of the thermal insulating legs. According to this option, the length of the legs are kept pretty the same, resulting in both high thermal insulation and high fill factor high absorption features. CEA/LETI – ULIS second generation microbolometer technology relies on this approach. Typical characteristics of this second generation technology for a 35µm pixel pitch are summarized in table 1, whereas typical microbolometer absorption spectra obtained from reflection experiments using an infrared integrating sphere are disclosed in figure 6. We can notice from these experimental data that for a given pixel pitch of 35µm:

- Thermal insulation (R_{th}) has been increased by a factor greater than 3.
- Similarly, the NETD figure has been improved by a factor of 5 due to the previous Rth enhancement and to an extra 60% 1/f noise reduction resulting from the optimization of the detector design (architecture improvement of the pixel as well as technological design rules shrink).
- As far as the time constant (T_{th}) of the 1st generation technology exhibits a tremendous margin regarding usual video frame rate, it has been possible to increase R_{th} keeping a fully usable time constant close to 12ms for a 35µm pixel pitch.
- Despite pitch reduction from 45 to 25µm and associated dramatic pixel area drop, fill factor larger than 80% and high optical efficiency in the 8 to 14µm wavelength range have been maintained as it can be noticed from spectra of figure 6.

320 x 240 IRFPA	Pitch (µm)	R_{th} 10^6 K/W	T_{th} (ms)	NEDT (mK)	Comments
1st generation (a-Si) bolometer	45	12	4	80	ULIS industrial process
1st generation (a-Si) bolometer	35	12	2	180	
2nd generation (a-Si) bolometer	35	42	12	36	Transferred to ULIS in 2003

Tab. 1. First and second generation (a-Si) technology comparison

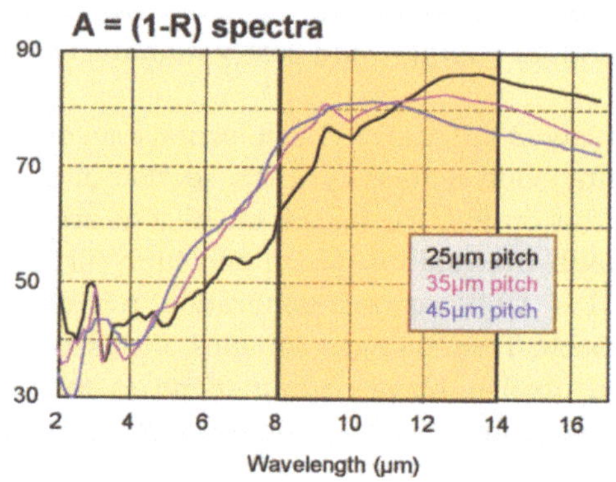

Fig. 6. Typical spectra obtained from various pixel ranging from 45 to 25µm pitch. 45µm spectrum = 1st generation ULIS industrial process. 35µm spectrum = 2nd generation process to be transferred to ULIS. 25µm spectrum = first lab demonstration at CEA/LETI.

Finally, CEA/LETIs basic option consisting of short and ultra thin design of suspended thermal legs results in high mechanical strength that could withstand high vibration rates and high mechanical shocks. This extra gain is particularly relevant for automotive application where microsystems devices must withstand an adverse mechanical environment.

2.5 Advanced Readout Development

Taking profit from the achievement and maturing of the second generation microbolometer technology ULIS and CEA/LETI have designed a 160x120 2D arrays in the scope of ICAR project with particular attention to the low cost automotive market. This new IRFPA is fed with a number of innovative on-chip features to simplify the use of this focal plane keeping a very small silicon

ROIC area down to $0.7cm^2$ for the 160x120 array, in order to reduce wafer-level processing costs per die. This new 160x120 is designed to fulfil low resolution, low cost applications. One of the most promising function is the possibility to adjust the skimming of the common mode current for each pixel by an automatic acquisition and in-pixel storage of non uniformity coefficients in a first step and readout pixel signal in a second step. At power on, the detector acquires its pixel compensation coefficients and stores them in on-chip memory for performing the current compensation during the following images acquisition and readout sequences. This automatic mode of operation could be changed to an external driving mode with non uniformity coefficients stored in an external memory (see figure7). The video output is available in analogic or digital format with an on-chip 12 bits (2x6) ADC. Most of the biases are generated inside of the ROIC for friendly user operation.

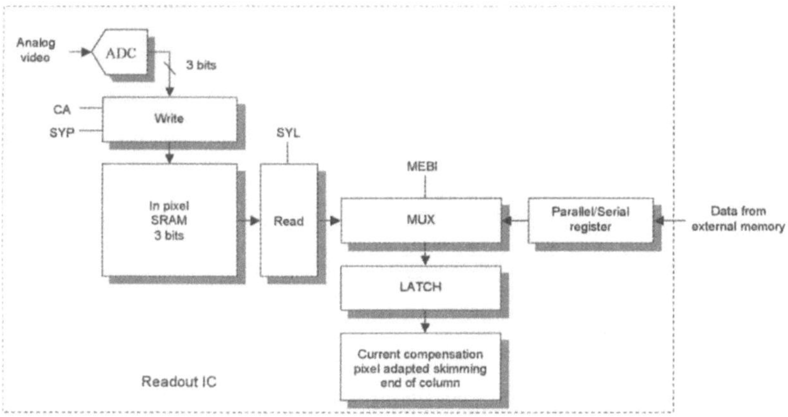

Fig. 7. Synoptic of 160x120 array on chip non uniformity compensating operation.

Several 160x120 microbolometer IRFPAs has been integrated under vacuum package and the usual electro-optical tests were performed under standard conditions including an operating temperature of 295K, a 100Hz frame rate and a flood illumination from a 300K blackbody through an f/1 limiting aperture. The resulting characteristics are summarized in table 2, whereas figure 8 shows a typical NETD histogram highlighting the weak dispersion of the IRFPA characteristics.

	Mean	Standard deviation (%)
Responsivity	16 mV/K	1.4 %
Total RMS noise	880 µV	10.4 %
NETD	56 mK	10.5 %
Operability	> 98%	

Tab. 2. Typical electro-optical characteristics of a 35µm pitch, 160x120 IRFPA.

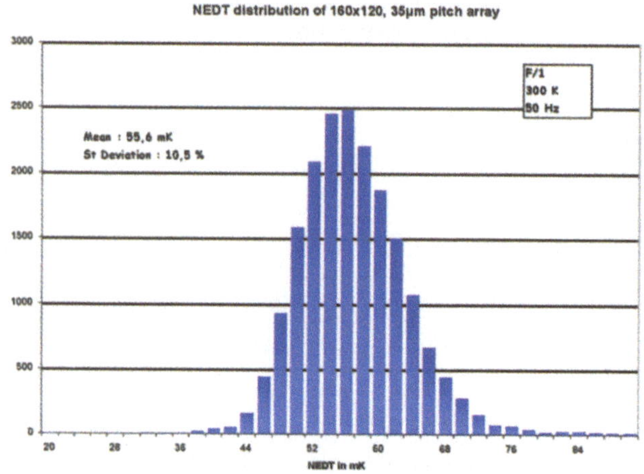

Fig. 8. NETD distribution and single frame obtained from a 35µm pitch, 160x120 IRFPA.

2.6 Packaging Development

Metallic Package

Metallic packages belong to the first generation of package used to integrate the microbolometer chip (figure 5), but their cost remains a large part of the total detector cost and this trend will be amplified in the near future as the pixel pitch will be reduced. As a consequence a less expensive package technology would be welcome and various developments are under progress in this field.

Ceramic Package

Ceramic packages (figure 9) are currently developed at ULIS. This technology is using available technologies developed for chips made in high volume production. Only the process used to assemble chip carrier and window carrier is adapted to take into account the required greater than 10 year lifetime under vacuum. These package constructions are compatible with automatic assembling machines that will contribute to decrease manufacturing cost.

Fig. 9. ULIS ceramic package developed for 160 x 120 microbolometer IRFPA.

Wafer Level Packaging

Beside these rather standard packaging techniques, CEA / LETI is working on the development of a wafer level packaging system in order to achieve the ultimate reduced manufacturing cost. This goal is completed in a 4 steps process. First, thanks to bulk micromachining techniques, a silicon micro-chip carrier is prepared on a silicon wafer. In a similar way, a second silicon wafer is processed in order to achieve an infrared window. These two substrates mainly consist of cavities and metallic thin film rings used for interconnecting and welding purpose. Then, individual microbolometer IRFPAs are positioned and wire bonded into each chip carrier cavity. Finally, the collective assembly (welding process) of the two wafers is carried out under vacuum leading to microbolometer IRFPAs shut under vacuum into silicon cavities. The major advantage of this technique in comparison to competitive option studied elsewhere [13], is that it does not require any extra soldering area on microbolometer die and consequently it contributes to IRFPAs cost reduction.

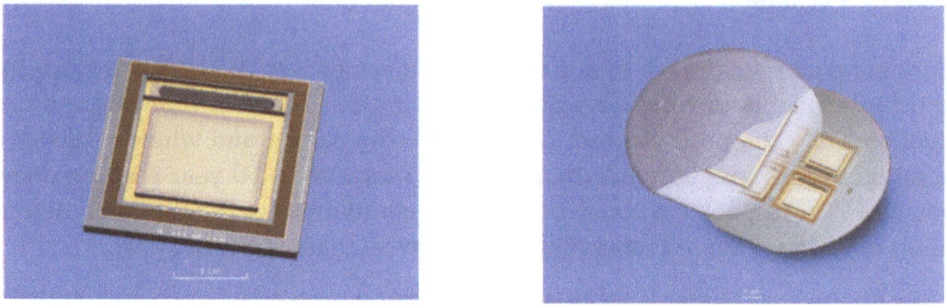

Fig. 10. Silicon chip carrier for microbolometer IRFPA (left) calls for a collective 4 microbolometer IRFPAs sealing (right) completely performed at the wafer level.

3 Application in the Automotive Safety

Traffic accidents are responsible for an unacceptable huge number of casualties all over the world. To improve this situation from the technological point of view, vision based systems for automotive safety represent one of the most promising development for driver assistance. But it is well known that in bad weather conditions, at night and usually every time the visibility is reduced the number of accident is increasing. Infrared video sensors, because they are not too much affected by such conditions, are well adapted to automotive applications. In the SAVE-U project, it has been chosen to use an infrared video sensor in addition to a visible wavelength video sensor and a radar to cope with difficult situations. The application reported in this paper was developed within the SAVE-U project. The objective was to detect obstacles coming in front of the vehicle to perform in a later stage their classification, which is not presented here. The final objective of SAVE-U is to develop a system able to protect vulnerable road users (pedestrians and cyclists).

Many current traffic accidents happen when unpredictable changes occur in the vehicle vicinity, therefore one of the main important task is to detect those changes. Looking at a scene taken by a camera mounted on a vehicle, it appears that main changes are due to vehicle global motion while minor changes are rather related to other moving objects such as cars, bicycles or pedestrians. If we then estimate the camera induced 2D motion field and use it to align two successive images, regions with secondary motions will be badly corrected and easily detectable (as illustrated below). The technique consists in estimating the camera motion and in finding regions whose motion is not consistent to it. This approach is based on image compensation techniques.

Fig. 11. Image at t with car and pedestrian motions (left).
Fig. 12. Image at t+1 (right).

Fig. 13. Contours of image t+1 on image t (left).
Fig. 14. Contours of compensated image t+1 on image t (right).

In the first section we will show how, using some restrictions, camera induced motion can be represented with a polynomial model. In a second part we will present the regression criterion used to compute the model parameters. As images present multiple motions simultaneously, the problem is reformulated in a robust framework such that secondary motions correspond to non conform data (outliers). These regions are then rejected from computation during the estimation process and will not corrupt the final solution. An iterative multi-resolution estimation scheme is also used to cope with high motion and reduce noise influence. Finally results and conclusions are given in the last section.

3.1 Camera Motion Model

In the real 3D world, the camera motion is described by two components: a translation T and a rotation Ω. Due to camera motion, a scene point P appears to be moving with rotation $-\Omega$ and translation -T. 3D velocity vector of P could be deduced from those remarks. After the modelling of the camera orientation and location, the projection of the real scene on the 2D sensor plane can be modelled using a perspective projection [17] but it leads to an expression of the 2D velocity field, which is not free from the Z-depth parameter always difficult to recover with a single camera. To get rid of the Z parameter, it is assumed that the scene can be approximated by a plane:

$$Z = a X + bY + c \quad (4) \tag{4}$$

Assuming this hypothesis (no strong depth variation in the scene) allows us to use a weak camera perspective [16]:

$$\begin{bmatrix} x \\ y \end{bmatrix} = \frac{f}{Z_{mean}} \begin{bmatrix} X \\ Y \end{bmatrix} = \lambda \begin{bmatrix} X \\ Y \end{bmatrix} \quad (5) \tag{5a}$$

where Z_{mean} denotes the average depth of the scene and f the focal length of the camera.

Then it is possible to deduced an affine 2d motion model, which is free from the depth parameter of the scene and which is able to handle translation, rotation and scaling:

$$\begin{bmatrix} \dot{x} \\ \dot{y} \end{bmatrix} = \begin{bmatrix} a_1 x + a_2 y + a_0 \\ a_4 x + a_5 by + a_3 \end{bmatrix} = m((x, y), (a_i)_i) \quad (5) \tag{5b}$$

3.2 Regression Criterion

Camera motion estimation can be performed as any common regression problem [15], finding the parameters $(a_i)_i$ of the model that best matches two images I (t) and I (t+1) i.e. such that

$$(a_i)_i = \arg\min \sum_{x \in I} \| I(X + \vec{m}(X, (a_i)_{i=1..6}), t+1) - I(X, t) \|? \tag{6}$$

The main drawback of the least square formulation is that the less conform the data are, the more influent they are, and they substantially biased the final solution. Robust estimation permits to overcome this drawback.

3.3 Robust Formulation

The field of robust statistics enables them to cope with huge errors which do not conform to the model assumptions [18,15,14]. The principle is to assign weights to data according to their adequation to the model and/or to their homogeneity with the rest of the dataset. This is achieved using a redescending function $\varphi(x,\sigma)$ for which influence of outliers tends to zero. σ is a scale parameter controlling the shape of the φ function and so the outliers selection process.

Then the regression criterion is reformulated as:

$$(a_i)_i = \operatorname{argmin} \sum_{x \in I} \varphi\left(\left\| I(X + \vec{m}(X, (a_i)_{i=1..6}), t+1) - I(X,t) \right\|, \sigma \right) \quad (7)$$

This can be easily converted into an equivalent iteratively re-weighed least square (IRLS) problem (see [14,18] for details)

$$(a_i)_i = \operatorname{argmin} \sum_{x \in I} w(r,\sigma) r? \quad (8)$$

with $r = \left\| I(X + \vec{m}(X, (a_i)_{i=1..6}), t+1) - I(X,t) \right\|$ the residuals and

$$w(r,\sigma) = \frac{\frac{\partial \varphi}{\partial r}(r,\sigma)}{r}$$

the weights at the current estimation step.

The general idea in IRLS techniques is to alternatively and iteratively: estimate a solution, assign weights to the residuals of this current solution and perform a new least square regression using these new weights.

3.4 Incremental Estimation

This criterion is non linear as referring to the parameters to estimate and might be non convex. To handle this problem, a common solution is to use an incremental minimisation scheme such as in [14,15]. The principle of those approaches is to suppose that an intermediate solution is known, then the full displacement u between first and second image can be decomposed in $u = u_k + du_k$. Using this remark the criterion can be linearized at point $X + u_k$ and the regression is done to find du_k. In practice this is achieved by constructing an intermediate image I^k warping the image at $t+1$ with the current motion estimation u_k and then re-estimating motion between I^k and the image at t.

3.5 Large Motion

To cope with large motions a coarse to fine strategy is used: a pyramid of low-pass filtered and sub-sampled images is constructed. At each level the image resolution is cut by half and low motion is first estimated at coarse level using the iterative scheme describe above. The computed flow field is then projected to the next level of the pyramid (rescaled as appropriate [15]) and the estimation begins again using this projection for initialisation at this level. The process is repeated until the flow has been computed at the full resolution. As scale and estimation evolve between each level and even at each step of the algorithm, the σ parameter of the φ function is reduced progressively during the process.

3.6 Results

To illustrate the algorithm results, several experiments have been performed in various situations. Each time the dominant motion between two successive frames is estimated, the first image is then warped and subtracted to the second to obtain the image of residuals. This one is then binarised [14] to build the detection map representing areas where non conform motions are detected. Intermediate results are detailed only for the first sequence, for the others, only the original image and the detection map are shown.

Experiment on a scene close to the camera:

Fig. 15. First image t (left).

Fig. 16. Second image t + 1 (right).

Fig. 17. Warped image, residuals and detection map.

Experiment with far and small targets:

Fig. 18. First image (left).

Fig. 19. Detection map (right).

Experiment with a low contrasted pedestrian (in the top left corner):

Fig. 20. First image (left).
Fig. 21. Detection map (right).

Experiment on a scene with multiple moving objects with different speeds and depths:

Fig. 22. First image (left).
Fig. 23. Detection map (right).

The main drawback of the algorithm is the planar hypothesis because large depth ranges in the scene will likely lead to false alarms. In fact in most of urban situations, the planar assumption is violated. In order to reduce these false alarms, some other approaches also use the outlier's map [21] or the residuals [22]. Otherwise, it is also possible to build a more complex motion model to improve the compensation [17,20].

4 Conclusion

This paper has put emphasis on the main features of CEA/LETI infrared microbolometer technology. One of the key points has been to elect a sensitive material made from amorphous silicon that features absolute compatibility with standard silicon processing. This basic option leads to high performance and low cost infrared imaging systems particularly suited for large market distribution such as automotive applications. This technology is now commercialised in mass production by the French company ULIS, while a brand new advanced technological arrangement has been demonstrated at CEA/LETI. The advent of this second generation of the technology results in a fivefold performance improvement compared to the current industrial process and NETD of 56mK obtained from 35µm pitch, 160x120 IRFPA has been demonstrated.

The results of detection by image processing techniques on infrared video sequences indicate that the approach gives better results compared to those obtained with images coming from video cameras working at visible wavelength. One reason seems to be that IR images present less detail but they are nevertheless textured enough for motion estimation. From that point they represent a very good alternative to visible wavelength sensors.

References

[1] F. Bertrand, J.L. Tissot, G. Destefanis, "Second generation cooled infrared detectors state of the art and prospects", 4th International workshop on advanced infrared technology and applications, Florence - Italie, 15 - 16 septembre 1997

[2] R.A. Wood, "Uncooled thermal imaging with monolithic silicon focal planes", Proceedings of SPIE Infrared Technology XIX, Vol. 2020, pp322-329, (1993)

[3] J.L. Tissot, F. Rothan, C. Vedel, M. Vilain, J.J. Yon, "LETI/LIR's amorphous silicon uncooled microbolometer development", Proceedings of SPIE Infrared Detectors and Focal Plane Arrays V, Vol. 3379, pp139-144, (1998)

[4] J.P. Chatard, X. Zhang, E. Mottin, J.L. Tissot, "Infrared Camera for cAR – ICAR a EURIMUS Project for Driver Vision Enhancement", AMAA yearbook 2002, Sven Krueger, Wolfgang Gessner Eds, pp141-145, ISBN 3-540-43232-9

[5] http://www.save-u.org

[6] C. Hanson, "Uncooled thermal imaging at Texas Instruments", Proceedings of SPIE Infrared Technology XIX, Vol. 2020, pp330-339 (1993)

[7] R. Watton, P.N.J. Dennis, J.P. Gilham, P.A. Manning, M.C.J. Perkins, M.A. Todd, "IR bolometer arrays; the route to uncooled, affordable thermal imaging", Proceedings of SPIE Infrared Technology XIX, Vol. 2020, pp379-390 (1993)

[8] T. Kanno et al, "Uncooled Infrared Focal Plane Array having 128x128 Thermopile Detector Elements", Proceedings of SPIE Infrared Technology XX, Vol. 2269, pp450-459, (1994)

[9] Tomohiro Ishikawa et al, "Low cost 320x240 uncooled IRFPA using conventional silicon IC process", Proceedings of SPIE Infrared Technology and applications XXV, Vol. 3698, pp556-564, (1999)

[10] P.I. Olden, E.A. Wachter, P.G. Datskos, T. Thundat, R.J. Warmack, "Optical and infrared detection using microcantilevers", Proceedings of SPIE Infrared

Technology and Applications XXII, Vol. 2744, pp345-354, (1996)

[11] J.L. Tissot, J.P. Chatard, E.Mottin, "Technical trends in amorphous silicon based uncooled IR focal plane arrays", Proceedings of SPIE Infrared Technology and Applications XXVIII, Vol. 4820, pp220-226, (2002)

[12] D. Murphy, M. Ray, R. Wyles, J. Asbrock, N. Lum, J. Wyles, C. Hewitt, A. Kennedy, D. Van Lue, "High sensitivity 25μm microbolometer FPAs", Proceedings of SPIE Infrared Technology and Applications XXVIII, Vol. 4820, pp208-219, (2002)

[13] R. Gooch, T. Schimert, "Low-cost Wafer-Level Vacuum Packaging for MEMS", MRS bulletin January 2003, pp55-59

[14] M.J Black.and P. Anandan, "The robust estimation of multiple motions: Parametric and piecewise-smooth flow fields", Computer Vision and Image Understanding, CVIU, 63(1), pp. 75-104, Jan. 1996.

[15] J.M. Odobez and P. Bouthemy, "Robust multi-resolution estimation of parametric motion models" Journal of Visual Communication and Image Representation, Vol. 6, No4, pages 348-365, décembre 1995.

[16] Subhashis Banerjee, "Camera models and affine Multiple Views Geometry" http://www.cse.iitd.ernet.in/~suban/vision/affine.pdf

[17] M. Irani, B. Rousso, S. Peleg , "Recovery of Ego-Motion Using Region Alignement" IEEE Trans. on Pattern Analysis and Machine Intelligence (PAMI), Vol. 19, No. 3, pp. 268 272, March 1997. http://www.wisdom.weizmann.ac.il/%7Eirani/abstracts/pami97.html

[18] P.J Rousseuw, A.M Leroy, "Robust regression and outlier detection" Wiley, New York, 1987.

[19] H. S. Sawhney, S. Ayer and M. Gorkani, "Model-based 2D&3D dominant motion estimation for mosaicing and video representation" In Proc. Intl. Conf on Computer Vision, pages 583-590, 1995 http://www.sarnoff.com/products_services/government_solutions/vision_technology/technical_papers/95sawh.pdf

[20] J. Y. A. Wang and E. H. Adelson, "Layered Representation for Motion Analysis", Proceedings of the IEEE Computer Society Conference on Computer Vision and Pattern Recognition 1993, pp. 361-366, New York, June 1993. http://persci.mit.edu/people/jyawang/wang.html#Publications

[21] G. Lefaix, E. Marchand, P. Bouthemy, "Motion-based Obstacle Detection and Tracking for Car Driving Assistance" IAPR Int. Conf. on Pattern Recognition, ICPR'2002,volume 4, pages 74–77, Quebec, Canada, August 2002 http://www.irisa.fr/vista/Papers/2002/2002_ICPR_lefaix.pdf

[22] M. Irani, B. Rousso, and S. Peleg, "Computing Occluding and transparent motion" International Journal of Computer Vision (IJCV), Vol. 12, No. 1, pp. 5-16, February 1994.http://www.wisdom.weizmann.ac.il/%7Eirani/abstracts/multipleMotions.html

J.J. Yon, E. Mottin
CEA – DRT – LETI/DOPT – CEA/GRE
17 Rue des Martyrs
38054 Grenoble Cedex 9
France
jyon@cea.fr, e.mottin@cea.fr

L. Biancardini, L. Letellier
CEA – LIST
91191 Gif/Yvette
France
loick.biancardini@cea.fr, laurent.letellier@cea.fr,

J.L. Tissot
PULIS – BP21
38113 Veurey-Voroize
France
jl.tissot@ulis-ir.com

Keywords: amorphous silicon, microbolometer, uncooled IR detector, IRFPA, image processing

Infrared Image Processing for Obstacle Avoidance

Y. Le Guilloux, J. Lonnoy, R. Moreira, SAGEM SA

Abstract

Most road accidents are due to a lack of perception of the driver about the road situation. PAROTO system's objective is to help the driver in detecting obstacles and thus allow him a better and earlier reaction in dangerous situations. Data coming from a Radar and an infrared camera are merged in order to take advantage of complementary information. Eventually the driver can be alerted after a filtering step. In this paper, we present the infrared channel, from the segmentation technique using the particular content of infrared images, to their exploitation in the interpretation of the road scene, through motion analysis.

1 Introduction

The goal of the PAROTO project is to develop a system, fitted on a standard car, capable of preventing collisions by anticipating the presence of hazardous obstacles ahead. PAROTO uses a Radar and an Infrared Camera to detect obstacles. Whenever it is needed, signals are emitted in due time through a dedicated interface in order to alert the driver in the most efficient way. The current phase of PAROTO, founded by the French Ministry of Research, will produce a demonstrator in order to evaluate the performance of the concept. The PAROTO consortium includes four partners: SAGEM SA (leadership, infrared, fusion), LASMEA (radar, fusion), INRETS (accident study, driver's interface management, system evaluation) and TEAM (driver's interface realisation).

In this presentation, we shall first recall the objective of the project. We briefly describe the overall architecture and the modules of the PAROTO system. Then we focus on infrared image processing, dedicated to obstacles detection. After presenting the chosen process strategy, we discuss in more details the different steps of the process. We conclude with some promising results and expectations.

2 General Description

2.1 Objective of the PAROTO Project

Detailed Accident Study (Etude Détaillée des Accidents) database from [1], which seems representative according to the French National global statistics, has shown that a large part of accidents are due to a lack of perception of the driver.

The objective of PAROTO system is to help the driver to better detect obstacles located on vehicle trajectory or being likely to cross it. The system is potentially able to accelerate the driver detection of the obstacles, so time could be saved and then used to cope with a possible dangerous situation.

Fig. 1. VELAC, the vehicle of LASMEA, equipped with an IR camera.

2.2 Functional Architecture of PAROTO System

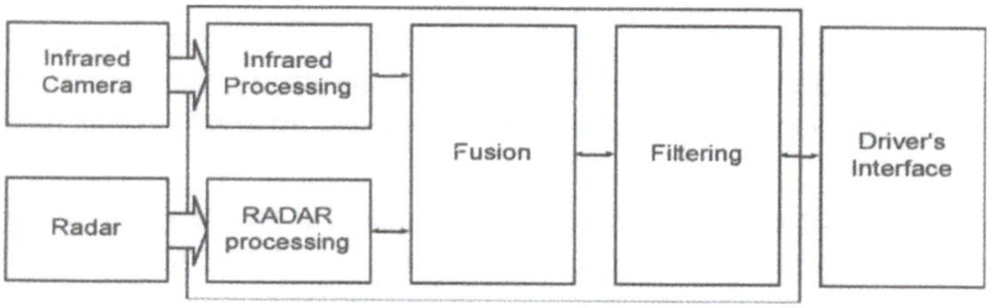

Fig. 2. Functional architecture of PAROTO System.

Both sensors feed respective processing capabilities, which in turn interact with a fusion module, the output of which is filtered in order to send only relevant signals to the driver's interface (Figure 2).

2.3 Radar

The key fact in the selection of the Radar technology for PAROTO is cost efficiency. The description of the Radar channel is out of the scope of this paper and will be discussed in later communications.

2.4 Fusion

Although fusion has clearly not reached the practical stage either, it seems important to recall how well the information coming from Radar and Infrared complement each other. Whereas Radar is able to tell at what distance it points echoes, the direction being somewhat fuzzy, Infrared can give the direction in which a relevant event is detected, but no direct access to the distance at which it occurs. Fusion of the sensors can thus be very fruitful, providing an extremely rich description of events.

2.5 Filtering

The system reliability (rate of false detection and of non-detection) at the fusion level is not sufficient to ensure the system acceptance by the driver. The driver awareness must be taken into account and can be measured by proprioceptive sensors. If the measured behaviour of the driver indicates that he copes with the situation, the filtering module does not have to distract him (although acceptance by the driver can benefit from agreeing with the system). The system must alarm the driver when he is not aware of the danger.

2.6 Driver's Interface

The way to warn the driver depends on the emergency of the situation, as explained in [2]. If the driver has quite a long time to react, i.e. the time to collision (TTC) is large enough, the filtering module commands the attention orienting mode of the Human-Machine Interface (HMI). In this mode, the warning is done through a set of LEDs which, when lighted, heads the driver's visual attention in the dangerous direction. The attention orienting mode allows the driver being conscious of a potential danger. If he has already perceived it, he can check the system's efficiency.

In the case of an imminent danger, the filtering module warns the driver using the alarm mode of the HMI. The alarm mode is used when the TTC is dangerously small and the driver doesn't seem to have perceived the danger. In such a case, the warning is both visual (red flashing LEDs) and acoustic (various sounds are tested).

3 Infrared Imagery: Goals and Means

3.1 Objectives of the Infrared Image Processing

The infrared image processing goes as far as possible in the interpretation of the road situation. The objectives are as follows:

- Detection of potential obstacles in image, also referred to as segmentation.
- Determination of their relative positions and velocities by tracking.
- Evaluation of the risk level (TTC).

3.2 Processing choices

The image positions of detected obstacles are given by segmentation, based on infrared image properties. From this information, we must recover the 3D positions with respect to the equipped vehicle (those positions are called relative positions). Having only one camera, we need an extra hypothesis to do so: we consider the road as a plane. The situation is described in figure 3.

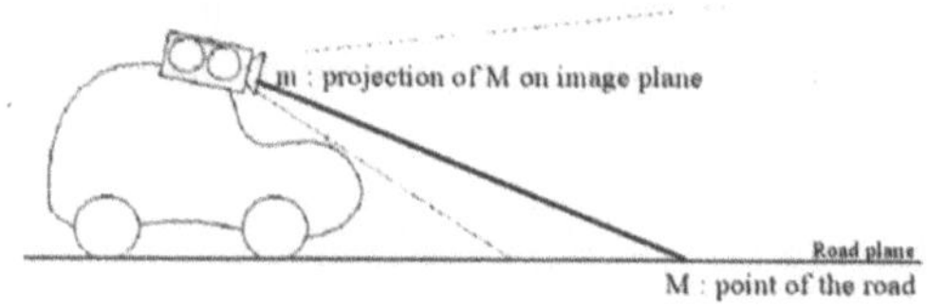

Fig. 3. Geometrical description of the acquisition system.

Under the planar road hypothesis, we can use the geometrical relations between road points and their projections in image to recover relative positions. Unfortunately, this is fragile due to vehicle's movements, especially pitch. Therefore we must estimate the ego-motion, through motion analysis in image, and compensate it in order to recover reliable 3D positions. It is a sort of stabilisation. Motion analysis is based on feature tracking.

3.3 Main Scheme of the Overall Process

As a consequence of the preceding remarks, segmentation and motion analysis are parallel channels whose results are used together to retrieve the most precise information about the situation. Figure 4 summarises this scheme.

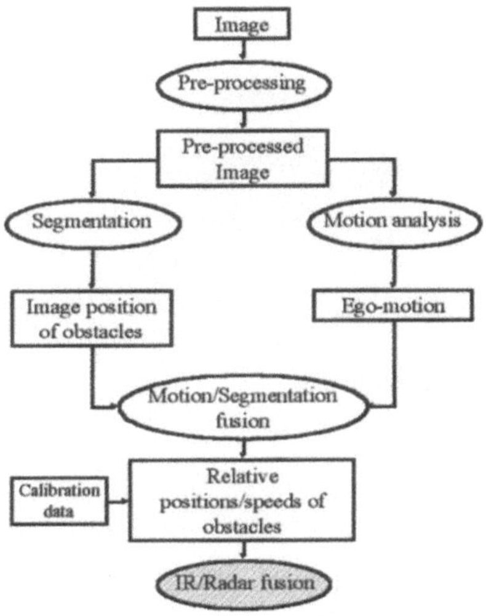

Fig. 4. Main scheme.

The results of the infrared image processing are relative positions, velocities and classes of obstacles. They are inputs to the fusion module of PAROTO system. This subject is not tackled in this paper. The next section describes segmentation. It is followed by a presentation of motion analysis.

4 Segmentation

This section discuss frame-by-frame obstacle detection. We first define the typical obstacles we will consider and show their IR aspects. Next, we describe the cues extracted to detect them. Finally, we outline the specific algorithms to segment a vehicle and a pedestrian from cues.

4.1 Obstacles Definition

In classical road situations, the sources of danger relate to humans. This leads us to focus the detection on vehicles, cycles and pedestrians. The relevant objects display a significant temperature difference with their environment: this is a major cue for their detection by an infrared camera. The hot parts of motorised vehicles are their engines, wheels and exhausts. In the case of a car, the engine heats the exhausts and the air that flows below it, producing a locally higher contrast. The human body radiates heat, so we are able to detect pedestrians as bright regions. However, in rare cases, high tech clothes can alter this thermal aspect.

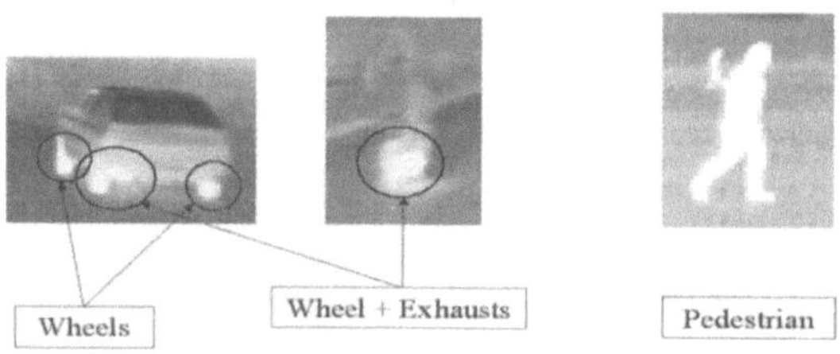

Fig. 5. IR aspects of road users.

4.2 Cues for Detection

Considering the obstacles definition, the first task is the extraction of hot regions, i.e. regional maxima of intensity with the following properties: strong contrasts with their neighbourhoods, thin frontiers and fair uniformity.

Various cues are used to segment the image.

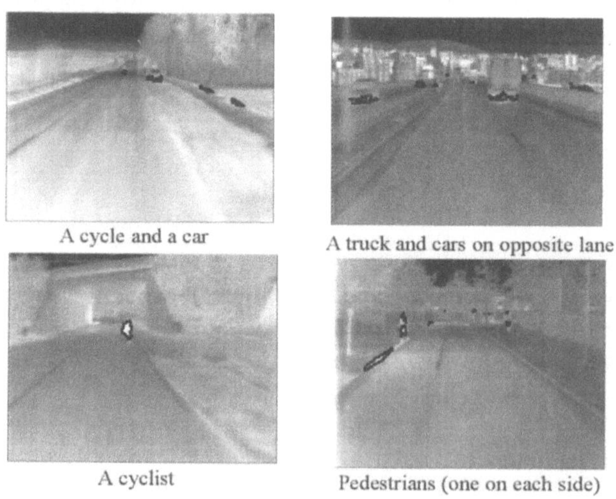

Fig. 6. Regional maxima (black regions) extracted on various obstacles.

Regional Maxima

In order to extract these regions, we use a morphological operator, basically a morphological aperture on a grey level image. Figure 6 illustrate some regional maxima. See [3] for details. Although regional maxima are significant cues, they are not sufficient to ensure the detection of obstacles:

- Some regional maxima doesn't correspond to obstacles.
- Obstacles often include more than one maximum.

Gradients based cues

Regional maxima give us the location of hot objects. To recognise them we need some information about their shapes. We construct this information from gradients, in three steps:

- Contours extraction from gradients norm.
- Estimation and quantification of orientations of these contours: this operation permits to handle the notions of left, right, top, bottom of an object.
- In the image of oriented contours, we look for particular shapes.

Fig. 7. Detected predefined shapes (black curves).

As illustrated above, we are able to detect:

- ▶ Wheels of a vehicle when it is close or not so hot (details appearing).
- ▶ Vehicles when they are very hot or far.
- ▶ Pedestrian feet, head, pelvis (usually the hottest parts).

This approach can be extended to other categories of objects by designing additional cues coping with different shapes.

Vertical Symmetries

Vertical symmetries are searched in the image of gradients' norm. This cue is mainly used to detect pedestrians.

Fig. 8. Detected vertical symmetries (black lines).

4.3 Obstacles Recognition

We combine the cues described above to find vehicles or pedestrians.

Vehicles

We use regional maxima and predefined shapes to detect vehicles. We score the combinations of these two cues and keep those matching some geometrical and textural criteria. Combinations are admitted or rejected depending on following criteria:

- ▶ Relative positions of cues.
- ▶ Dimensions of the bounding box versus image position.
- ▶ Textural contents of the rectangle.

Those fuzzy criteria are computed based on a model of a standard vehicle. Figure 9 illustrates some results. An admitted combination is represented by a bounding box including the supposed vehicle

Fig. 9. Black boxes are detected vehicles.

Some explanations are needed to completely understand results in figure 9. In the left image, the motorcycle is detected because the geometrical criterion is flexible. In the right image, the occluding pole near the left border causes the vehicle not to be detected. We recall that segmentation is done frame-by-frame: a simple temporal filtering will solve the problem of occlusions. We note some false detection near right border, due to the properties of these areas.

Pedestrians

The approach used to detect pedestrians is inspired from the technique just described: we combine regional maxima and vertical symmetries. Other shapes, more specific to pedestrians, are studied. Criteria of admission or rejection are calculated based on a pedestrian model. The textural contents criterion consists here in the comparison of the candidate with a probabilistic template, as in [4]. A template characterises the IR aspect of a pedestrian. As we search for pedestrians at various distances, an unique template is not sufficient. We have constructed multiple templates depending on the image position, that is, in turn, on distance. Some results are shown in figure 10. Cyclists are currently detected the same way as pedestrians. Trees and poles can be classified as pedestrians in some cases.

Fig. 10. Black boxes are detected pedestrians.

4.4 Conclusion on Segmentation

Results provided by frame-by-frame segmentation algorithm are satisfying. Some improvement has to be achieved in urban context, where more artefacts pop up.

5 Motion Analysis

We explain here the way we stabilise our image sequences (in order to retrieve reliable 3D positions from image).

5.1 Feature Tracking

Motion analysis is done through the analysis of the displacements of carefully chosen and tracked features. We used a feature tracker.

Fig. 11. Features tracked over five frames.

5.2 Ego-Motion Recovery

Velocity and rotation describe the motion of the vehicle. The velocity is known through an odometer. Thus, we have to estimate pitch, yaw and roll. In the case of a pure rotation, the displacement of an image point is approximated by:

$$\begin{cases} u = -f.\Omega_Y + y.\Omega_Z \\ v = f.\Omega_X - x.\Omega_Z \end{cases} \tag{1}$$

where f is the focal length and Ω the rotation vector. With a sufficient number of features, Ω can be estimated using equation 1.

5.3 Sequence Stabilisation

Stabilising an image sequence is done by integration of rotation components. A low-pass filtering is necessary to cope with changes of direction of the vehicle.

6 Motion / Segmentation Fusion

6.1 Relative Positions and Speeds of Obstacles

At this point, we have the image positions of obstacles, given by the segmentation module, and the ego-motion information, from the motion analysis. We can correct the positions of obstacles in image and project them in the road plane. We obtain the relative positions of obstacles.

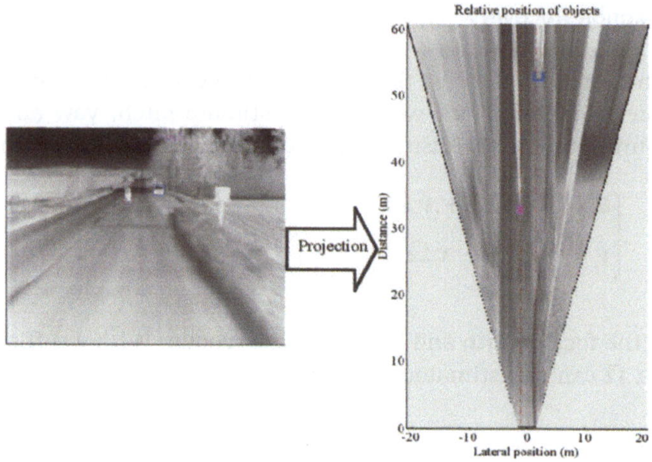

Fig. 12. Image positions to relative positions.

Figure 13 shows the impact of stabilisation on estimated obstacles trajectories, projected onto the road plane, with respect to the PAROTO vehicle (section 3.2). Reconstructed trajectories are much smoother and more reliable in the stabilised case.

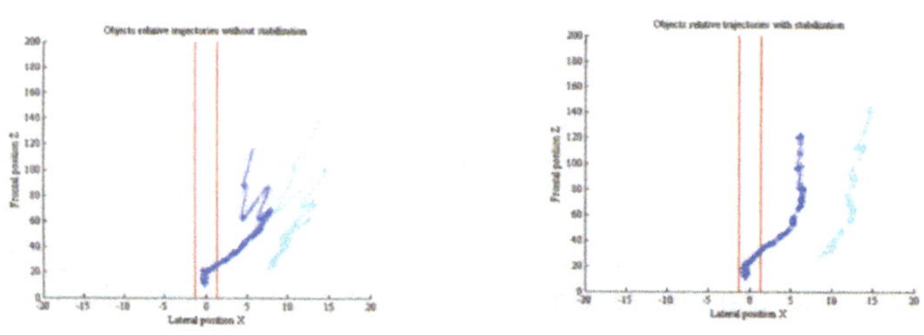

Fig. 13. Reconstructed vehicles (left) trajectories without and with sequence stabilisation.

Finally, the obstacles are tracked, using a Kalman filter (with a constant acceleration model), to estimate relative speeds.

6.2 Dealing with Artefacts

Manhole covers or thermal shadows on the road are particular "objects". In some configurations, they can lead us to detect false obstacles (figure 14). To reject those false obstacles, we have to estimate their height above the road plane to test them against the clearance of the vehicle. We use tracked features once more.

Fig. 14. Left, hot area on the road surrounded by thermal shadows. Right, a false detection.

Displacements in the image of points belonging to the same object depend on the relative velocity of this object and their height with respect to the ground. This relationship yields an estimate of height above the road plane, which allows to reject false obstacles.

7 Elements for Risk Level Evaluation Using IR Sensor Only

The information provided by the IR module, ie height and velocity, can serve as a basis for the classification of objects as dangerous or not. For example, the following diagram illustrates the analysis of objects found ahead the PAROTO vehicle.

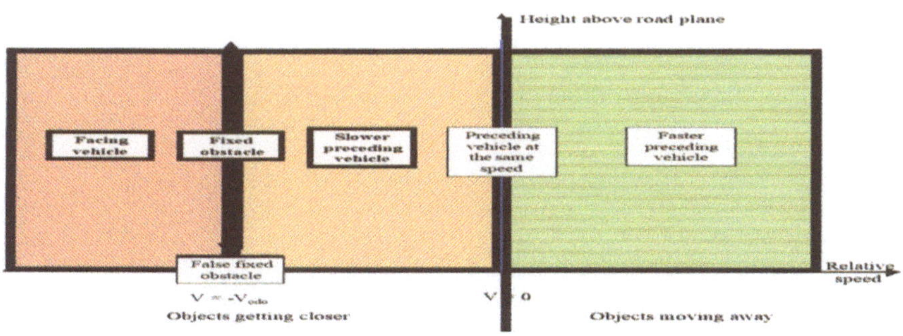

Fig. 15. Situation interpretation depending on relative speed and height of the obstacle.

8 Conclusion

We have presented the IR channel of the PAROTO system. It consists in frame-by-frame detection of potential obstacles and a motion analysis to recover relevant information about the road situation. Motion analysis and reconstruction of obstacle trajectories are fixed parts: no major evolution will occur on these algorithms, with an unique sensor. The quality of information provided by IR channel depends directly on the quality of segmentation. Our segmentation algorithm is efficient in motorways and sub-urban contexts. It needs improvement in urban contexts, where sources of false detection are numerous. Next work on PAROTO project concern IR/Radar fusion and on-board tests.

Acknowledgements

This study is being conducted within the PAROTO project. Although a few authors were involved in the redaction of this article, it builds upon the results of a collection of efforts led by all partners of the consortium, whose dynamic involvement is gratefully acknowledged.

References

[1] P. Van Eslande, L. Alberton, "Scénarios types de production de l'erreur humaine dans l'accident de la route: problématique et analyse qualitative", rapport INRETS N°218, 1997.

[2] Y. Le Guilloux, J. Lonnoy,, R. Moreira, M.-P. Bruyas, A. Chapon, H. Tattegrain-Veste, "PAROTO project: the Benefit of Infrared Imagery for Obstacle Avoidance", IEEE Intelligent Vehicle Symposium IV'2002, June 2002.

[3] L. Vincent, "Morphological Grayscale Reconstruction in Image Analysis: Applications and Efficient Algorithms", IEEE Transactions on Image Processing, Vol.2, No. 2, pp. 176-201, April 1993.

[4] H. Nanda, L. Davis, "Probabilistic Template Based Pedestrian in Infrared Videos", IEEE Intelligent Vehicle Symposium IV'2002, June 2002.

Yann Le Guilloux, Jacques Lonnoy, Raphaël Moreira
SAGEM SA
Division Défense et Sécurité
95100 Argenteuil
France.
yann.leguilloux@sagem.com

Keywords: road safety, infrared image processing, obstacle avoidance.

A Compact 77GHz Transceiver Module Using G3D Diode Technology for Automotive Applications

N. Priestley, B. Prime, e2v technologies Ltd

Abstract

This paper looks at a new 77GHz adaptive cruise control (ACC) radar transceiver design, which has placed considerable emphasis on low parts count, ease of assembly and test, excellent performance over temperature and ease of system integration. Unlike the currently available 77GHz transceivers, which use either a cavity Gunn diode voltage controlled oscillator (VCO) or a monolithic microwave integrated circuit (MMIC) chip set for the signal source, this unit uses a novel graded gap Gunn diode (G3D) planar VCO technology. This approach is inherently low cost as the 77GHz oscillator is assembled into the module as a small (10x10x4mm), hermetically packaged device. The millimetre wave section of the module was built on low dielectric constant softboard, which fits into a metal housing, which is 50x45x15mm excluding the waveguide feed. The operation and fabrication of the Gunn diode is discussed as well as a description of the physics behind a frequency modulated continuous wave (FMCW) radar sensor.

1 Introduction

The use of radar technology in automotive applications has been growing steadily over the last ten years. Hence many of the microwave component suppliers of adaptive cruise control parts have had to learn how to address the automotive market from the design process through to production line operation. The market for ACC is now set for rapid growth over the next five years; from less than 100,000 units in 2002 to a forecast of over 8 million units in 2006. Over seventeen percent of all European vehicles will be fitted with ACC at this time and of these 50% will be radar based and the balance will use lidar which uses light rather than microwaves to sense the road ahead. The challenge to the designers of radar based next generation ACC components is to reduce cost, size and parts per million failure rates while increasing production volumes and capacity.

This is being achieved by the introduction of a range of technology improvements which span from high frequency semiconductor devices to the antenna design.

2 Sensor Design

The principle of radar is to use echo and doppler shift to measure distance and relative speed. If a microwave signal is transmitted from a transmitter as a short burst, it moves away at the speed of light until it hits a target and it is then reflected back to the receiver (sited next to the transmitter). If the time taken for this to happen is measured, then the distance between the source and the target can be calculated. A change in frequency of the reflected signal ,called Doppler shift, enables us to measure the relative velocity between the site of the radar and a target. If the two are closing together then the reflected signal will rise in frequency because the electromagnetic wave will be stacked up and therefore compressed. Many ACC systems use a form of radar known as FMCW (Frequency Modulated Continuous Wave). This relies on the frequency of the transmitted signal being constantly ramped to give the trace of a saw tooth. By measuring the difference in frequency between the out-going signal and the incoming signal a very accurate measurement of the distance can be made. For this to work well enough at 77GHz, and to be able to measure the distance between the radar and the target at a distance of 150 metres to an accuracy of better than 0.5 metre, the ramp of the saw tooth must be very linear. The schematic of this FMCW radar transceiver, as shown in figure 1, is comprised of a number of key components which may be built either from microwave monolithic integrated circuits (MMICs) or using discrete microwave semiconductor devices. Some of these components will now be treated in more detail especially those which have a strong link to the cost and performance of the sensor.

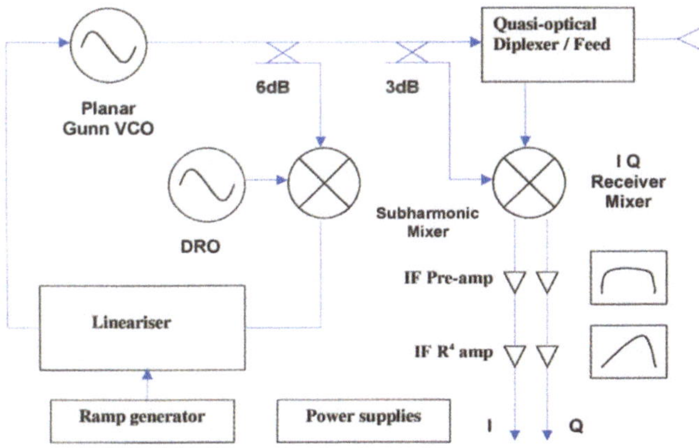

Fig. 1. FMCW Radar Schematic.

2.1 Voltage Controlled Oscillator

Currently there are two technology alternatives available to make the VCO; Gunn diode or MMIC, both of which are made using Gallium Arsenide (GaAs) rather than silicon due to the high frequency of operation. The Gunn diode has been used in radars since soon after it was invented in 1962 and currently it is used in about half of the ACC systems being installed in cars today. MMIC sources use pHEMT (pseudo-morphic high electron mobility transistor) technology and generate the signal typically at the 4th or 5th sub-harmonic. This is followed by a succession of multiplication and amplification stages to reach 77GHz with the output power required.

One of the strengths in being a designer and manufacturer in both compound semiconductors and microwave components is that in-house technology may be tailored and harnessed to provide the optimum solution. Gunn diode technology was therefore selected as the best technology for the VCO.

2.2 Gunn Diode Technology

Gunn diodes which operate at millimetre wave frequencies are made by manufacturers in the UK, USA and Japan. A particularly important innovation was introduced in the early 1980's [1, 2] which is now known as the 'graded gap' Gunn diode which overcame the limitations which inhibited the use of these devices at frequencies over 60GHz. To understand more about this patented structure and indeed the Gunn-effect we need to understand some of the principles of GaAs and the epitaxial structures which are used.

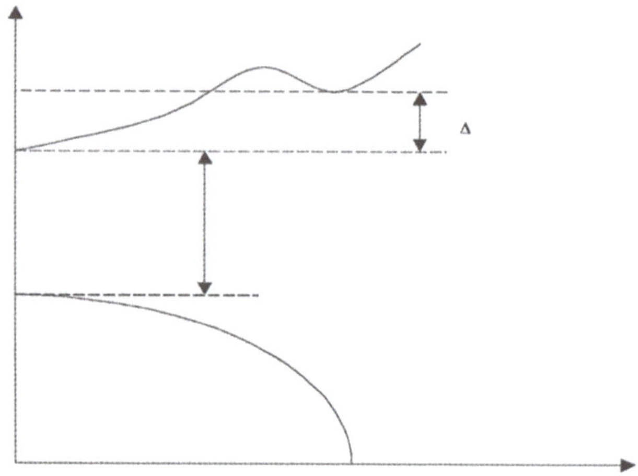

Fig. 2. Gallium Arsenide Band Structure.

A schematic representation of the band structure for GaAs is shown in figure 2, and this is only a small part of a very complex energy band structure. Only the conduction bands need to be considered for the study of electron dynamics, since electrons in the valence bands are stationary.

There are two energy levels which we need to consider; A (central valley) and B (satellite valley). In the lower central valley the electrons have a smaller effective mass and a high mobility; however in the satellite valley electrons have a larger effective mass and a lower mobility; futhermore it can be seen that the two valleys are separated by a small energy gap DE. At moderate electric fields both valleys will be populated by electrons, but at high fields most electrons will be in the upper valley. When this happens the device will produce phenomena known as Negative Differential Mobility i.e. decreasing mobility with increasing field.

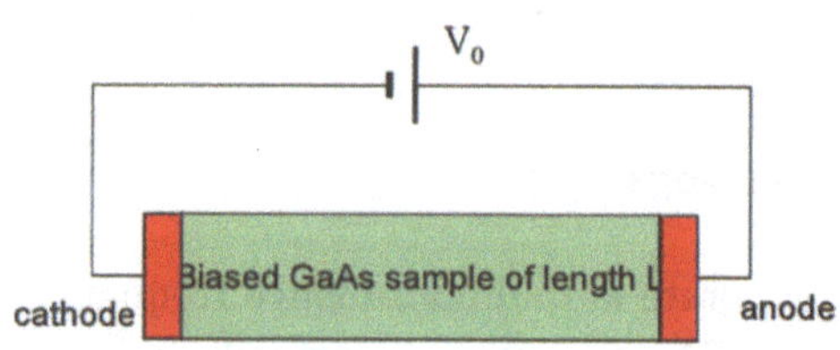

Fig. 3. DC Biased GaAs Sample.

If we take a sample of uniformly doped n-type GaAs of length L and apply a constant voltage V_0 to it as in figure 3, electrons will flow from the cathode to the anode with constant velocity. However if we assume that a small perturbation in the net charge arises at a point between the cathode and anode, electrons will experience a higher electric field. If this takes them into the satellite band (V_0 = threshold voltage), then they will drift towards the anode with a smaller velocity. Consequently a pile up of electrons will occur and this will increase the negative charge in that region. The region on the anode side will become progressively more depleted of electrons due to their higher drift velocity towards the anode. This initial charge perturbation therefore grows into a dipole domain commonly known as a Gunn-domain which will grow while propagating towards the anode until a stable domain has been formed. As soon as the domain is absorbed by the contact region, the average electric field in the sample rises and domain formation can again take place. The successive formation and drift of Gunn-domains leads to high frequency current oscillations at the contacts, and the frequency is largely dictated by the distance the domains must travel before being annihilated at the anode. Hence the frequency of a Gunn diode is determined by the doping and epitaxial thickness of the n-type GaAs (drift region).

A major problem with the operation of such conventional devices over the temperature range -55°C to +25°C is that as the temperature reduces, the 'turn-on' voltage, the point above threshold at which coherent RF power is obtained, increases to a point where it equals the peak power voltage. This forces the diode to be operated at a higher voltage with much reduced power, efficiency, poor noise performance and the increased possibility of device failure. A solution to this is to tailor the electric field at the cathode which can be done by using a hot electron injection technique whereby very large electric fields are created inside the semiconductor by using built in potentials. e2v use a graded AlGaAS injector; the structure for which is shown in figure 4.

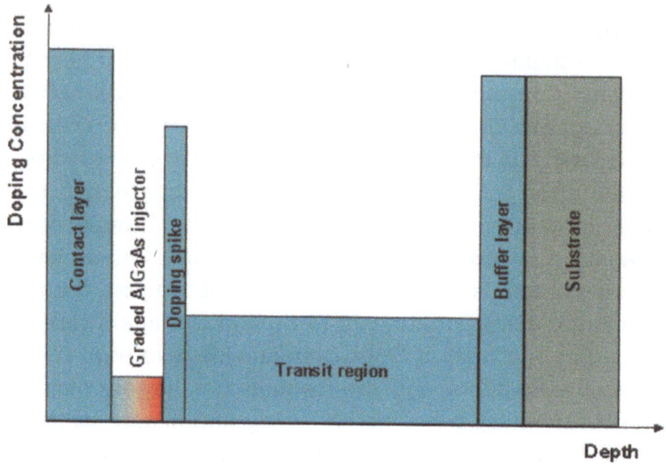

Fig. 4. Structure of Graded AlGaAs Injector.

The growth of the epitaxial layers, which demand very close control of the doping level and the thickness of each component, is performed by molecular beam epitaxy, MBE. Semi-insulating GaAs wafers, 3 or 4 inches in diameter are grown with layers as shown in figure 4 and then contacted with gold contacts and etched to form Gunn diodes with a carefully controlled diameter so as to achieve the oscillation characteristics to make a 77GHz oscillator. This results in the so-called integrated heat sink (IHS) structure which is 300 microns in overall diameter and is shown in figure 5. The gold heat sink (cathode) is attached to a copper heat sink in the oscillator by either solder, thermo-sonic weld or epoxy and the top contact (anode) is connected by using a Maltese cross which is also made in gold.

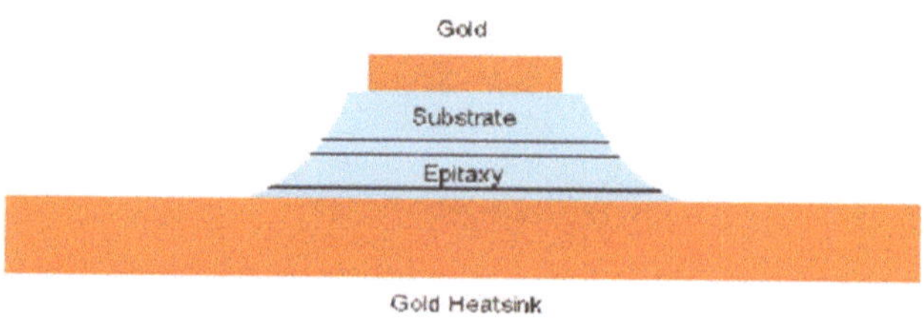

Fig. 5. Integrated Heat Sink Gunn Diode Structure.

2.3 VCO Design and Construction

The oscillator circuit [3] is in microstrip and is fabricated on fused quartz using thin film technology. Fused silica has a dielectric constant, $E_r = 3.81$, a dielectric loss tanδ = 0.0004 at 30GHz and a coefficient of thermal expansion of $0.5 \cdot 10^{-6}K^{-1}$. It is readily available and is fully compatible with thin film processing. The thin film circuit is approximately 5mm x 7mm in size and contains the elements. A surface mount package was designed to provide the housing for the VCO circuit and to give excellent heat dissipation for the Gunn diode. A high thermal conductivity metal composite was chosen for the base and the Gunn diode chip is mounted directly to this base which in turn is designed to be mounted directly with two screws to the rest of the radar module.

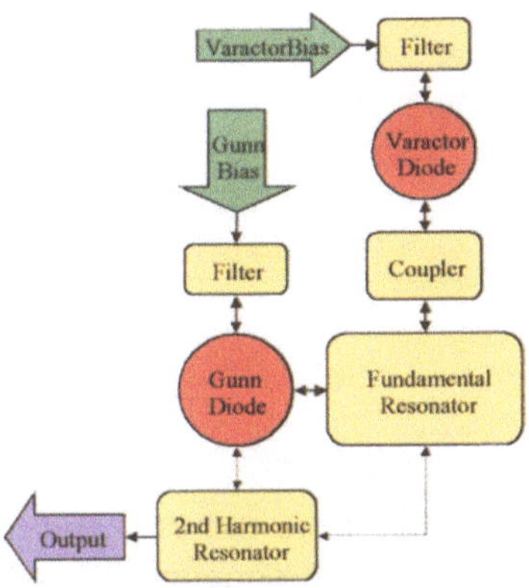

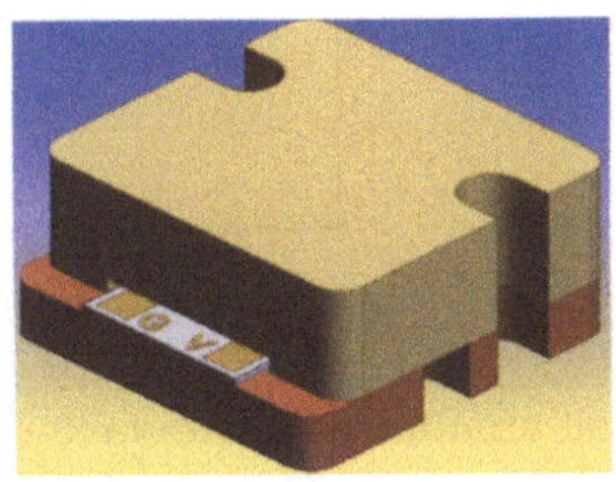

Fig. 6. VCO schematic circuit diagram (left).

Fig. 7. Solid model illustration of VCO package (right).

The Gunn diode is thermosonically mounted to the gold plated package floor and epoxy is used to mount the varactor diode chip, also on to the floor of the package. Interconnection to the top contact on the Gunn diode is made using a solid gold Maltese cross structure which is bonded using a thermosonic wedge bonder. Gold wire is used to make a similar connection to the varactor. The hermetic package (figure 7) for volume manufacture is fitted with ceramic feed-throughs and the lid is seam-seal welded.

2.4 Loop Electronics

FMCW radars [4] require the microwave source to be frequency modulated in a very linear way. This is usually in the form of a frequency ramp such as a triangle or sawtooth waveform. Because the microwave source will have a tuning characteristic, which is not usually linear, some form of correction circuit is required. Figure 8 shows the block diagram for the circuit used to linearise the VCO.

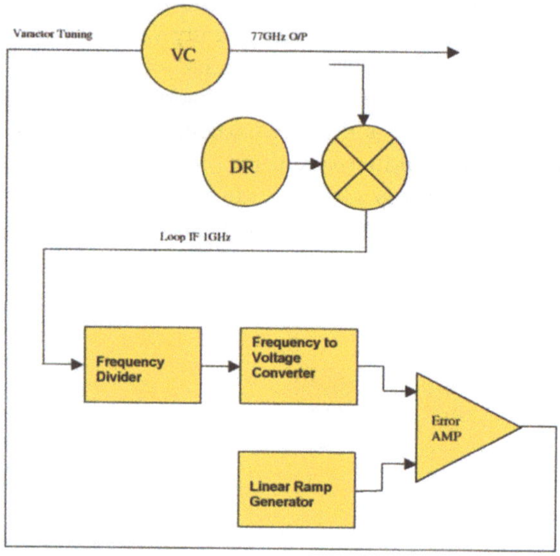

Fig. 8. Lineariser block diagram.

A small amount of microwave power is sampled from the VCO and fed to the harmonic mixer. This produces an Intermediate Frequency (IF) which is a faithful representation at approximately 1GHz of the microwave frequency being generated by the VCO. At this IF frequency low cost standard electronic parts can be used. The IF is then divided by 160 using logic gates. This very low frequency is converted to a DC voltage using pulse integration of the divided down frequency. Voltage proportional to the VCO microwave frequency is a process called frequency discrimination. The change in this voltage is proportional to the frequency deviation of the VCO.

The ramp generator produces a voltage characteristic the shape of which represents the frequency sweep plan. If a sawtooth frequency sweep is required, the ramp generator will produce a voltage sawtooth the amplitude of which will set the frequency deviation required from the VCO.

The voltage from the ramp generator and the voltage from the frequency discriminator are compared in an error amplifier. This tries to make the voltage from the frequency discriminator the same as that of the ramp generator. The difference between these voltages is the error voltage which in turn is proportional to the frequency error.

The error voltage is fed back to the VCO to correct the frequency and hence make the VCO lock to the sweep plan set by the ramp generator.

2.5 Receiver Mixer

A mixer is a device for comparing the difference between two frequencies. A reference frequency called the Local Oscillator (LO) is fed into one port of the device. Another frequency to be compared against the LO is fed into the Receive port (Rx). The difference between these two frequencies will appear at the Intermediate Frequency (IF) port.

In the radar the mixer is being used to compare the transmitted signal with the received signal to produce an IF which contains range and relative speed information.

An in-phase and quadrature mixer (IQ) is used in the radar sensor. This is comprised of two mixers, the LO of one being fed 90 degrees out of phase to the other. The returned signal, Rx, is fed to both mixers in phase. By comparing the relative phase of these two Ifs, any phase modulation placed on the carrier by the relative movement of an object in front of the radar, can be retrieved.

In the FMCW radar a sample of the transmitted frequency is fed to the LO of the mixer. This transmitted frequency is continually changing with time (ramping). It then travels out from the radar and if it hits an object, some of the energy will return to the radar receiver. The instantaneous transmitting frequency and the returned echo arrive at the LO and Rx ports of the mixer. Because the returned signal took time to travel to the object and back the frequency at each port is different. This difference is proportional to the distance travelled by the carrier to the object and back again. If the object is moving the return will also contain Doppler frequency which can be used to measure direction and relative speeds.

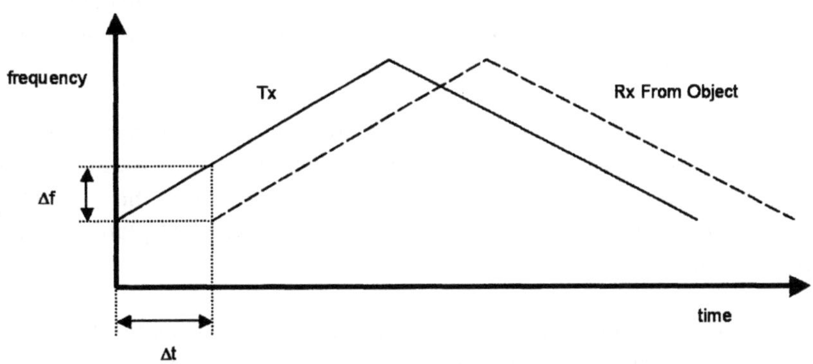

Fig. 9. Δt = time taken for carrier to travel to object and back
Δf = difference freqeuncy between the transmit freqeuncy and the receive freqeuncy

$$\Delta f = \Delta c \cdot 2(d/C)$$

where

Δf = intermediate frequency from driver (Hz/m)
Δc = modulation rate of the carrier (GHz/s)
d = distance to object (m)
C = propagation velocity ($3 \cdot 10^8$m/s)

For example, a carrier ramped at a frequency of 150MHz in 1mS, reflecting off an object 10M away will result in an output frequency from the radar of 10KHz. This equates to 1KHz per meter range.

2.6 Diplexer Feed

A single antenna is used to both transmit the carrier (Tx) and to receive the reflected signal from an object (Rx). This requires a diplexer to route the transmitted carrier from the VCO out through the antenna and the incoming reflected signal routed to the receiver.

Diplexing is achieved in the waveguide antenna feed converting the linearly polarised carrier generated at the VCO into a circularly polarised carrier to be transmitted from the unit. For the purposes of an explanation, lets assume that the Tx carrier is linearly vertically polarised. This enters the waveguide antenna feed system in the vertical plane and is converted into circular polarisation which then leaves the unit via the antenna. When the circularly polarised return signal Rx, enters the antenna it will pass back through the feed system which will convert it back into linear polarisation. After passing through the feed back into the unit, this returned signal will be linear and in the horizontal plane.

At the point where the signals leave the microstrip circuit and enter the waveguide output we now have the Tx in the vertical plane and the Rx in the horizontal plane. These can be transformed from the waveguide to the microstrip circuit by the use of two probe transitions, one in the vertical plane and the other in the horizontal.

Both the diplexing and antenna feed are achieved in one part using the dielectric properties of a cross-linked polystyrene material. A dielectric vane is used within circular waveguide angled at 45 degrees to the plane of the linear polarised Tx and Rx feeds. This same vane protrudes out of the waveguide to form the antenna fed.

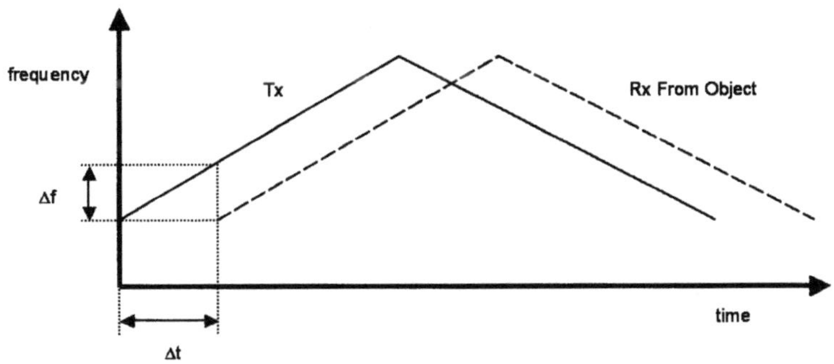

Fig. 10. Transceiver electronics board assembly.

3 Transceiver Module Construction

Two circuit boards contain the transceiver electronics; each board being approximately 45x50mm in size and these are placed either side of a chromated aluminium core. One board, figure 10, contains all the low frequency circuitry; power supply, ramp generator, loop lineariser, IF pre-amplification and final IF amplification.

Microwave softboard copper clad material is used for the other board which contains the microwave functions of the module. Most components are surface mount solder attached the only exceptions being the flip-chip mixer diodes and the DRO resonator. The VCO is fixed directly to the aluminium base so as to give a good thermal path. The 77GHz output is delivered through the wall of the VCO package via a high frequency performance glass to metal seal. The microwave board is mounted directly to the aluminium base and an aluminium lid encloses this circuit as well as providing partitioning between the various circuit functions. The overall size of the module excluding the connector interfaces is 50x45x15mm^3 (figure 11).

Fig. 11. Transceiver module.

4 Performance Measurement

An initial performance demonstration has been performed in the laboratory. For the purposes of this demonstration the transceiver module was fitted in a box (100mm cube) with a dielectric feed being used to illuminate a quasi-optic lens made of high density polyethylene (HDPE). An output power from the transceiver of 10dBm was obtained and the conversion loss of the mixers was 8.5dB. A 25dB isolation was achieved between the right hand and left hand polarisations. Using a corner reflector, targets have been detected at a range of 100 metres with the output frequency equating, as expected, to 1kHz per metre.

5 Conclusion

The challenge of achieving low cost and small size with the performance required for a long range (150m) adaptive cruise control sensor has been demonstrated. The performance of the planar VCO, which is regarded as the critical component in the sensor, was shown to be adequate during initial measurements which have been made. Standard surface mount device (SMD) technology was used in the construction of the microwave board and a very simple low cost design for the mechanical housing was achieved. The next important step is the design of the antenna arrangement which must have the ability of being able to receive reflected signals from several known angles.

References

[1] N. R. Couch, H. Spooner, P. H. Beton, M. J. Kelly, M. E. Lee, P. K. Rees, and T. M. Kerr, "High-Performance, Graded AlGaAs Injector, GaAs Gunn Diodes at 94 GHz.", IEEE Electron Device Letters., vol. 10, no. 7, pp. 288-290, July 1989.

[2] I. Dale, J. Stephens, and J. Bird, "Fundamental mode graded gap Gunn operation at 77 and 84 GHz,"Microwaves 1994 Confπerence Proceedings, October 1994, London, p 248

[3] N. Priestley, K. Newsome, I. Dale and P. Norton", A Gunn Diode Based Surface Mount Oscillator for Automotive Applications", IEEE MTT-S International Microwave Symposium Digest, June 2002, Seattle, pp1863-1866.

[4] P. L. Lowbridge, "Low cost 94GHz FMCW Radar for Precision Range Measurement", Military Microwaves Conference Proceedings, 1990

Nigel Priestley, Brian Prime
e2v technologies Ltd
Carholme Road
Lincoln
UK
nigel.priestley@e2vtechnologies.com

Keywords: millimeter-wave, transceiver, ACC, MMIC, gunn diode, sensor, radar, signal processing

Development of Image Processing Camera (IPCAM)

Y. Fujii, K. Sakata, H. Tsukaoka, H. Ishikura, T. Yamamoto, K. Yasui, Mitsubishi Electric Corporation, Automotive Electronics Development Center

Abstract

This paper outlines methods of cost reduction and downsizing of our Image Processing Camera (IPCAM) intended for use in lane departure warning system or lane keeping system. These systems have potential needs in the society because they may be abele to reduce fatal traffic accidents. But we faced two issues to solve before the popularisation of these systems. First one was how to reduce the cost of image-processing camera currently consisting of many expensive devices. Second one was how to improve the lane detection performance and the accuracy. Our attempt to provide solutions to the above-described problems began with innovation of hardware architecture for image processing, based on our unique algorism for lane detection. We realized that pre-processor for image processing and external memory ware eliminated by the innovation. In addition, an adoption of 1 chip CMOS imager was realized by long exposure within those systems delay. Moreover we could materialize conversion from glass lens into plastic one. Then we successfully devised the way to operate an image processing only with a general-purpose microcomputer and an image sensor. Now the volume of the camera is just $50cm^3$. Compared with costly conventional ones, our image-processing camera is far smaller and low cost. In this paper, we are going to describe our revolutionary image-processing camera aimed for wide diffusion.

1 Introduction

The lane departure warning system (LDWS) and the lane keeping system (LKS) have gradually attract attention from automotive industry as one of the means to prevent inadvertent lane departure, traffic accidents and/or crises, and to relieve drivers' stress. As a matter of fact, these systems are beginning to be installed from luxury segment and commercial vehicles [1-4]. The International Standards Organization (ISO) has been working on the standardization draft of LDWS, knowing that the LDWS and the LKS will be increasingly popular in a foreseeable future. However, current option prices of these systems remain well over US$3,000, which is deemed the main inhibiting cause for its wider diffusion. It is the cause of expensive systems that lane detection method base on real-time image processing technology, because the method needs many expensive hardware resources. And consequently, the systems contained the image-processing camera are very expensive. In other words, a bottleneck for the cost reduction of image-processing camera is its massive hardware volume. This fact clearly signifies some kind of breakthrough is indispensable to reduce the hardware volume, namely, the cost. However, elimination of hardware components simply results in degradation of the performance, e.g. detection ability, accuracy, speed calculation, and so forth. In the course of our struggle against this contradiction, Mitsubishi's know-how accumulated from our long manufacturing history since 1995, when we pioneered the lane detection system, has been fully applied [5]. Below are the details of our approaches toward overcoming such contradictory problems through renewing the conventional hardware architecture and optimising its algorithm ability.

2 Problems of the Image Processing Hardware

2.1 Conventional Image Processing Camera

Figure1 depicts a typical hardware architecture of conventional image-processing camera that constitutes chiefly of image capturing block and image processing block [6].

An image-capturing block was mainly composed of a CCD and some peripheral ICs, A/D converter, and a unit of power supply and ground for the CCD. In the image-processing block, a pre-processor used for image processing (a customized ASIC may also operate the task), an image memory (with a control ASIC), and a microprocessor were believed minimal composition to achieve real-time processing. If heavy load is anticipated, these pre-processors and MPU's may be multiplied. This kind of architecture is deemed fairly normal. Confined in this architecture, it can readily be understood how difficult it would be to reduce hardware volume and cost. Thus we thoroughly reviewed each processing for lane recognition and concluded that not in all cases had we used up the total potential of such massive hardware, despite having uniquely small algorithm feature in our product. A part of our review that resulted in the findings of the barely necessary hardware ability is shown in the Table 1.

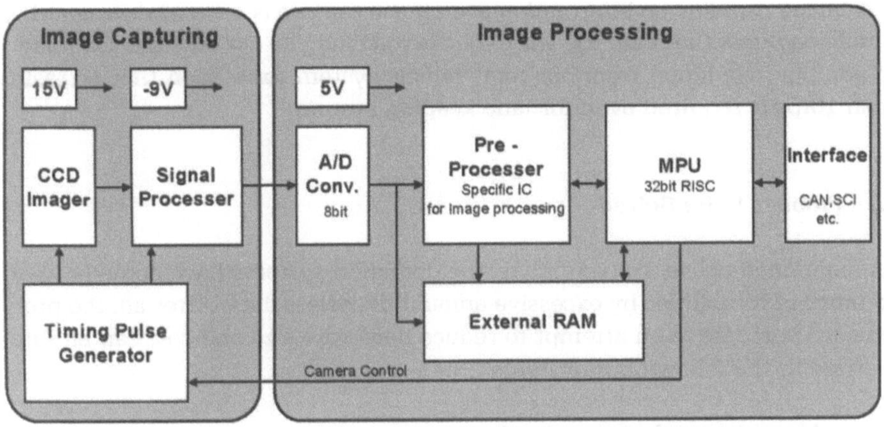

Fig. 1. Typical hardware structure of conventional image-processing camera.

Component	Conventional	Minimal Specification	Remarks
MPU	32 bit RISC or higher	32 bit RISC	Obviously, it needs high processing power.
Pre-Processer	Necessary	Omissible	This can be shrinked if algrithm successfully devised.
External RAM +ASIC	1MB or more	50KB	Only an area barely necessary for lane marking detection should be processed.
Image Sensor	640(H) × 480(V)	640(H) × 100(V)	Only an area barely necessary for lane marking detection should be processed.
Frame Rate	30fps	10fps	Processing cycle time of 100ms sufficient for LKS.

Tab. 1. Comparison of architecture between the conventional hardware and the minimal specification for lane detection.

As shown in table 1, we found all hardware items except the MPU were redundantly massive. The pre-processors for image processing could also be omitted without sacrificing the performance, if we could somehow devise the algorithm. In table 1, presumed data is in a VGA format. The reason to set the vertical pixels at 100 is because the detection object (road surface) has around 100-pixel worth of vertical dimensions in the image. On the contrary, the following areas in the image are unnecessary: sky, road surface without markings, and areas where accurate detection cannot be guaranteed due to dispersions. The area of "road surface without markings" could be neglected because this image contains no information even if the camera is swayed downward so it fully captures the road, the width of the road may be cut off from the image. In addition, we found from our total vehicle system simulation that no faster than 10fps is required even for lane keeping system.

2.2 Problem to Be Solved

As mentioned above, conventional lane detection systems have actually wasted much of its abilities by excessive amount of useless data. After all, the problems to overcome in an attempt to reduce hardware size and cost can be summarized in the following four items.

1. Algorithm to downsize hardware
2. Solutions to reflect small capacity of required memory to cost reduction
3. How to utilize CMOS imager with worse sensibility
4. Improvement on detection accuracy

The first problem listed above points out that the characteristics of algorithm are the determining factor of hardware size; that is, cost. In fact, it is better to adopt an algorithm that best matches its detection object than to have an image-processing library, which is dependent upon the specific pre-processor in the framework of hardware size reduction. This is a very critical point for the development of image-processing camera; downsizing hardware is vital, where degradation of detection performance is not allowed.

The second problem states that the cost for memory IC cannot be simply reduced even if the capacity of required memory is small. That is because the number of chips is independent of the capacity of memory. Even worse, the cost will be higher if the memory moves away from its trend of the time. In addition to that, when dealing with image processing, it is still necessary to use a memory-access control IC to operate read-and-write despite that the capacity of required memory is reduced. This is because a control function of image data cannot be eliminated, image data from CMOS imager are written onto the memory in an order along the scanning line, but on the other hand, the MPU reads out the data in a random order. Therefore the memory even if its capacity is reduced still needs 2 chips and does not contribute to any cost reduction.

The third problem is a concern over the basic performance of imager chip. To realize imaging functions listed on the table 1, CMOS imager that can manage random access is more suitable than CCD. CMOS imager is also attractive in the light that it only needs a single power supply and can operate in 1 chip, although current CMOS imagers still have problems in night driving situations due to its low sensibility.

The fourth problem deals with detection accuracy. "Lane detection" basically detects a lateral position of itself within a drive lane, if used in LKS and LDWS. The accuracy highly relies on the number of pixels from the imager chip, but it should be exaggerated that the performance of lens unit, which provides 2-dimensional pictures, plays an important role in deciding hardware potential on detection accuracy of the camera. Needless to say that a lens unit is an analogue component and its characteristics (distortion, ghost, dispersion of focal point etc.) actually affect the accuracy. In the following section, we look into how to solve these four problems we confronted.

3 Solutions

We were not confined to a general, conventional architecture of image processing hardware upon searching solutions to the foresaid problems. Firstly, the pre-processor was eliminated as a result of applying our unique compact algorithm. Secondly, an external memory was omitted as a result of optimising the usage of internal memory of the MPU. Additionally, the sensibility problem of the CMOS was confirmed avoidable with adoption of extended frame rate. Finally, the adoption of aspheric lens realized high resolution and low distortion; even adoption of plastic lens seemed possible. We will be explaining detailed contents from the next section onward.

3.1 Algorithm to Downsize Hardware

Typical image-processing camera as shown in figure1 includes specific pre-processor because algorithm of image processing generally consists of repetition of processing, which is simple but requires calculation cost, so it makes it impossible to perform in real time without specific pre-processor. In developing an image-processing algorithm for lane recognition, we have been concentrated on simplifying the processing, that is performed repeatedly, and requires calculation cost. One of the outstanding results is called the Top Hat Filter (THF), which is a simple but effective filtering method. Figure 2 shows how the THF works.

By using this filtering method, we can obtain accurate position of lane markings with little calculating cost, even in the situation of different luminance at right and left as shown in figure 2. Moreover, this filtering method matches general imager because the process is closed in one line. This match occurs because the process is closed in one line. General methods, such as Huff conversion for instance, store original picture in the 2-dimensional RAM area. On the contrary, this method finishes its process only in 1 line. After the process, the MPU stores only detected lateral position data, not the original picture anymore, and can free the rest of its RAM area. It is for this reason that we can efficiently utilize available hardware resources. As described in detail, this algorithm doesn't only match the characteristics of CMOS, but also compact, high speed, and light in hardware. If this filtering method is processed by software with 32bit RISC MPU, the computing cost is estimated around 15ms. This fact means we can even eliminate the expensive pre-processor for image processing in conventional image sensors with this algorithm. As described above, we succeeded in performing all image processing only by software with a general purpose MPU.

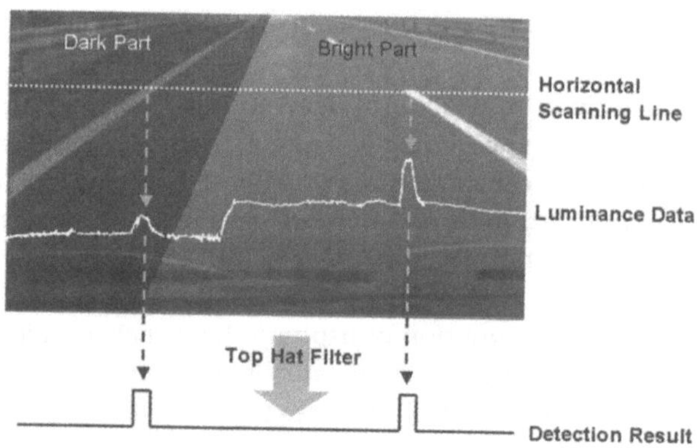

Fig. 2. The filtering, we call "Top Hat Filter".

3.2 Solutions to Reflect Small Capacity of Required Memory to Cost Reduction

In a general image-processing camera as shown in figure 1, whole image data are stored in a huge external RAM at first. Then MPU accesses the image data, and performs image processing, also using specific IC for image processing. Though, the algorithm we have developed does not require the whole image data. The memory capacity required in order to guarantee proper operation of our algorithm is less than 50 Kbytes. In this algorithm, the processing is made possible only with an internal RAM in the MPU. Therefore, we decided to eliminate this external RAM, for instance, we limited image data required for image processing carefully and built the structure, which could transmit the image data output from the imager to the internal RAM of MPU directly. The reason why we made it to realize the structure was that we had been developing the simple recognition algorithm that would requires little memory resource and little calculation costs. Thus the elimination of the external image memory, which is unthinkable in conventional image processing, is finally carried out, and as a result, our advantage of compact memory is fully linked with cost reduction.

3.3 How to Utilize CMOS Imager With Worse Sensibility

By carefully selecting a CMOS imager, we can possibly obtain various merits: i.e., single power supply voltage and integration of peripheral ICs. However, there are also some demerits, such as low sensitivity, and a lot of noise at high temperature. Generally, a CMOS imager is said that sensitivity is lower than CCD imager. Supposing a situation to obtain an image of dark road only with headlights using CMOS imager, for example, the image would be almost black at a normal frame rate. We then investigated how long the exposure time could be extended at maximum. Assuming an application for the LKS, we obtained an acceptable delay time of image processing by total vehicle system simulation, and optimised the processing cycle time. According to the results, the vehicle motion remains stable up to 100ms cycle time (on expressways in Japan at 140km/h). It was also found from these results that the faster the processing was, the better the performance would be. Repetition of investigation and experiments finally told us that the exposure time should be set at 80ms to capture lane markings with both headlamps. The frame rate was also reduced from normal video level of 30fps down to 12.5fps (with 80ms exposure time, F4), and hence we overcame the low sensibility problem of CMOS chips.

As a result, we succeeded in adopting a CMOS imager. In addition, we selected the imager of 640 pixels in horizontal, because we judged that it was enough resolution for detecting lane markings to 40m ahead. It is significant to note that we made it to maintain the same performance level with such a dark lens as F4, because a brighter lens would be very costly. This result greatly influences the design of the lens unit described in the next section.

3.4 Improvement of Detection Accuracy

Conventional lenses for video cameras have appreciated brightness and resolution more than distortion. However, those lenses used for the detection of lateral position must value distortion most. The lens unit that satisfies these needs generally becomes very expensive, for it combines many into one. That is why we cut necessary performances off from unnecessary ones and developed an optical lens unit suitable for lane marking detection processing. In other words, the distortion performance was maximized, while the lens unit brightness was reduced because sufficient sensibility was successfully obtained with F4, as described above. For the plastic lens, we paid attention to its temperature-dependent refractive index upon designing so it can obtain the barely necessary resolution at +85℃Celsius temperature. As a result of the above efforts, an optical system composed of only 2 lenses was successfully developed as shown in figure 3. The significant distortion performance is measured under 0.15% even at the peripheral area. This high fidelity made distortion compensation by software unnecessary, and alleviated load of the MPU. Hence a more accurate lane marking detection was realized. Figure 3 shows an outline of the designed lens unit.

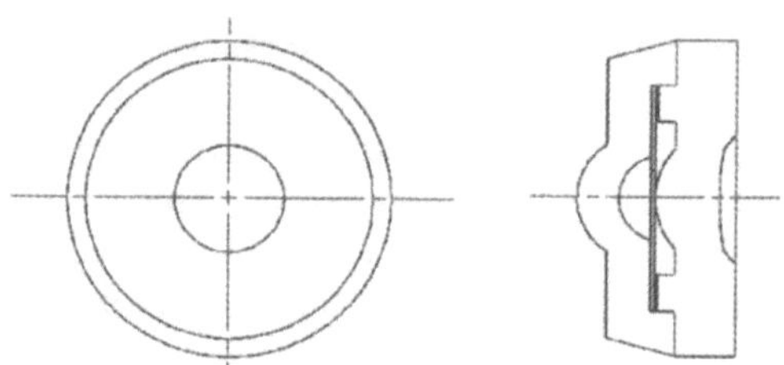

Fig. 3. Designed plastic lens unit.

4 Result

4.1 New Hardware Structure and Specifications

As a result of the solutions we found, we succeeded in building the lane recognition system mainly in 2 chips composition: a CMOS imager and a MPU. Figure 4 shows the hardware structure of our image-processing camera for detecting lane markings (we call "2 chips IPCAM"). And Fig.5 is a photograph of the prototype.

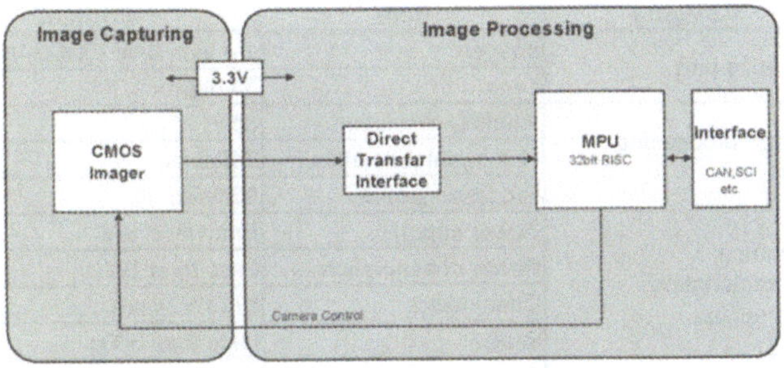

Fig. 4. Hardware structure of our "2 chips IPCAM".

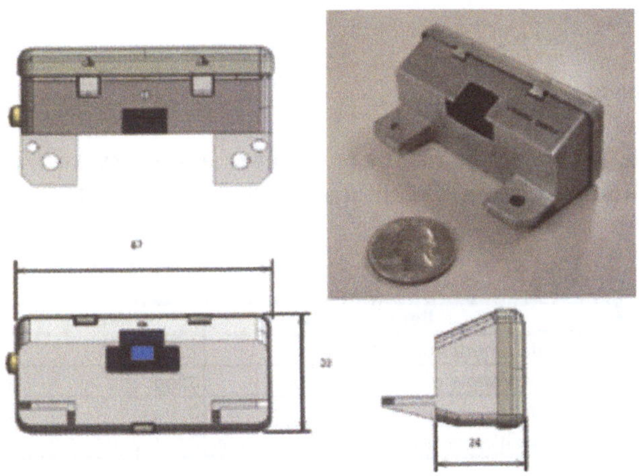

Fig. 5. Prototype 2 chips IPCAM.

As shown in figure 5, it can readily be understood how compact the structure is with only 67W x 32H x 24D dimension. Since the volume of this prototype is approximately 50 cm3, it would not hinder driver's visibility if it were mounted on the front windshield. The fact that the camera is far smaller than a rear-view mirror definitely facilitates vehicle interior design. In addition, table 2 shows main specifications of the prototype, 2 chips IPCAM. We will describe the performance in detail in the following section.

Category	Item	Content
Imaging part	Imager	1/4 inch B/W CMOS VGA
	Lens	33 degrees, F4
Image processing part	Intensity resolution	8bits
	Horizontal resolution	640pixels
	Processing cycle	80msec
Electrical characteristics and outline	Power supply	From 9 to 18V
	Power consumption	Less than 1W
	Dimensions	67x32x24mm
	Weight	Less than 300g
I/O	Communication	SCI or CAN

Tab. 2. Main specifications.

4.2 Performance

In this section, we show the results of our evaluation tests on static detecting accuracy of lateral position. Conditions of the evaluation tests are shown in table 3.

Item	Value
Height of camera from the ground	1.2m
Depression angle of camera	0deg
Lane width	3.6m
Lateral position of camera from lane center	0.0, 0.2, 0.8, 1.2m

Tab. 3. Conditions of the evaluation test about detecting accuracy.

Detection errors stated in this section means lateral deviation divided by detected distance (that is, the $\tan\theta$, if the lateral deviation is represented with a camera angle θ to detected point). Considering possible applications such as the LDWS or the LKS, we judged that acceptable detection error of lateral position should be less than 0.5% (5cm at 10m ahead). Table 4 shows the result of evaluation tests with a prototype unit.

As shown in table 4, our target specification of less than 0.5% is satisfied. In addition, the prototype has a function that calculates a radius at 40m ahead.

Moreover, we must mention the recognition performance. Figure 6 shows sample image of processed result.

Lateral position of camera from lane center [m]	0	0.2	0.4	0.8	1.2
Lateral Error Rate [%]	0.082	0.155	0.163	0.221	0.312

Tab. 4. The result of evaluation test for detecting accuracy of lateral position.

The result of lane recognition is shown as white dots in the image. Although, there are sunny part and shady part in the image, lane recognition is performed properly. Table 5 shows the lane marking recognition rate during actual road tests. The rate is shown in 1 - error time /Total drive time. The data are the accumulation of 50 hours of road tests on Sanyo, Chugoku and Meishin expressways in Japan, including day and night. Our ultimate goal is to heighten the rate to 100%, but for the time being it remains at 96%, which can be considered acceptable for the actual usage because test drive included adverse condition such as no lane marking area, situation of cutting in vehicle, and sudden rain, etc.

Fig. 6. Sample image of processed result.

Location	Recognition Rate 1 - (Error Time /Total Drive Time) [%]	
	Day time	Night time
Sanyo Expressway	97	97
Tyugoku Expressway	97	96
Meisin Expressway	96	96

Table 5. Lane Marking Detection Rate during Actual Road Test.

6 Conclusion

After all, we made successful approaches to the development of an image-processing camera with substantial reduction in cost and size, high performance in detection, and high accuracy. As a result, a reasonably low cost image-processing camera, i.e. 2 chips IPCAM, we originally pursued was spawned compared with very expensive conventional ones. A very reasonably priced system structure can be realized with the IPCAM we developed. This IPCAM is expected to widen the range of vehicle segments equipped with LDWS or LKS from luxury and commercial vehicle to all vehicle including compact and subcompact segments. Penetration of these systems into the market promotes more safety and convenience for drivers at reasonable price. We are now looking into the possibilities of applying the same hardware to detection of preceding and cut-in vehicles, and later to the so-called "Automatic Stop-and-Go" function. Our relentless pursuit for the provision of drivers' safety and comfort πthrough adding more values will ever continue.

Reference

[1] Yoshiki Miichi, Takashi Ohta, Akihiro Shin, Hisao Arita; "Development of Mitsubishi Driver Support System", Mitsubishi Motors Corporation Technical Review, 2000, No.12, pp.81-84.

[2] On Sadano, Minoru Tamura, Shigeki Satou, Yoshinori Yamamura, Takayuki Watanabe, Kiyomi Hirasago, Minoru Kanehira, Hitoshi Satou; "Development of the Lane Keep Support System and Brake Assist System with a Preview Function", Nissan Technical Review, February 2001, No.48, pp. 13-17.

[3] Makoto Nishida, Hisashi Satonaka, Takashi Hiramaki, Arata Takahashi, Satoru Masuda; "Development of Lane Monitor Using Rear view Camera", Society of Automotive Engineers of Japan, Inc. Annual Congress, 2002, No.114-02, pp.1-4

[4] Makoto Kawai, Shinnosuke Ishida, Takayuki Tsuji; "Intelligent Vehicle and Advanced Safety Technology", Journal of Society of Automotive Engineers of Japan, February 2003, Vol.57, No.2, pp.44-49.

[5] Kazumichi Tsutsumi, Yuhei Akasu, Kazuhiro Ikebuchi, Shigekazu Muraoka, Hiromichi Kushizaki and Minoru Nishida, "Headway Distance Control Technology", Mitsubishi Electric Technical Review, September 1996, Vol.70, No.9, pp.13-16.

[6] Yoshiyuki Fujii, Hideki Tsukaoka; "Technology to Integrate a Vehicle-Mounted Camera and Image-Processing Unit", Mitsubishi Electric Advance, June 2001, Vol.94, pp. 14-17.

Yoshiyuki Fujii, Kazuki Sakata, Hideki Tsukaoka, Hisashi Ishikura, Takayuki Yamamoto, Katsuaki Yasui
Mitsubishi Electric Corporation
Automotive Electronics Development Center
840 Chiyoda-Machi
Himeji, Hyogo, 670-8677
Japan
fujiiy@hime.melco.co.jp, sakatak@hime.melco.co.jp,
tsukaoka@hime.melco.co.jp,ishi@hime.melco.co.jp
yamataka@hime.melco.co.jp, yasui@hime.melco.co.jp

Keywords: image processing camera, lane recognition, lane departure warning system, lane keeping system

VisiTyre: A TPMS Solution Employing Directly Connected, 2-Wire Communication Channel to Achieve Highly Predictable and Reliable Performance

P. Cohen, ETV Corporation Limited

Abstract

The problem of interfacing automotive body mounted electronics to sensor elements mounted on a vehicle's rotating wheels presents a serious challenge. The main requirement is to reliably pass continuous real time pressure, temperature and force vector data from sensors that are mounted inside a vehicle's rotating wheels to chassis mounted receiver electronics.

This paper focuses on ETV's VisiTyre, a direct measurement Tyre Pressure Monitoring System (TPMS) incorporating sensor elements which are complete "system on silicon" subsystems having a simple yet robust, directly connected, 2-wire interface and communication channel to chassis mounted receiver electronics.

VisiTyre obviates the need for sensor batteries and RF transmitters inside the tyre and introduces the concept of standardised "TPMS Enabled Tyre Rims" and "TPMS Enabled Wheel Hubs" that allow easy tyre and wheel removal, replacement and transparent wheel rotation using existing tyre industry equipment and facilities. The VisiTyre standard is proposed for adoption as a universal standard for direct connect tyre pressure measurement systems.

1 Introduction

With the introduction of mandatory TPMS on new passenger vehicles sold in the USA [1], the issues of accuracy, reliability and cost of proposed solutions for this important safety system are receiving prominent attention [2].

There are two categories of TPMS currently being considered for adoption which alert a vehicle's driver to abnormal tyre pressure conditions:

- Direct measurement systems; and
- Indirect measurement systems.

A direct measurement system measures tyre pressure directly with physical pressure sensors. Indirect measurement systems measure something other than actual tyre pressure, such as relative wheel angular velocities or axle to road height changes.

Presently, there are serious performance and limitation issues inherent in all available systems. These shortcomings, which are summarised as follows, contra-indicate their use for such an important safety system.

2 Limitations of Current Systems

To date, direct measurement solutions have typically involved battery powered Radio Frequency (RF) transmitter modules mounted inside the tyre/rim cavity or chassis mounted antennae systems that energise a wheel's passive transponder sensors. The major competitors offering Direct Measurement systems are Beru, Johnson Controls, Schrader and SmarTire.

The most common Indirect measurement system is based on Adaptive Braking System technology (also called Anti-lock Braking System and known as ABS). These systems rely on rotational wheel speed sensors that are implemented using a rotating toothed metal wheel to modulate a magnetic field. The magnetic field induces a signal with a frequency proportional to angular velocity in an adjacent coil. Electronics is then used to detect small changes in one wheel's speed relative to another. This approach is based on the tyre's radius decreasing when a tyre loses pressure. Indirect TPMS based on ABS technology is implemented by TRW, Delphi Automotive Systems and BMW (M Series & Mini).

All of these approaches continue to suffer from system limitations and reliability problems. These limitations are summarised as follows [2]:

2.1 Limitations of RF Based Direct Measurement TPMS

- The receiver's signal and/or electronics is subject to deafening and interference by spurious energy fields from various higher power external sources sharing the air interface medium (e.g, television transmitters, cell phones, Bluetooth radio devices, LAN wireless devices, garage door openers and CB radios etc.).

- Cross-talk between other adjacent vehicles with active transmitters and receivers.
- Being affected adversely by the influence of metallic vehicle parts on energy field density patterns in the vicinity of the receiver's antennae. The performance of these solutions varies greatly between vehicle models because every vehicle model has a different geometry that may interfere adversely with energy field densities.
- Dependence on battery power sources inside the tyre. Once batteries are depleted, operation is compromised and subsequent serviceability is a costly ongoing maintenance factor.
- In order to conserve battery power continuous monitoring is not possible. Transmission takes place in concert with "sleep mode" algorithms and then only when pressure activation floor is passed (this is a non-deterministic transmission mechanism and as such is deemed to be not ideal for such an important safety system).
- Difficulties identifying tyre positions after tyres undergo rotation schedules. Usually, tyre positions must be manually reprogrammed by skilled individuals to avoid consumer set-up errors.
- The battery & RF transmitter wheel modules add significant weight, which affects wheel balance.

2.2 Limitations of ABS Based Indirect Measurement TPMS

- In order for ABS based TPMS systems to work, the vehicle must be moving above a minimum speed, ~24kph and below a maximum speed, ~112kph.
- Cannot measure tyre temperature (Abnormal temperature is a common precursor to tyre failure).
- Cannot be easily calibrated for change in atmospheric pressure (e.g., with change in altitude).
- Poor resolution: Can only detect a 30% pressure loss differential between tyres (i.e., if all tyres were inflated to say 28lb/in^2 one tyre would have to deflate to 19.6lb/in^2 before the system would detect it).
- Cannot tell which tyre is under-inflated.
- Cannot detect some combinations of 2 under-inflated tyres. This is a result of the measure of relative wheel rotation speeds.
- Cannot monitor spare tyre.
- Significant time (tens of minutes) to detect pressure loss.
- Only suitable for use with a "Tell Tale" alarm.

3 Impact of Legislative Requirements

The mandating of TPMS as a safety feature in passenger vehicles sold in the US market[1] has important impacts on manufacturer's responsibilities as well as their profits and is concomitant with product performance expectations of the consumer.

Whereas previously, manufactures could include TPMS in their options cost model they must now pass on the cost of a compulsory safety system to all consumers. Consumers have the rightful expectation that a tyre safety warning system fitted to their vehicle under legislative requirements will be accurate, reliable and function correctly over the life of the vehicle. Any failure of these systems could lead to loss of life or property. In addition manufacturers could be embroiled in litigation or expensive product recall scenarios.

For these reasons it is vital that any TPMS solution used should have the highest possible quality, reliability and performance within acceptable cost frameworks of the automotive industry and the consumer. It should not have any of the limitations previously summarised.

This is the rationale leading to the need for and design of VisiTyre.

4 General VisiTyre Model

At a system level the VisiTyre TPMS solution comprises eight distinct sections. Referring to Fig. 1, these are:

1. Sensor Elements
2. Communication Channels Between Sensor Elements and Electronic Control Module
3. Electronic Control Module
4. Communication Channel Between Electronic Control Module and Human User Interface
5. Human User Interface
6. Control Software/Algorithms
7. Human User
8. Power Supply

The VisiTyre implementation of sections 1,2,3 and 8 will constitute the remainder of this paper.

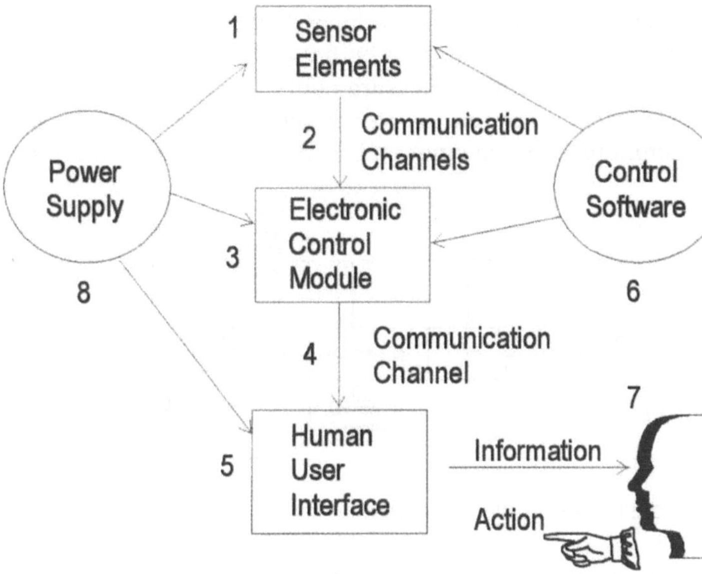

Fig. 1. Generalised VisiTyre System Components.

5 VisiTyre System Components

5.1 Sensor Elements

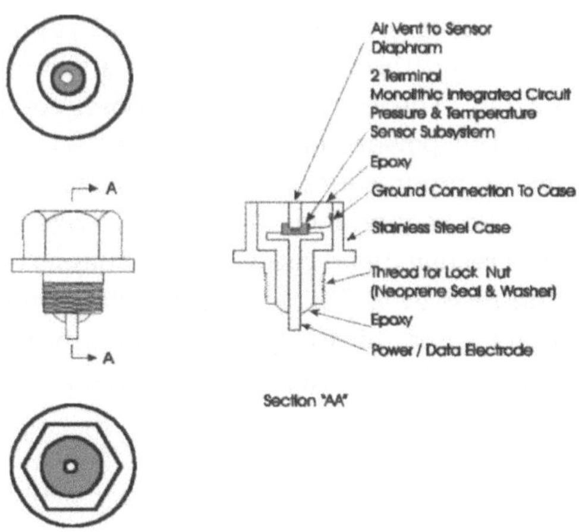

Fig. 2. ETV DCS01 Sensor Package.

The VisiTyre solution is a Direct Measurement TPMS incorporating proprietary ETV DCS01 sensor elements (figure 2) that are complete "system on silicon" subsystems, and performing measurements of pressure, temperature and force vector. They have a simple yet robust 2-wire interface for direct connection to the communication channel and to the chassis mounted receiver electronics.

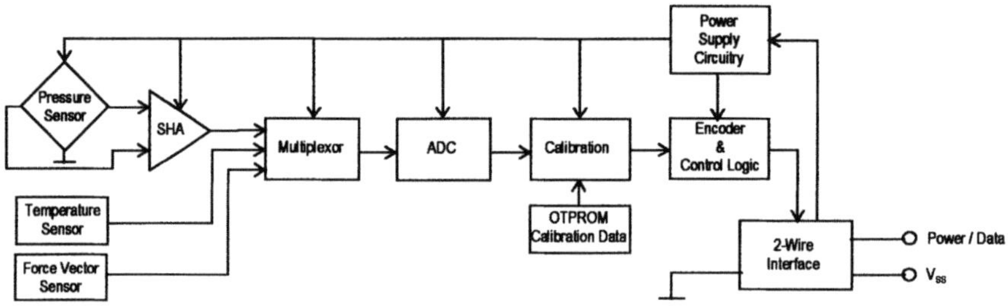

Fig. 3. ETV DCS01 Sensor Block Diagram.

The sensor subsystem leverages SensoNor SPxx family technology [3] with a sealed vacuum cavity which is micro machined with a single-crystal silicon diaphragm fabricated to contain four environmentally protected piezoresistors. The piezoresistors are aligned to the diaphragm's stress field resulting in an extremely low zero point and low sensitivity drift with change in temperature. The subsystem is designed around a digital core which is pre-calibrated at time of manufacture to ensure high accuracy over the automotive temperature range of -40°C to +125°C.

The ETV DCS01 sensor's 2-wire electronics interface comprises a Ground, and Power/Data connection. This interface is capable of supplying power to the sensor subsystem, initiating the measurement of a tyre's pressure temperature and force vector data. The interface superimposes the transmission of the sensor's tyre data on the power connection as a serial datagram. VisiTyre assumes an unreliable communication channel with the serial datagram being transmitted with a Cyclic Redundancy Check (CRC) code following the CCITT norm. The datagram is received and recovered by chassis mounted Electronics Control Module. If the recovered CRC is good the datagram is made available for hand-off to CAN bus communication channel for subsequent display on the driver's instrument cluster. The overall communications pathway is represented in figure 4.

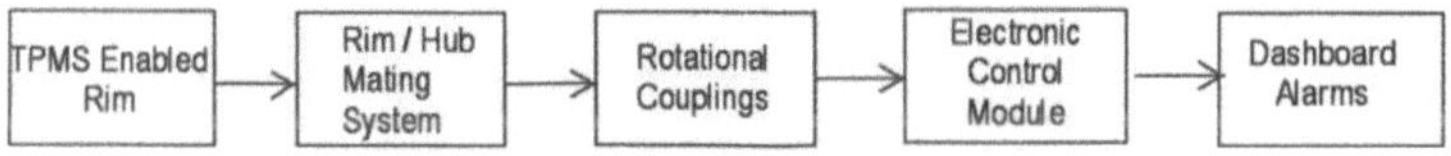

Fig. 4. Overall Communication Path Way.

It is important to note that upon the application of power, the VisiTyre interface initiates a deterministic measurement of a tyre's pressure, temperature and force vector data by the ETV DCS01 sensor. Importantly, this solution obviates the need for sensor batteries inside the tyre, uses the minimum number of physical conductors (2), is economical and has a highly predictable and reliable performance.

5.2 TPMS Enabled Rims

A "TPMS Enabled Rim" is a steel or alloy rim that is equipped with a 2-Terminal ETV DCS01 sensor and a contact pad to facilitate automatic mating and de-mating of the sensor's electrical connection to the vehicle's "TPMS Enabled Hub" during wheel removal and replacement operations.

Fig. 5. TPMS Enabled Steel Rim (Rear View).

5.3 TPMS Enabled Hubs

A "TPMS Enabled Hub" is a drive or stub axle wheel hub that is equipped with a contact pin (shown in red on figure 6) to facilitate automatic mating and de-mating of the rim mounted sensor's electrical connection to the chassis mounted Electronic Control Module during wheel removal and replacement operations.

Fig. 6. TPMS enabled hub (contact pin surrounded by red insulator).

5.4 Rotational Coupling

The function of the rotary couplings is to make two continuous electrical connections from points on a stationary chassis to points on rotating wheels. Electrical connections are for both ground reference and power, with the sensor subsystem's transmitted data signal being superimposed on the power line connection. Coupling rotation speeds vary from zero (stationary) through a slow creep to high revolutions per minute (e.g., 2000 RPM).

Fig. 7. Stub axle hub & bearing assembly fitted with rotational coupling.

Rotational couplings may be based on Electro Magnetic Induction (EMI) designs where changing magnetic flux in a driving coil induces current to flow in another coil in close proximity or they may consist of electrodes in contact. The successful use of electrode type rotational couplings to implement VisiTyre's 2-wire communication channels is dependent on achieving:

- Low ohmic contact resistance (less than 1mΩ).
- Low electrical noise insertion.
- Electrical isolation between circuits.
- Low maintenance and high reliability at continuous operating speeds of up to 2000rpm.
- Compactness and the arrangement of their physical installation and connector wiring pathway.
- Low cost.

VisiTyre has been successfully tested with different rotational coupling contact arrangements for noise sensitive data signals. These include: fully "wetted" electrodes connecting through a shielded mercury pool and low ohmic contact electrodes riding on slip rings coated with advanced materials such as titanium nitride. Multiple contacts per connection have been used to keep electrical noise extremely low using the contact materials indicated. The rotational couplings used are selected to be cost effective and to provide a long, highly reliable service life in the wheel area environment where water, dirt and both high and low temperatures are prevalent.

The use of a Cyclic Redundancy Checking (CRC) algorithm applied to the sensor sub system's serial datagram further enhances the ability of the overall system to cope with any noise introduced into the communication channel.

5.5 Electronic Control Module

VisiTyre's receiver interface electronics in the chassis mounted electronics control module extracts the power line's superimposed serial data. A microprocessor system in the electronics module analyses the recovered data for abnormal conditions. The receiver electronic control module has further interface options that are suitable for connection to various manufacturer's motor vehicle body electronics systems. These interface options include the Bosch Controller Area Network (CAN) bus, GM's LAN bus, RS232 serial port and "Tell Tale" warning light with audible alarm.

6 Conclusion

It is vital that mandatory TPMS solutions should have the highest possible quality, reliability and performance within acceptable cost frameworks of the automotive industry and the consumer. This is the rationale leading to the need for and design of VisiTyre. The VisiTyre design leverages two technologies to provide a completely new technology platform for Tyre Pressure Measurement Systems:

- Advanced Sensor Element with a 2-Terminal Interface.
- Direct Connect, 2-Wire Communication Channel.

The Direct Connect 2-Wire Communication Channel employs two important components; rotational couplings and a unique mating contact system based on standardised "TPMS Enabled Rims & Hubs". In combination, these components:

- Supply power to the sensor element, thus obviating any battery requirement.
- Facilitate deterministic transmission of tyre data directly to chassis mounted Electronic Control Module.
- Accommodate transparent tyre removal, replacement & rotation.

VisiTyre's use of rotational coupling technology represents a change in attitude – instead of treating couplings as something TPMS engineers should mitigate VisiTyre treats them as a technology which is exploitable with the use of modern advanced materials.

VisiTyre introduces the concept of standardised TPMS Enabled Tyre Rims and Hub Mating System. This standard is proposed for adoption as a universal standard for two-conductor Tyre Pressure Measurement Systems. The VisiTyre standard allows easy tyre and wheel removal, replacement and transparent wheel rotation using existing tyre industry equipment and facilities.

Advantages of the VisiTyre system over other systems being developed or in the market place are:

- Measures pressure, temperature and force vector data continuously in real time.
- Transmits data as a digital datagram with an embedded cyclic redundancy check (CRC).
- Transparent tyre rotation without the need to reprogram tyre position;
- No requirement for sensor subsystem batteries.
- Not subject to being deafened by radio frequency sources.
- Highly predictable and reliable system performance over the vehicle's life.
- Sensor packaging with negligible added weight to affect wheel's balance.
- Reports normal and abnormal conditions simultaneously on all tyres

including spare tyre.

- High reliability of a micro machined silicon fabrication processes.
- High yield of sensor subsystem in mass production.

References

[1] H.R. 5161 TREAD Act Sec 13 USA 106th Congress 2D Session November 1 2000.

[2] NHTSA DOT Tire Pressure Monitoring Systems; Controls & Displays Docket No. NHTSA 2000-8572

[3] SensoNor asa, A Low Cost Fully Signal Conditioned Pressure Sensor Microsystem With Excellent Media Compatibility, SensoNor asa Horten, Norway. 1999

Phil Cohen
BE (Elec Eng) UNSW
Technical Director
ETV Corporation Limited
Level 18 201 Elizabeth Street
Sydney, NSW 2000
Australia
P.Cohen@usa.net

Keywords: tyre pressure measurement systems, TPMS, sensor interfaces, 2-wire communication channel, direct connection, TPMS enabled rims, tyre safety, tire

Short Range Radar System for Automotive Applications

G. Rollmann, V. Schmid, M. Mekhaiel, DaimlerChrysler AG
P. M. Knoll, Robert Bosch GmbH

Abstract

The number of cars in Europe is increasing, leading to a higher traffic density. The average age of drivers is increasing consistent with demographics of the total European population. Every second accident involving vehicles is related to traffic situations in which faster reaction of the driver could have mitigate crash consequences. So, there is an increased need and appreciation for obstacle detection that operate at day and night.

The industry has responded to the European Commission programs and has developed new, efficient 24GHz Short Range Radar (SRR) solutions for Road Safety and Intelligent Transport Systems. Further the new 24GHz Radar system is an essential "building block" of the EU Project RESPONSE. Herein included is a new generation of active safety systems and Advanced Driver Assistance Systems as well as passive safety. These systems will take into account not only the driver and the vehicle but also the environment of the vehicle. The task to finalize the regulatory issues soon is important to get the technology started.

1 Introduction

Today, triggering of restraint systems is based on vehicle deceleration signals during an impact. Airbags are deployed by sensing the crash event. They are typically activated after 5ms. There is no doubt, that airbag and windowbag systems have significantly reduced the number of severe and fatal accidents. Accident analysis, however, shows that in two thirds of all accidents a fairly long phase with a critical driving situation precedes the actual impact. A first step toward making use of this phase is realised for the first time with PRE-SAFE. PRE-SAFE has now seen its first series application in the 2003 S-Class, and combines active and passive safety. The system recognizesrecognises critical driving situations with increased accident possibility. It triggers preventive measures to prepare the occupants and the vehicle for possible crash by eavaluating the sensors of the Electronic Stability Program (ESP) and the Brake Assist (BAS). Reversible belt-pretensioners for occupant fixation, passenger seat positioning and sunroof closure are activated.

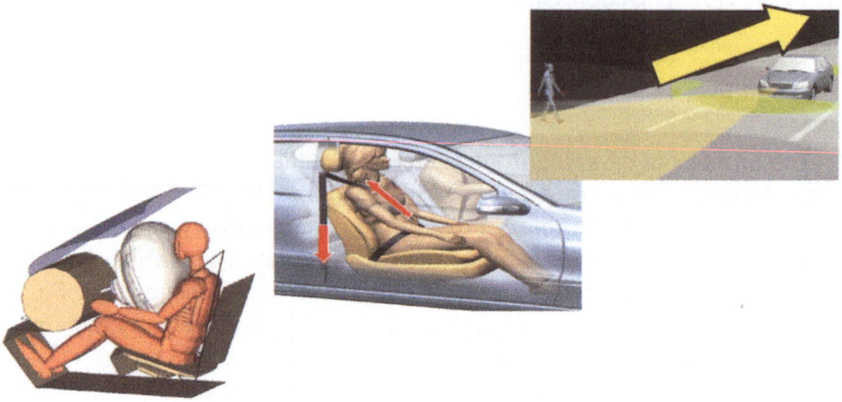

Fig. 1. Saftey system evolution: Airbag – Pretensioning of seat belts – object detection.

However, these are merely first steps in the new and evolving safety technology. Next generation vehicles will additionally take the near-range environment into consideration. Short range radar is thus expected to be the key enabling technology. Placing 24GHz Short Range Radar sensors around the car leads to a "virtual safety belt" allowing to explore several safety and assistance functions, such as collision warning, Precrash sensing, Prefiring Preconditioning of restraints and airbags, Lane change aid, Stop, & follow, Stop & and roll, Recognition of Traffic members, Blind spot detection, Parking aid and Pedestrian protection. Figures 1 and 2 show some examples of possible functions based on Short Range Radar and its sensor data fusion in the future.

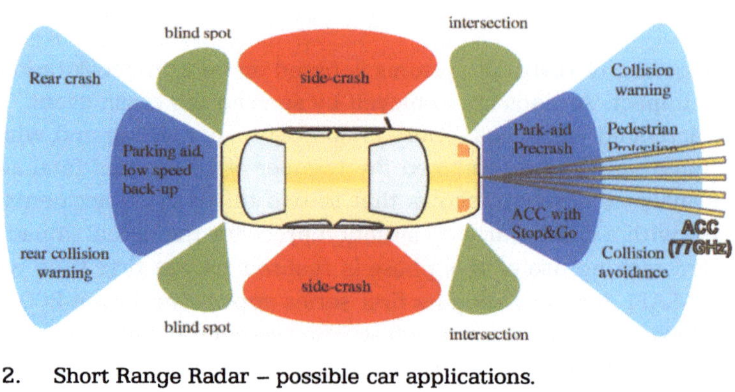

Fig. 2. Short Range Radar – possible car applications.

2 Why 24GHz UWB for Short Range Radar?

Several items have to be taken into account to determine the appropriate frequency for SRR's:

- The **unit size**, which is a consequence of the antenna size resulting from antenna gain and the carrier frequency. The higher the frequency the smaller can be made the antennae. The automotive market accepts an area coverage of about $50cm^2$.
- The **unit cost**. Automotive products may not exceed a certain cost level to be accepted also for mid size and compact cars. The availability of discrete RF components parts (e.g. transistors with sufficient gain) limits the maximum carrier frequency choice up to 30GHz, today. Higher frequencies will cause use of MMICs with increasing costs and high development efforts in the next years.
- **Required Bandwidth**. Radar range separation is proportional to the occupied bandwidth. Short range applications detecting objects of complex shape need a considerable high separation of a few centimeters. This results in a required bandwidth of 5GHz.
- **Application capability**. Automotive electronics needs to find mounting space at the car under conditions which allow for full functionality. Standard bumper material must be used. It suffers from absorption loss, which is proportional to the frequency. The reflection loss coming from the metallic particles in the bumper varnish is proportional to the 4th power of the frequency. Comparing 77GHz versus 24GHz shows an entire transmission loss ratio of more then 20dB for the 77GHz frequency range. New bumper materials have to be studied allowing a placement of 77GHz units behind bumper material.
- **Capability for mass production**. This implies the reliable availability of components and a stable and reproducible production process.
- **Available frequency band**. Global marketing and use or deployment possibility is a very important issue as well as the compatibility to established services. A bandwidth of 1GHz is available for automotive applications (76.5$\pm$0.5GHz). A 5GHz UWB in the 24GHz range (24.125 $\pm$2.5GHz) is available in USA explicitely also for automotive applications. It is not yet available in Europe and is matter of intensive discussions in the EC.

24GHz SRR technology allows a low-cost design at reasonable cost already today and can keep the product size small enough to fit in the space available while providing useful range resolution and object separation which is needed for Cartesian object tracking.

Such SRR functions as mentioned above, today cannot be covered by other means or systems operating e.g. at 77GHz or 5.8GHz . This is because of of size, manufacturability, and cost constraints. Attenuation of the bumper material increases with frequency. At 24GHz, it is at a value of around 3dB, which is tolerable. compared to 20dB at 77GHz. Only cost-efficient systems will be accepted by the market for widespread use, which in turn is a prerequisite for effective public benefit. Currently, the 77GHz ACC sensor of DaimlerChrysler is offered for more than 1000 Euro despite efforts within the last four years since introduction in 1999. A minimum of 8 SRR sensors as required for environmental coverage would therefore cost over 8000 Euro!

For automotive applications, it is essentially to measure radial distances and velocity of objects relative to the own vehicle with an appropriate high resolution (<3cm). Especially detection of pedestrians in front of biglarge objects is jeopardized by low resolution. Precise measurements of object movements are essential for the prediction of trajectories - high resolution helps to prevent false alarms. SRR's with high resolution require a bandwidth of approxapprox. ± 2.5GHz,. On the other hand, only extremely low but at very low emission levels. These emissions are at or below the spurious limit of EN 300 440, fulfilling FCC Part 15 regulations (-41.3dBm in 1MHz) according to the defined emission mask. The emissions outside the proposed emission mask are reduced below -50 dBm in 1MHz.

Extensive compatibility studies are neededhave been carried out under CEPT ECC WGPT SE-24 group, responsible for UWB coexistence studies, in order to determine emission levels, antenna characteristics, and other mitigation factors that will help to avoid harmful emissions to services operating in the relevant frequency bands. Theseat sharing studies have demonstrated that SRR systems can operate on a compatible basis with existing services if 10 percent or fewer of automobiles in the total population in service in Europe are equipped with SRRs. This, which percentage of operation would be achieved in 2014. A package solution is in discussion to start with a first sensor generation at 24GHz and a 2nd generation at 77GHz. The 77GHz technology, being today on the market for ACC (adaptive cruise control) systems today, is not yet mature for SRR :

- High volume, low cost series production at 77GHz for safety and conveniencemfort systems is not feasible within the next 10 to15 years due to significant commercial and manufacturing issues. These issues span the full range from semi-conductor techniques to manufacturing processes and materials.
- Significant development in the area of MMW circuit boards is ongoing but much work is requested to identify new materials that meet both,

the cost and material properties required to support low cost production.

- MMIC technologies require special SMD technologies and packing techniques which are not yet available today.
- The development of these semi-conductor technologies takes about 5-8 years. To bring the technology to production, and to achieve the required reliability for the automotive application of the final products another 5 years are necessary.

3 Locating the Surrounding Objects

A " virtual safety belt" consisting of about 8 to 12 SRR units mainly mounted in the front and rear bumper can supply all the safety applications as well as the conveniencemfort functions with all the information about the surrounding objects. Each application would have its own controller, which evaluates these information according to its concrete function. A stop and go system for example, would only extract the information about the object moving in front.

Since a we need to track targets tracking of moving obstacles at widely dissimilar speeds is required, the radar range resolution must be better than 310cm, as indicated above. Target aquisitionacquisition time should be as low as 20-30ms. The maximum detectable range should be at least 20m. A typical car environment consists generally of more than one object. The system must therefore be multi-target capable. Figure 3 illustrates the main blocks of the SRR-system. It mainly consists of about 8 to 12 front-end units. Theyse have an interface to the CAN-bus. Each unit generates independently of the others a list of up to 10 targets. For each target the corresponding relative velocity component towards the car direction is proccessed by using Doppler frequency.

The Radar Decision Unit (RDU) has access to the front-end target lists via an interfacing processor, which mainly handles all I/O-routines. The RDU unit contains the core signal processing algorithm and has the following tasks:

- Combines all target lists together, generates one list containing validated object information.
- Corrects measurement errors by applying robust tracking techniques.
- Estimates the lateral size of the objects, their central point and their speed vector.

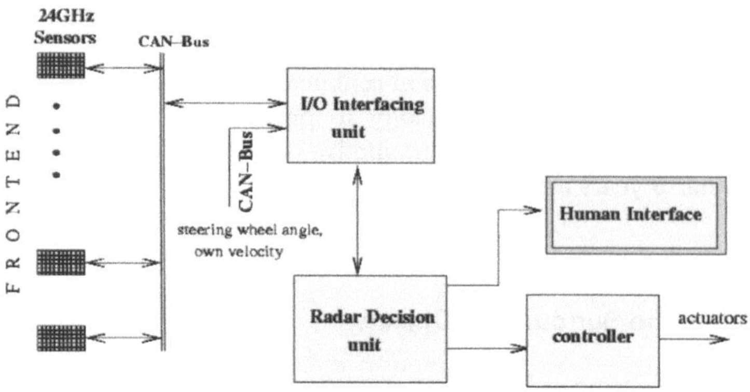

Fig. 3. Block diagramm of the 24 SRR system.

Summary

The first generation SRR at 24GHz is ready for introduction depending on the degree of required reliability. 77GHz still needs about 10 years of further development, and is, therefore, immature for the first generation of SRR units. This is supported by the development results and the past years experience with immense funding investments. The cost as well as manufacturing structure (MMIC, Packaging, Test, Yield,) are most critical factors and they are not expected to be solved at a short term basis.

So only the 24GHz SRR can give a significant contribution to the EU goal of reduction of traffic fatalities until 2010.

References

[1] Ronald K. Jurgen (editor), "Object Detection, Collision Warning and Avoidance Systems", Society of Automotive Engineers Inc. (SAE), USA, 1998

[2] Rodolfo Schöneburg, Karl-Heinz Baumann, Rainer Justen, "PRE-SAFE- the next step in the enhancement of vehicle safety", ESV, Nagoya, Japan, 2003 (sub.)

Gerhard Rollmann, Volker Schmid, Moheb Mekhaiel
DaimlerChrysler AG
Calwer Str 1
71059 Sindelfingen
Germany
gerhard.rollmann@daimlerchrysler.com
volker.schmid@daimlerchrysler.com
moheb.mekhaiel@daimlerchrysler.com

Peter M. Knoll
Robert Bosch GmbH
Leonberg
Germany
peter.knoll@de.bosch.com

Keywords: short range radar, collision-mitigation, precrash, presafe, collision warning

Laserscanner Innovations for Detection of Obstacles and Road

U. Lages, IBEO Automobile Sensor GmbH
K. Ch. Fuerstenberg, K. C. J. Dietmayer, University of Ulm

Abstract

This paper presents innovations for detection of obstacles and road using a Multilayer Laserscanner. The classification of pedestrians and other road users require robust object tracking algorithms. Their reliability strongly depends on a precise ego-motion compensation. A new method of ego-motion estimation based on relative velocities of tracked objects will be introduced. This method operates especially under undefined driving conditions, such as skidding of the ego vehicle, where algorithms based on the vehicle's INS data fail.

High performance object tracking and classification algorithms for real-time data processing will be introduced. Especially the results of a new pedestrian recognition technique will be presented. Methods for road detection and road type classification based on Laserscanner data will be explained in a separated chapter. The fusion of road detection and object recognition will enable a better interpretation of driving situations.

1 Introduction

There are more than 1,200,000 accidents in the European traffic with about 1,600,000 injured persons and 42,000 deaths every year [1-4]. Pedestrian accidents - about 160,000 with more than 6,000 killed pedestrians - represents the second largest source of traffic related injuries.

Last year object detection and tracking in the nearfield of the ego vehicle using Laserscanners were presented at the AMAA conference [5]. In order to obtain a further improvement of accident prevention and a reduction of both injuries and deaths, innovative safety systems must be integrated into future cars. Laserscanners and their algorithms are able to deliver data relevant for such active safety systems [6,7].

2 Multilayer Laserscanner

The MultiLayer Laserscanner used for environment detection is the latest member of IBEO Automobile Sensor GmbH's Automotive Laserscanner family (figure 1).

Fig. 1. Multilayer laserscanner.

Based on the experiences with the previous LD Automotive model, the LD ML uses 4 scan planes instead of one for the observation of the vehicles environment, obtaining a vertical opening angle of 3.2 degree to the front. This allows a compensation for the vehicles pitch angle, as shown in figure 2.

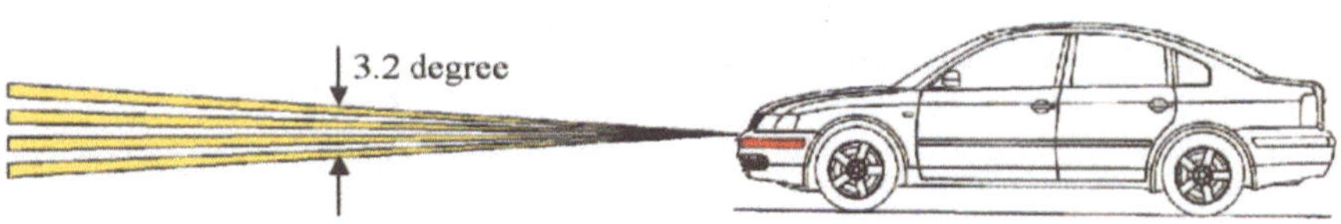

Fig. 2. Multilayer Laserscanner with four individual scanning planes.

The Laserscanner provide a variable scan frequency from 10 to 40Hz. The angular resolution varies with the scan frequency, at 10Hz the angular resolution is 0.25 degree. The horizontal scan area is up to 270 degree and the object range up to 100m. More Laserscanner details and a description of the working principle can be found in [7,9].

3 Ego-Motion Estimation

The Laserscanner measures relative distances to objects in the scan area. Consequently, the Laserscanner is able to calculate only the relative speed of all objects. However, based on the known movement of the test vehicle, it is possible to calculate the absolute movement of all objects.

In order to estimate the ego-motion under defined driving conditions a vehicle model, based on the vehicle's INS-data (Inertial navigation system), can be used. A reliable ego-motion estimation under undefined driving conditions, such as skidding of the ego vehicle, can be determined by using the relative velocities of static objects in the vehicle's environment such as posts, reflection posts and trees. Therefore a distinction between moving and static objects is necessary in order to apply this new method.

3.1 Ego-Motion Estimation using INS-Data

A vehicle model was developed for calculating the movement of the test vehicle. Therefore, INS-information, such as the steering angle and the speed of the test vehicle is used. Neglecting pitch and rolling of the vehicle as well as movement on the z-Axis, three remaining parameters Δx, Δy (vehicle translation) and $\Delta\psi$ (vehicle rotation) describe the movement of the vehicle between two scans of the Laserscanner in a time period T.

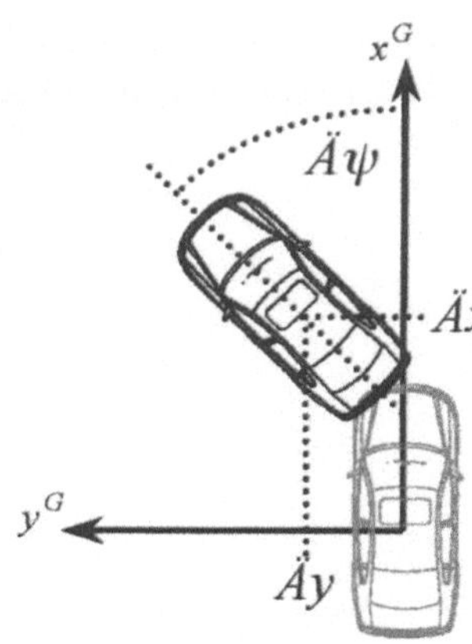

Fig. 3. Three parameters describe the movement of the vehicle in the global co-ordinates x^G, y^G.

$$\Delta x = v \cdot T \cdot \cos(\beta) \quad (1)$$

$$\Delta y = v \cdot T \cdot \sin(\beta) \quad (2)$$

$$\Delta \psi = \frac{v \cdot T}{l_h} \cdot \beta \quad (3)$$

v: Velocity of ego vehicle

T: Samplingrate (typ. 100 ms)

β: Free floating position angle

l_h: Distance from the centre of gravity to the rear axis of the car (see figure 4)

The vehicle's INS-sensors provides its velocity v. The calculation of the free floating position angle β is done by so called vehicle models using INS data as well.

A simple model is based on the assumptions, that there are no changes in wheel load, no cross acceleration, no accelerating or braking forces and that aerodynamic forces are neglected.

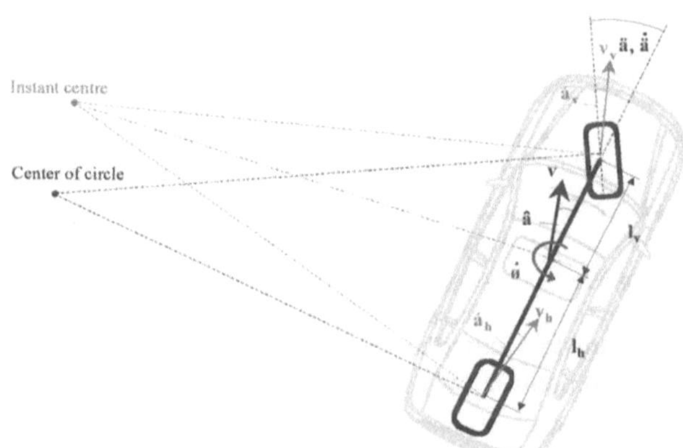

Fig. 4. Vehicle model.

Using the simple model the free floating position angle b is only dependant on the steering angle δ.

$$\beta = \frac{l_h}{l_v + l_h} \cdot \delta \tag{4}$$

$l_{v/h}$ Distance from the centre of gravity to the front/rear axis of the car

δ Front wheel steering angle

A more complex model considers the slip angles α_v, α_h modelling a cross acceleration up to 4m/s [8].

This model uses a Kalman-Filter to provide the free floating position angle β.

$$\dot{\mathbf{x}} = \mathbf{A} \cdot \mathbf{x} \tag{5}$$

In equation 5 the state vector x and the state transition matrix A are calculated as follows:

$$\mathbf{x} = \begin{pmatrix} \beta & \dot{\psi} & \delta & \dot{\delta} \end{pmatrix} \tag{6}$$

$$\mathbf{A} = \begin{pmatrix} -\frac{C_v + C_h}{m_g \cdot v} & \frac{C_h \cdot l_h - C_v \cdot l_v - m_g \cdot v^2}{m_g \cdot v^2} & \frac{C_v}{m_g \cdot v} & 0 \\ \frac{C_h \cdot l_h - C_v \cdot l_v}{J_z} & -\frac{C_h \cdot l_h^2 - C_v \cdot l_v^2}{J_z \cdot v} & \frac{C_v \cdot l_v}{J_z} & 0 \\ 0 & 0 & 0 & 1 \\ 0 & 0 & 0 & 0 \end{pmatrix} \tag{7}$$

$C_{v/h}$: cornering power of the front/rear tires

m_g: mass of the vehicle

J_z: yaw moment of inertia of the z-axis

ψ: Yaw rate

Both models were evaluated in several scenarios with almost the same results. A closed loop scenario is chosen in Fig. 5 to present the performance [9].

Tracklength 256 m	Calculated position	Deviation absolute	Deviation relative
Global X	1.53 m	1.53 m	0.8 %
Global Y	-1.28 m	-1.28 m	
Global ψ	-361.7 degree	-1.7 degree	0.5 %

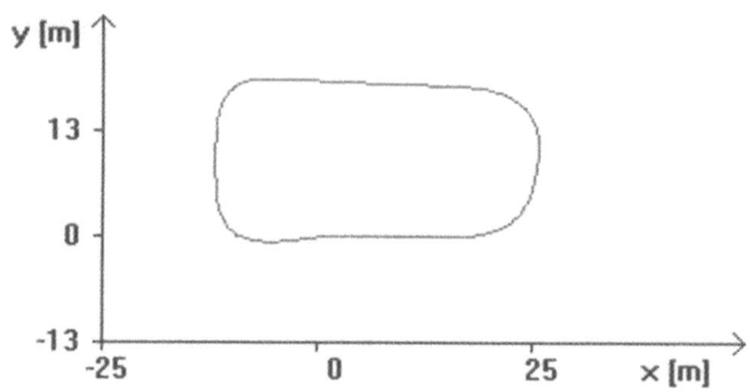

Fig. 5. The trajectory of the calculated global movement (left) and the calculated position (right) shows, that the calculated start and end points are almost the same, when driving once around a parking place.

3.2 Ego-Motion Estimation Using Velocities of Objects

Using the relative velocity of static objects in the vehicle's environment such as posts, reflection posts and trees the ego motion in a global co-ordinate system can be determined. Therefore moving and static objects in the vehicles surrounding have to be distinguished. The proposed method offers both, a classification of static objects and an ego-motion estimation.

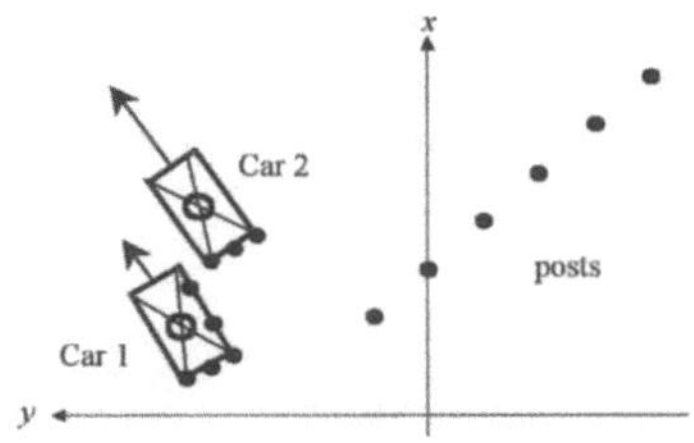

Fig. 6. Two moving cars and six posts detected by a Laserscanner displayed in a bird's-eye view.

Consider the scenario shown in figure 6. Eight objects, two moving cars and six posts, which might be detected by a Laserscanner fixed at the front bumper of a moving vehicle, drawn in the moving Laserscanner's co-ordinate system.

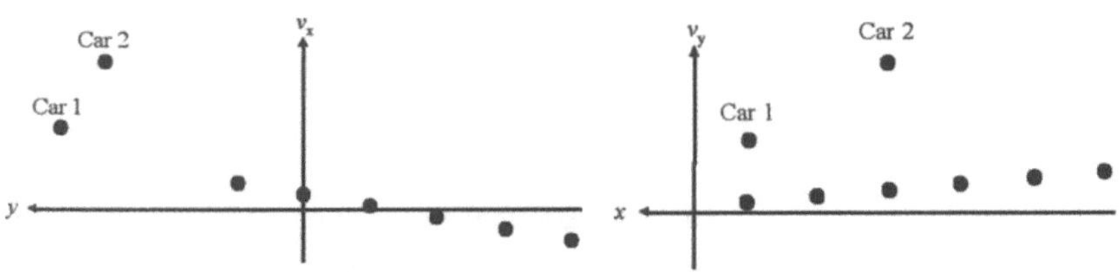

Fig. 7: Graph of the v_x-component over y and the v_y-component over **x** for each tracked object.

Drawing the relative velocity components v_x, v_y for each object over y and x respectively, as shown in figure 7, it can be noticed that all dots assigned to static objects can be interconnected roughly by a straight line. Using this characteristic the ego motion can be derived. Now the theoretical description is discussed and explained by means of the scenario shown in figure 6.

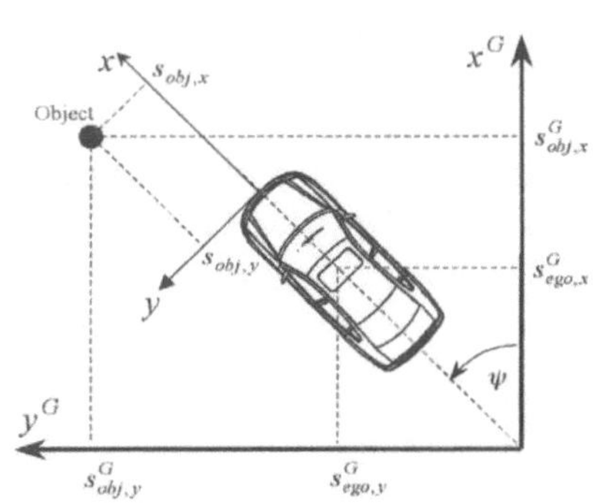

Fig. 8: Position of ego vehicle and an object in a global (x^G, y^G) and the Laserscanner's (**x**, **y**) co-ordinate system.

The relative position $S_{obj,x/y}$ and velocity $V_{bj,x/y}$ of objects surrounding the ego vehicle, which are detected and determined using the moving Laserscanner's co-ordinate system, as shown in Fig. 8, can be described in the global co-ordinate system as follows

$$\begin{pmatrix} s_{obj,x} \\ s_{obj,y} \end{pmatrix} = \begin{pmatrix} \cos(\psi(t)) & \sin(\psi(t)) \\ -\sin(\psi(t)) & \cos(\psi(t)) \end{pmatrix} \begin{pmatrix} s^G_{obj,x} - s^G_{ego,x} \\ s^G_{obj,y} - s^G_{ego,y} \end{pmatrix} \tag{8}$$

$$\begin{pmatrix} v_{obj,x} \\ v_{obj,y} \end{pmatrix} = \begin{pmatrix} \cos(\psi(t)) & \sin(\psi(t)) \\ -\sin(\psi(t)) & \cos(\psi(t)) \end{pmatrix} \begin{pmatrix} v^G_{obj,x} - v^G_{ego,x} \\ v^G_{obj,y} - v^G_{ego,y} \end{pmatrix} + \dot{\psi} \begin{pmatrix} -s^G_{obj,y} \\ s^G_{obj,x} \end{pmatrix} \tag{9}$$

with

$$\dot{\psi} = \frac{d\psi(t)}{dt} \tag{10}$$

The relative velocity of objects in the Laserscanner's co-ordinate system can be separated into the translational and rotational components.

$$\begin{pmatrix} v_{obj,x} \\ v_{obj,y} \end{pmatrix} = \begin{pmatrix} u^{trans}_{obj,x} \\ u^{trans}_{obj,y} \end{pmatrix} + \dot{\psi} \begin{pmatrix} -s^G_{obj,y} \\ s^G_{obj,x} \end{pmatrix} \tag{11}$$

Looking at static objects for which holds, the translational component can be calculated as follows

$$\begin{pmatrix} U^{trans}_{obj,x} \\ U^{trans}_{obj,y} \end{pmatrix} = \begin{pmatrix} \cos(\psi(t)) & \sin(\psi(t)) \\ -\sin(\psi(t)) & \cos(\psi(t)) \end{pmatrix} \begin{pmatrix} -v^G_{ego,x} \\ -v^G_{ego,y} \end{pmatrix} \tag{12}$$

This results in linear equations which describes the components of the relative velocities for static objects. Using Equation 11 these components can be calculated according to

$$\begin{pmatrix} v_{obj,x} \\ v_{obj,y} \end{pmatrix} = \begin{pmatrix} U^{trans}_{obj,x} \\ U^{trans}_{obj,y} \end{pmatrix} + \dot{\psi} \begin{pmatrix} -s^G_{obj,y} \\ s^G_{obj,x} \end{pmatrix} \tag{13}$$

The linear equations, outlined in equation 13, describe the two straight lines interconnecting the static objects as mentioned above (see figure 7).

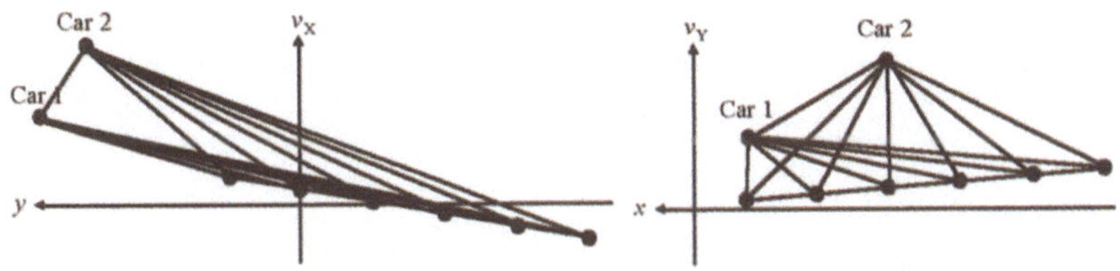

Fig. 9. Interconnections made in the graph of the v_X-component over y and the v_Y-component over x of all objects.

In order to distinguish between static and dynamic objects, the following method is proposed. All dots assigned to objects in the $v_{x/y}$ and the $v_{y/x}$-diagram respectively are interconnected by straight lines, as shown in figure 9. Then the parameters of each straight line, intersection point with the v_x-axis, v_y-axis respectively and the gradient ψ, are calculated and drawn in $\psi/u^{trans}_{obj,x\,/\,y}$-diagrams, as shown in figure 10. The dots assigned to connections between static objects accumulate around one point in each diagram.

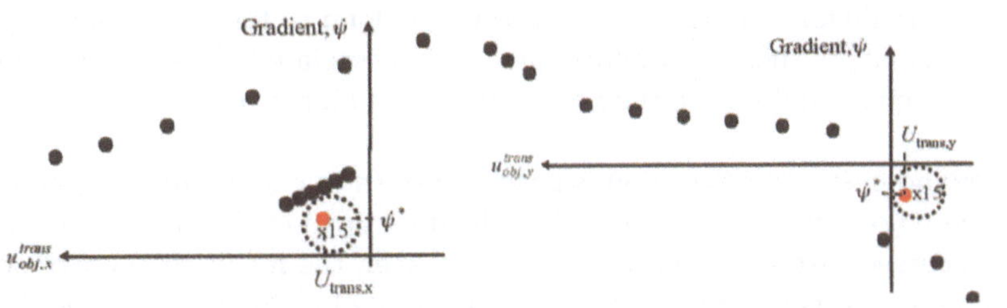

Fig. 10. Graph of the line parameters Gradient ψ over $u^{trans}_{obj,x}$ and $u^{trans}_{obj,y}$ respectively.

The two points of accumulation, as shown in figure 10, comprise all dots assigned to static objects. Using these points of accumulation the translational components $U_{trans,x}$, $U_{trans,y}$ and the rotational component ψ^*, which is calculated twice, of the static objects are determined. As static objects have no velocity in the global co-ordinate system itself, the translational and rotational components from above describe the movement of the car.

The movement Δx, Δy and $\Delta\psi$ between two scans of the ego vehicle, due to the translational and rotational movement of the vehicle, in a time period T, can be determined as follows

$$\Delta x = U_{trans,x} \cdot T \tag{14}$$

$$\Delta y = U_{trans,y} \cdot T \qquad (15)$$

$$\Delta \psi = -\dot{\psi}^{*} \cdot T \qquad (16)$$

Therefore both ego-motion Δx, Δy (vehicle translation) and Δy (Vehicle rotation) and classification of static objects is carried out.

4 Object Recognition

The first step in object recognition is the segmentation, which clusters the Laserscanners distance measurements, which belong to the same real object. The ego-motion of these segments and their associated distance measurements is compensated before they are assigned to object tracks.

The following object-classification is based on the static and dynamic parameters of the object to find a suitable classification. Road detection is done with the unprocessed distance measurements. However, the measurements of classified vehicles are removed, to obtain a reliable road estimation. An overview of the data processing is shown in Fig. 11.

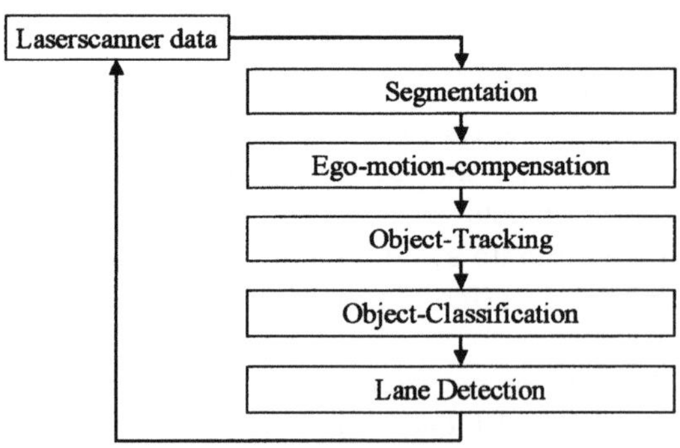

Fig. 11. Overview of the object recognition process.

4.1 Segmentation

The first step in the data processing is the segmentation of the distance measurements. The determination of a suitable segmentation criteria is quite crucial for the subsequent data processing.

The distance measurements generated by one revolution of the rotating head (scan) are divided into segments, which are assumed to belong to the same real object. These segments are represented by a rectangular box around the measurements, providing length and width of the segment. The centre of geometry and the centre of gravity of the segment is calculated.

4.2 Tracking

Comparing the segment parameters of a scan with predicted parameters of known objects from the previous scan(s), established objects are recognised. Unrecognised segments are instantiated as new objects, initialised with default dynamic parameters.

The Tracking process is usually divided into three sub-processing steps, as shown in figure 12.

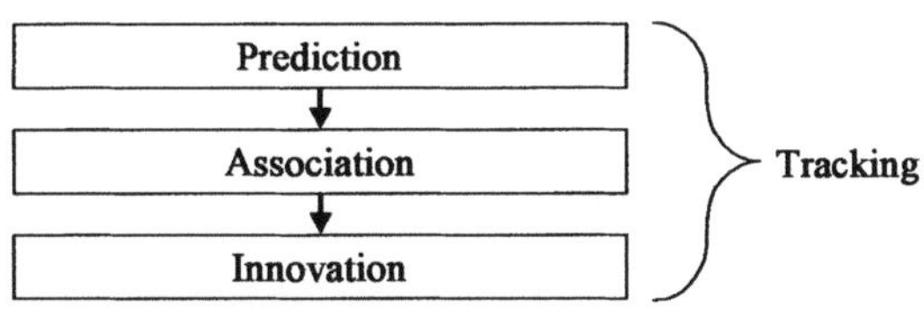

Fig. 12. The tracking process.

In order to estimate the object state parameters a Kalman filter is well known in the literature and used in various applications, as an optimal linear estimator [10]. A simplified Kalman filter, the alpha-beta (α–β) tracker is used often instead [11]. The Kalman filter was chosen, as it allows for more complex dynamic models, which are necessary for a precise object tracking. We evaluated our data on different association methods, such as the nearest neighbour and the global nearest neighbour method [12].

5 Object Classification

Object classification is based on object-outlines (static data) of typical road users, such as cars, trucks/busses, poles/trees, motorcycles/bicycles and pedestrians. Additionally the history of the object classification and the dynamics of the tracked object are used in order to support the classification performance [13,14].

5.1 Object Classification

In a simple algorithm road users are classified by their typical angular-outline using only the geometric data [15]. Additional the objects history and its dynamic data are necessary to enable a robust classification [16].

In case there is not enough information for an object classification a hypothesis is generated, based on the objects current appearance. The temporary assignment is valid, as long as there is no violation of a limiting parameter. However the classification is checked every scan to verify the assignment of the specified class [8,17].

An environmental model supports the selection of a suitable class. The understanding of the traffic situation may further improve the classification results.

5.2 Pedestrian Classification

If an object is detected the first time, the classification is based solely on the outline of this object. As a small object can be either a pedestrian or a small static object, such as a pole or a lamppost, the classification 'ped?' is determined. As long as a small object is not moving towards the road, there is no hazard implied neither for the vehicle nor the object itself. If a small object is moving, it is immediately classified as a pedestrian, based on the previous classification 'ped?' and its absolute velocity [16].

In order to classify a pedestrian more robust its typical movement can be used to differentiate between pedestrians and static objects. The range of velocity, the typical appearance of the moving legs and the objects history are additional criteria's for the classification of a pedestrian.

Figure 13 shows a pedestrian stepping onto the road from behind a parking car. All other object classifications are suppressed in order to highlight the pedestrian. The object number, absolute velocity and the estimated classification are shown close to the object. The pedestrian is classified two scans after first detection in a distance of 25.2m (x=25.0m, y=3.5m). This first classification is performed in less than 300ms at 10Hz scan frequency. The same classification result can be obtained in less than 75ms at 40Hz scan frequency. The update rate of tracking and classification in the following scans is 100ms at 10Hz scan frequency, 25ms at 40Hz scan frequency respectively.

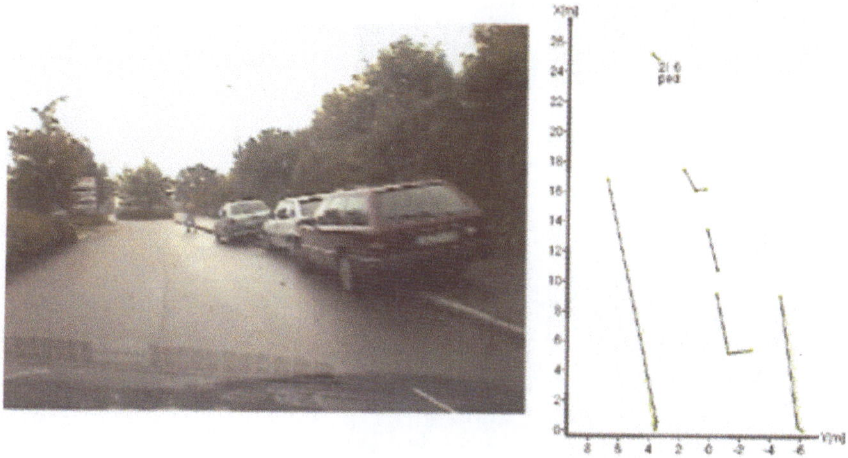

Fig. 13. A video image and object data of a pedestrian entering the road behind a car.

Pedestrians can be detected, tracked and classified even in large distances, as shown in Fig. 14, which enables the robust detection of potentially harmful situations. In Fig. 14a the pedestrian is detected and classified in the Laserscanners data in a distance of 49.0m (x=48.0m, y=-14.1m) in front of the ego vehicle, which can't be verified in the reference video image due to bad weather conditions. This pedestrian, with the object ID 0, is tracked from scan to scan. It can hardly be recognised 3.3s later in the video image, as shown in Fig. 14b, in a distance of 29.7m (x=28.9m, y=-7.0m) measured by the Laserscanner. After 5.4s in a distance of 14.1m (x=14.0m, y=-1.9m) the pedestrian can be identified in the video image, as shown in figure 14c.

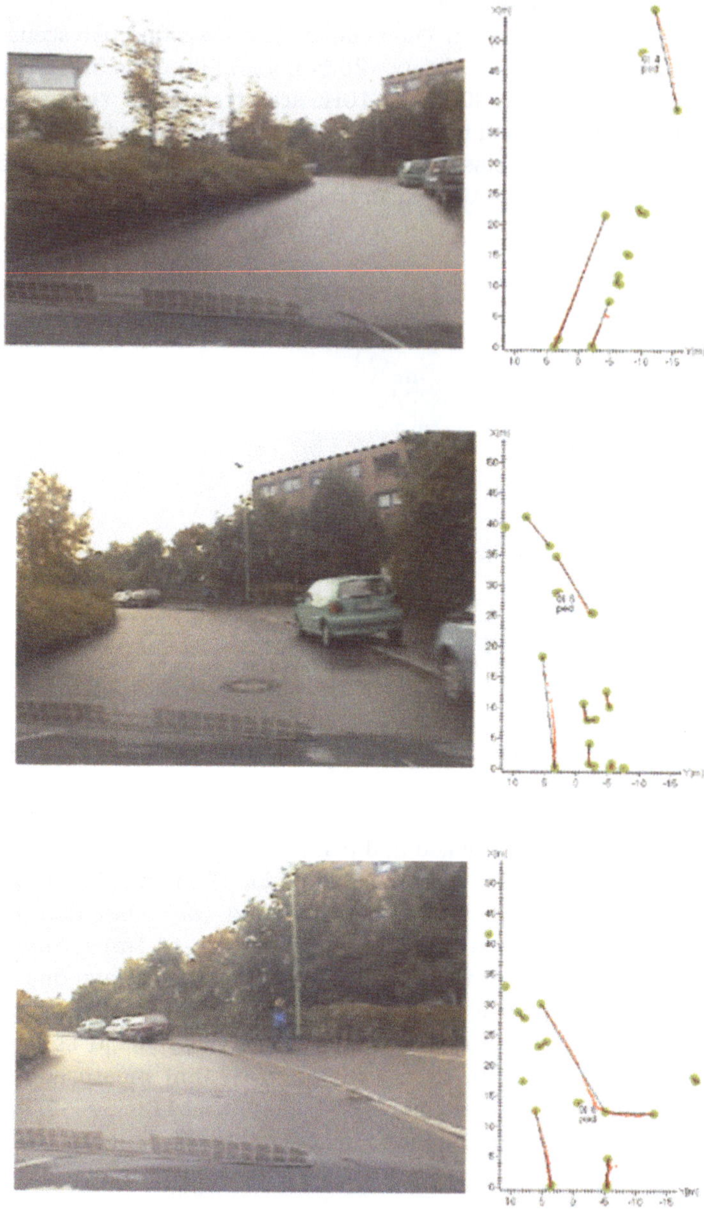

Fig. 14. A video image and object data (right) of a pedestrian walking along the road. Its object ID is 0 and the velocity 4km/h and 6km/h respectively.

6 Road Detection with Non-linear Filtering

The algorithm used for the road detection is based on unprocessed distance measurements from the Laserscanner [18]. However, all distance measurements stemming from vehicles on the road have to be removed from the scan in a pre-processing step.

A road is described by a set of two parallel parabolas. The road are characterised by the width, w, the offset a_0 of the car to the road boundary and the parameter a_2 for the curvature. Additionally, an adaptive range of view is used that limits the range of the parabolas.

The middle, the left, and the right side of the road are therefore described by

$$\begin{aligned} f_{middle}(x) &= \bar{a}_0 + \bar{a}_2 \cdot x^2 \\ f_{left}(x) &= \bar{a}_0 + \bar{w}/2 + \bar{a}_2 \cdot x^2 \\ f_{right}(x) &= \bar{a}_0 - \bar{w}/2 + \bar{a}_2 \cdot x^2 \end{aligned} \tag{17}$$

where x is the co-ordinate axis in driving direction. The over-lines indicate that all parameters are filtered according to the algorithm described below.

6.1 Range of View Adaptation

The range of view used for the curvature estimation, X_c , which is the maximum range of measurements that are taken into consideration, is determined depending on the maximum distance with significant information and the present curvature itself. The maximum distance with significant information is determined by the maximum distance in front that contains most valid measurements as shown in figure. 15.

Let $P_{n,x}$ and $P_{n,y}$ be the Laserscanner data in Cartesian co-ordinates,

$$\begin{aligned} P_{n,x} &= r_n \cos\alpha_n \\ P_{n,y} &= r_n \sin\alpha_n \end{aligned} \tag{18}$$

then measurements not used, P_{out}, are

$$P_{out} = \{P_n \in P | P_{n,x} \in [X_c, X_{c,\max}]\}$$ (19)

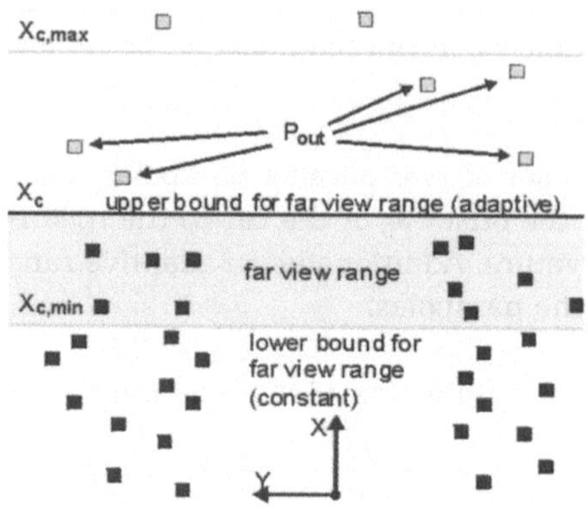

Fig. 15. View range adaptation.

and the number of these measurements is N_{out}. In Equation 19, $X_{c,max}$ is constant however the range of view for the curvature estimation X_c is adapted in every time step k depending on N_{out} as follows

$$X_c(k) = \begin{cases} X_c(k-1) + \Delta X_{incr} & N_{out} \geq N_{out,\max} \\ X_c(k-1) - \Delta X_{decr} & N_{out} < N_{out,\max} \end{cases},$$ (20)

Additionally the range of view is adapted to the present curvature using the heuristic function

$$f_c(\bar{a}_2) = X_{c,\max} \cdot e^{-b_c|\bar{a}_2|}$$ (21)

The range of view is adapted with

$$X_{\max} = \min\{X_{c,\max}, f_c(\bar{a}_2)\}$$ (22)

$$X_c(k) = \begin{cases} X_{\max} & X_c(k) > X_{\max} \\ X_{c,\min} & X_c(k) < X_{c,\min} \\ X_c(k) & \text{otherwise} \end{cases} \qquad (23)$$

Due to this adaptation scattered measurements lying far away and therefore being unreliable for the determination of the curvature can be excluded. The reduction of the range of view with an increasing curvature is done, because the sight into curves is reduced by occlusions that are caused from objects at the inner side of the curve.

6.2 Determination of Width and Position

The width and relative position of the road is determined by observing the measurements at both sides of the road within a lower range of view. This is visualised in figure 16.

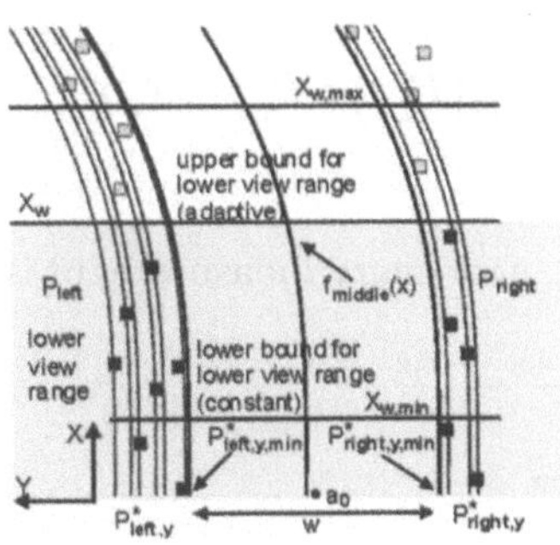

Fig. 16. Calculating the roads middle and width.

The lower range of view X^w is adapted similarly to X_c. Additionally, a dependence from Xc exists.

The dependence on the present curvature is now described by the heuristic function

$$f_w(\bar{a}_2) = X_{w,\max} \cdot e^{-b_w|\bar{a}_2|} \qquad (24)$$

The current limit is determined by evaluation of Equation 25 and Equation 26, where $X_{w,\max}$ and $X_{w,\min}$ are constant maximum and minimum limits.

$$X_{\max} = \min\left\{X_{w,\max}, f_w\left(\bar{a}_2\right)\right\} \tag{25}$$

$$X_w(k) = \begin{cases} X_{\max} & X_c(k)/2 > X_{\max} \\ X_{w,\min} & X_c(k)/2 < X_{w,\min} \\ X_c(k)/2 & \text{otherwise} \end{cases} \tag{26}$$

The middle line of the road divides the current measurement scan into measurements lying on the right side and the left side of the road. Considering the view range adaptation described above, this can be formulated mathematically as follows

$$P_{left}\left(k\right) = \left\{P_n \in P \;\middle|\; P_{n,y} > f_{middle}\left(P_{n,x}\right) \wedge P_{n,x} < x_w\right\} \tag{27}$$

$$P_{right}\left(k\right) = \left\{P_n \in P \;\middle|\; P_{n,y} \leq f_{middle}\left(P_{n,x}\right) \wedge P_{n,x} < x_w\right\} \tag{28}$$

Within the lower range of view X_w, now every measurement is transformed to the y-axis.

$$P^*_{n,left,y}(k) = P_{n,left,y} - \bar{a}_2 \cdot P^2_{n,left,x} \tag{29}$$

$$P^*_{n,right,y}(k) = P_{n,right,y} - \bar{a}_2 \cdot P^2_{n,right,x} \tag{30}$$

Afterwards, the two measurements closest to the present middle line of the road are searched according to

$$P^*_{left,y,\min}\left(k\right) = \min_n\left\{P^*_{n,left,y} - \bar{a}_0\right\} \tag{31}$$

$$P^*_{right,y,\min}(k) = \min_n \left\{ \bar{a}_0 - P^*_{n,right,y} \right\} \tag{32}$$

These two measurements define new values for the smallest width w and the relative position a_0 within the lower range of view in the current scan

$$\begin{aligned} w(k) &= P^*_{left,y,\min}(k) - P^*_{right,y,\min}(k) \\ a_0(k) &= \left(P^*_{left,y,\min}(k) + P^*_{right,y,\min}(k)\right)/2 \end{aligned} \tag{33}$$

Unfortunately, these values do give a correct value for the actual width of the road and the relative position of the car on the road in every measurement scan. For noise reduction and additional robustness, a history of road width and position values are filtered by calculation of a weighted mean value. The lengths of the filters differ, because of the slower changes of the width but faster changes of the relative position of the car to the middle of the road. The highest weighting is done when the current estimation of the width is close to the mean estimation. A medium weighting is performed for width values that are smaller than the mean estimation. They appear rarely, i.e. on appearance of reflection posts. A small weighting is useful for width values significantly larger than the mean estimation, which occur very often and lead to a wrong result. This method reflects the strong non-linearity of measurements compared to the model parameter. The weighting factors are determined as follows

$$\begin{aligned} g_{w,i} &= \begin{cases} 1.0 & \left|\overline{w} - w_i\right| < \varepsilon_w \\ g_{w,medium} & \overline{w} - w_i \geq \varepsilon_w \\ g_{w,small} & w_i - \overline{w} \geq \varepsilon_w \end{cases} \\ g_{p,i} &= \begin{cases} 1.0 & \left|\overline{w} - w_i\right| < \varepsilon_w \\ g_{p,medium} & \overline{w} - w_i \geq \varepsilon_w \\ g_{p,small} & w_i - \overline{w} \geq \varepsilon_w \end{cases} \end{aligned} \tag{34}$$

Additionally, the weighting is adapted depending on the closest measurements found at the right and left side of the middle line of the road. If these two points build a goal, for example because two reflection posts are found, the weighting factor $g_{\Delta d,i}$ is chosen high. The weighting factor $g_{d,i}$ is chosen high, if the distance of the closest point to the y-axis is small, which means a relative secure estimation in the close range was made. This can be described mathematically as follows

$$g_{d,i} = \left(1 + 0.2 \cdot \min\left\{P^*_{left,x,\min}, P^*_{right,x,\min}\right\}\right)^{-1}$$
$$g_{\Delta d,i} = \left(1 + 0.2 \cdot \left|P^*_{left,x,\min} - P^*_{right,x,\min}\right|\right)^{-1} \tag{35}$$

Let F_w comprise the last N_w width values w and F_p comprise the last N_p position values a_0. Then, the weighted mean values are calculated with the respective weights according to

$$G_w = \sum_{i=0}^{N_w} \left(g_{w,i} \cdot g_{d,i} \cdot g_{\Delta d,i}\right) \tag{36}$$

$$G_p = \sum_{i=0}^{N_p} \left(g_{p,i} \cdot g_{d,i} \cdot g_{\Delta d,i}\right) \tag{37}$$

$$\overline{w} = \frac{1}{G_w} \sum_{i=0}^{N_w} g_{w,i} \cdot F_{w,i} \tag{38}$$

$$\overline{a}_0 = \frac{1}{G_p} \sum_{i=0}^{N_p} g_{p,i} \cdot F_{w,i} \tag{39}$$

These filtered values describe the width of the road and position of the car relative to the road very robust. For robustness these values are additionally checked for plausibility. If the filtered right bound of the road is lying on the left of the sensor or vice versa, this means the sensor is located outside the current road estimation, the algorithm will be initialised. Moreover, if the road is smaller than the vehicle also an initialisation process is carried out.

6.3 Determination of the Curvature

Hereafter, the curvature of the road has to be determined. The parameters used are visualised in figure 17.

First, a set of curvatures A_2 is determined, which ly around the current curvature. The number of curvatures in the set, $2N_s+1$, depends on the recent classification for the road type. The higher the curvature of a road type may be, the larger the number of curvatures in the set is.

$$A_2 = \left\{a_{2,-N_S}, \ldots, a_{2,-1}, a_{2,0}, a_{2,1}, \ldots, a_{2,N_S}\right\} \qquad (40)$$

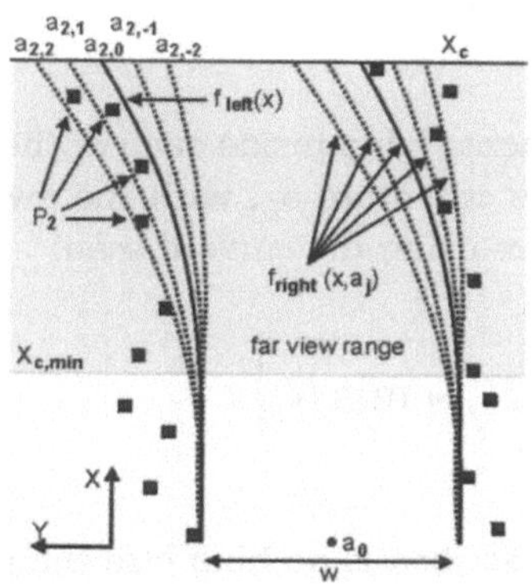

Fig. 17. Determination of the curvature.

For each curvature of the set the measurements of the Laserscanner lying inside the respective road, P_j, is determined. The functions $f_{left}(x,a_j)$ and $f_{right}(x,a_j)$ describe the road for a specific curvature $a_{2,j}$ out of the set A_2 according to

$$\begin{aligned} f_{left}\left(x,a_j\right) &= \bar{a}_0 + \bar{w}/2 + a_{2,j} \cdot x^2 \\ f_{right}\left(x,a_j\right) &= \bar{a}_0 - \bar{w}/2 + a_{2,j} \cdot x^2 \end{aligned} \qquad (41)$$

Using these functions, it can be determined whether measurements lie inside or outside a road hypothesis, given by a specific curvature parameter out of the set A_2.

$$P_j = \left\{ P_n \in P \middle| \begin{matrix} P_y < f_{left}\left(P_x, a_{2,j}\right) \wedge \\ P_y > f_{right}\left(P_x, a_{2,j}\right) \end{matrix} \right\} \tag{42}$$

Now the number of measurements lying inside each of these roads is counted and set to the variable N_j. The curvature $a_{2,j}$ with the lowest count N_j is the best estimation of the curvature using the current scan.

$$j_{\min} = \left\{ j \in \left[-N_s, N_s\right] \middle| N_j = \min_l N_l \right\} \tag{43}$$

In other words, the method described is nothing else than a matching of possible roads with discrete curvatures to given measurements. A following mean filtering with F_c comprising the last N_c values of a_{2,j_min} increases the robustness and reduces noise.

$$\bar{a}_2 = \frac{1}{N_c} \sum_{i=1}^{N_c} F_c(i) \tag{44}$$

6.4 Road Type Classification

The run of the road is not the entire useful information about a road. The performance of the road detection can be increased significantly by additionally estimating the road type. On the one hand, static limits for the width and curvature of the road may exclude wrong values from calculations. Dynamic limits for the change of the curvature may work in the same manner. On the other hand, the road type can reduce the number of possible object types in the environment, which stabilises the object detection and this removal from the scan.

Using the mean value for the width of the road, a simple road type classification can be implemented, because different road types show a different width. However, this classification will not be unequivocal, as of course no detailed information like the number of lanes on the road can be determined.

The width detection used for a road type classification is very similar to the road detection described above. The main difference is a larger observation time, because the width of a road may change quickly, but road types change very rarely. Therefore, the minimum values are stored over a longer period and the mean value calculation is done over a longer period of time.

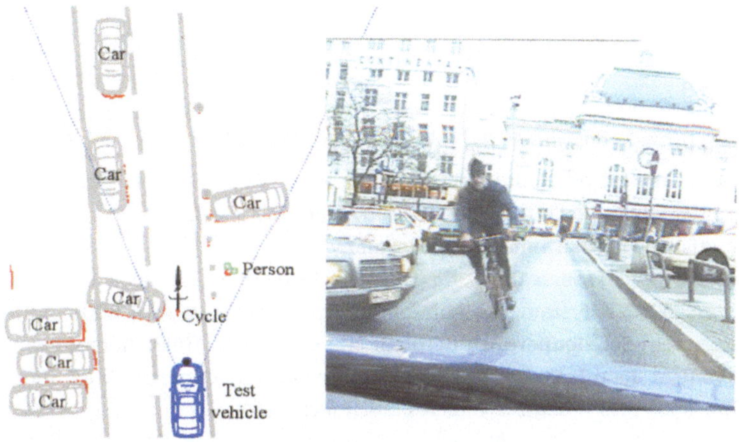

Fig. 18. Urban scenario in the Laserscanners view using all knowledge of vehicle, road and object modules is summarised at the left hand side. The video picture is shown as a reference.

7 Conclusions

The Laserscanner LD ML is able to observe the vehicles environment. A reliable detection and tracking of the objects around is performed. This paper introduced a new method for ego-motion estimation using the velocities of objects tracked by a Laserscanner. The proposed method is very accurate under both, defined and undefined driving conditions, the latter being most relevant for active safety systems. The classification of objects based on tracking and ego-motion estimation, especially the results of the pedestrian classification were shown. The road detection and road type classification, which was described in detail in the chapter before, is a new approach to determine the road parameters using Laserscanner data. The algorithms are developed and tested offline on recorded data of real traffic scenarios and evaluated online in a test vehicle.

Using the object tracking and classification together with the road detection, it is possible to provide data for future active safety systems. Fig. 18 outlines the aim of a complete knowledge based environmental model of the surrounding of a driving vehicle. The introduced vehicle, road and object modules are a essential step on the way of increasing safety due to active safety systems.

References

[1] Promoting road safety in the EU - COM (97) 131 fin.

[2] Toward fair and efficient pricing in Transport – COM (95) 691 fin.

[3] Statistica degli incidenti stradali – 1997 Italy ACI ISTAT.

[4] Statistisches Bundesamt – 1996 Germany.

[5] Lages, Ulrich: Laserscanner for Obstacle Detection. Yearbook of AMAA 2002, Advanced Microsystems for Automotive Applications, AMAA 2002, Berlin, 2002.

[6] Weisser, H.; Schulenberg, P.; Bergholz, R.; Lages, U. (1998) Autonomous Driving on Vehicle Test Tracks: Overview, Motivation and Concept. IEEE International Conference on Intelligent Vehicles, Stuttgart, 1998.

[7] Willhoeft, V.; Ewald, A. (2000) Laserscanners for Obstacle Detection in Automotive Applications. Proceedings of IV 2001, IEEE Intelligent Vehicles Symposium 2000, IV 2000 Dearborn, Paper IVS-008.

[8] Streller, Daniel; Fuerstenberg, Kay Ch.; Dietmayer, Klaus C.J.: Vehicle and Object Models for Robust Tracking in Traffic Scenes using Laser Range Images. Proceedings of ITSC 2002, IEEE 5th International Conference on Intelligent Transport Systems, ITSC 2002 Singapore.

[9] Fuerstenberg, K. Ch.; Dietmayer, K.; Eisenlauer, S.; Willhoeft, V.: Multilayer Laserscanner for robust Object Tracking and Classification in Urban Traffic Scenes. Proceedings of ITS 2002, 9th World Congress on Intelligent Transport Systems, ITS 2002 Chicago, Paper 2054.

[10] Welch, Greg; Bishop, Gary: An Introduction to the Kalman Filter. http://www.cs.unc.edu, 2001.

[11] Gavrila, Dariu M.; Giebel, Jan; Shape based Pedestrian Detection and Tracking. Proceedings of IV 2002, IEEE Intelligent Vehicles Symposium, IV 2002 Versailles.

[12] Blackman, Samuel: Design and Analysis of Modern Tracking Systems. Artech House, London, 1999.

[13] Willhoeft, V.; Fuerstenberg, K. Ch.; Dietmayer, K. C. J.: New Sensor for 360° Vehicle Surveillance. Proceedings of IV 2001, IEEE Intelligent Vehicles Symposium, IV 2001 Tokyo.

[14] Fuerstenberg, K. Ch.; Willhoeft, V.: Pedestrian Recognition in urban traffic using Laserscanners. Proceedings of ITS 2001, 8th World Congress on Intelligent Transport Systems, ITS 2001 Sidney, Paper 551.

[15] Fuerstenberg, K. Ch.; Hipp, J.; Liebram, A. (2000) A Laserscanner for detailed traffic data collection and traffic control. Proceedings of ITS 2000, 7th World Congress on Intelligent Transport Systems, ITS 2000 Turin, Paper 2335.

[16] Fuerstenberg, K. Ch.; Dietmayer, K. C. J.; Willhoeft, V.: Pedestrian Recognition in Urban Traffic using a vehicle based Multilayer Laserscanner. Proceedings of IV 2002, IEEE Intelligent Vehicles Symposium, IV 2002 Versailles, Paper IV-80.

[17] Dietmayer, K. C. J.; Sparbert J.; Streller, D.: Model based Object Classification and Object Tracking in Traffic scenes from Range Images. Proceedings of IV 2001, IEEE Intelligent Vehicles Symposium, IV 2001 Tokyo, Paper 2-1.

[18] Sparbert, J.; Dietmayer, K. C. J; Streller, D.: Lane Detection and Street Type Classification using Laser range Images. Proceedings of ITSC 2001, IEEE 4th International Conference on Intelligent Transport Systems, ITSC 2001 Oakland, page 454-459.

Kay Ch. Fuerstenberg, Klaus C. J. Dietmayer
University of Ulm
Department of Measurement, Control and Microtechnology
Albert-Einstein-Allee 41
89081 Ulm
Germany
kay.fuerstenberg@e-technik.uni-ulm.de

Ulrich Lages
IBEO Automobile Sensor GmbH
Fahrenkroen 125
22179 Hamburg
Germany
UL@ibeo.de

Keywords: laserscanner, obstacle detection, ego-motion, tracking, classification, pedestrian recognition, road detection

High Resolution Piezo Film Sensor Systems for Automotive Applications

P. Mirow, Mirow Systemtechnik GmbH

Abstract

This paper deals with applications and concepts for PVDF film sensors in automotive safety applications. Due to their extraordinary dynamic range (286dB) piezoelectric PVDF-plastic film may be used as an active electro-mechanical converter system for highly sensitive as well as for highly loaded measurement tasks. Strain measurements on metal structures are possible with forces high enough to result in plastic deformations.

Furthermore, the high flexible piezo film may be applied to moderately 3D shaped surfaces. A novel, by Mirow Systemtechnik GmbH developed analogue-digital signal amplification process enables a significant dynamic gain for slow signals, exceeding the amplifier dynamic range of 120dB.

Accordingly measurement systems may be realized that use the sensor signal in multiple ways. Application examples are the activation of pedestrian airbags, the measurement of the force distributions in the front bumper for the optimisation of passenger hold-back systems.

1 Introduction

Piezoelectric film made from PVDF is a very thin (standard 28µm) electromechanical converter, which convert 12-15% of a mechanical into an electrical energy, or vice versa. Thereby they may be used as active sensors as well as actuators and for energy generation. The high strength, flexibility and electrical sensitivity of the material allows the use in an enormous dynamic and frequency range, as an example in transverse direction for strains of 10^{-8} - 10^{5}N/cm^{2} with frequencies from 0,001Hz up to GHz. These ranges are used in parts, depending on the amplifier layout.

Starting point for the activities of Mirow Systemtechnik GmbH had been flow investigations with multi-sensor arrays at the aerospace faculty (ILR) of the technical university Berlin (TUB). Following projects dealt with several measurement task from fluid dynamics, medical and dental systems, vibration analysis, manufacturing, automotive as well as other disciplines.

Recently an increase of the dynamic range was achieved for several solutions, using a novel, patent applied analogue/digital signal amplification and correction process. This was done without the need to switch rates for the amplification electronics. Some results are presented with this paper.

2 Measurement Principle

Used for signal measurement the piezo film is an capacitor, whose current state between the surface electrodes changes in proportion to the applied mechanical stress (normal, longitudinal or transversal deformations) on the foil. By principle the piezo film, as well as all other piezoelectric transducers (quartz, ceramics), are only suitable for the recording of dynamic forces.

Suitable amplifiers may be stabilized by means of a shortening resistance, which drains the signal with a defined time constant ($T = R*C$) to the zero level. With a suitable selection of the R/C-combination an overload of the amplifier is avoided while still retaining a high sensitivity. All signals at or below the cut-off frequency are damped amplitude-wise and the static part of the signal (0Hz) is totally suppressed. Within the mechanical capabilities of the sensor material any stress state may be defined as the zero level.

3 Correction Process

Using modern data acquisition and signal processing methods the physical set amplitude constrains (for example +5V) may be extended to almost any numerical voltage value.

For the realization of successful signal corrections all disturbances are to be minimized carefully, leaving only the signal to be processed. This is especially important for very low frequent, quasi-static processes, where temperature effects will become increasing disturbing.

When recording the signal any amplifier overload situation, as shown in figure 1, is to be avoided in any case, this means the signal amplitude of the amplifier has be within the supply voltage range.

Signal overload would lead to an undefined step in the correction function, leaving the sensor signal useless for the considered time interval. A new definition of the zero level would become necessary.

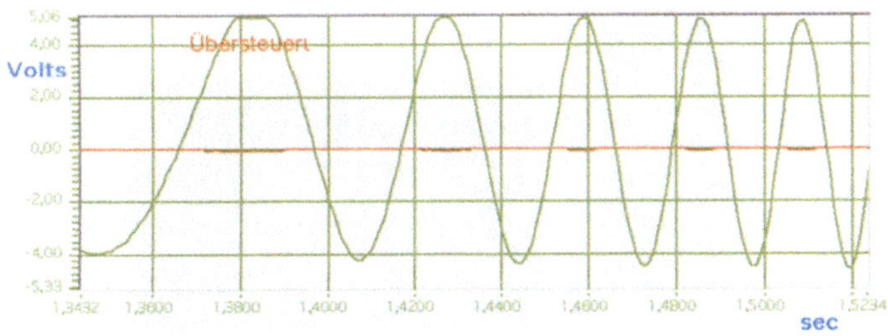

Fig. 1. Example for signal overload.

The numerical post-processing of correctly digitised signals gives the following advantages:

- ▶ Calculation and correction of the amplifier offset at a free defined rest level and time instance.
- ▶ Optimisation of the electrical range and sensitivity by a problem-suited R/C-combination.
- ▶ Recalculation of the cut-off frequency of the analogue amplifier to any chosen value including 0Hz (quasi-static measurement).
- ▶ Increased dynamic gain for comparably slow signal components, by means of a numerical extension of the supply voltage range, in theory going up to the maximum numbers used in the data processing environment.

In real applications of cause there will always be some level of signal disturbance, which will limit the degree of precision attainable in a measurement.

4 Sensor Film Manufacturing Enhancements

To lessen the realization time and increase the quality for the fabrication of the sensor build-ups, which may be quite complex in design, it was necessary to introduce computer aided methods.

To aid the design of the sensors novel computer-based layout and simulation tools were developed, which enabled an optimisation of the manufacturing process and lead to enhanced sensor build-ups. In addition to the electronic sensor layout the size and arrangement of the sensors are designed by computer. Based on these data the sensors are machined from the raw material in a computer-controlled process. Figure 2 shows an example for flow measurements on a turbine blade, created using these techniques.

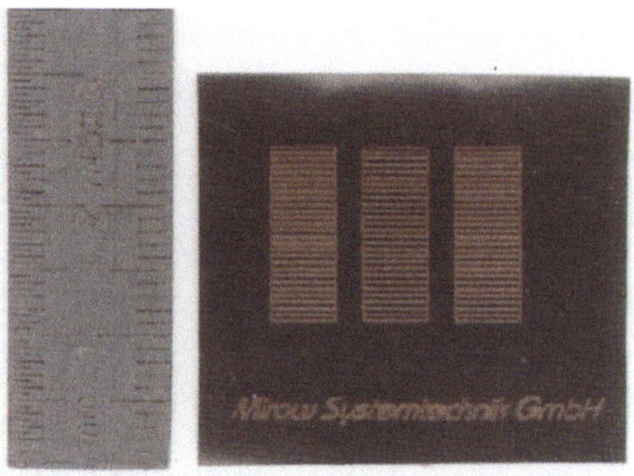

Fig. 2. High-resolution flow measurement sensor array with 3x27 active sensor areas.

Fig. 3. Meniscus disk with sensor array.

To apply the piezo film on 3D-curved surfaces, special coating and bonding techniques where developed. For the display, processing and storage of the recorded signals mainly PC-based hardware (slot cards) in combination with object oriented software (virtual instruments) are used. Due to the ever increasing performance of computers and acquisition hardware increasingly complicated and more precise processing becomes possible (on-line filtering, FFT). Figure 8 shows a application with an artificial meniscus, which was used for the measurement of the pressure distribution of a stainless steel ball joint and meniscus disk.

5 Integrated Sensor-Amplifier Systems

Some flexible sensor and amplifier systems where developed (figure 4) for the measurement of vibrations and deformations at buildings, vehicles and machinery as well as in safety systems

Fig. 4. Some flexible transducers.

By varying the electronic components the frequency behaviour as well as the sensitivity may be matched for a broad range of different measurement tasks, figure 5 shows two output signals with different sensitivity which are available from one flexible sensor system.

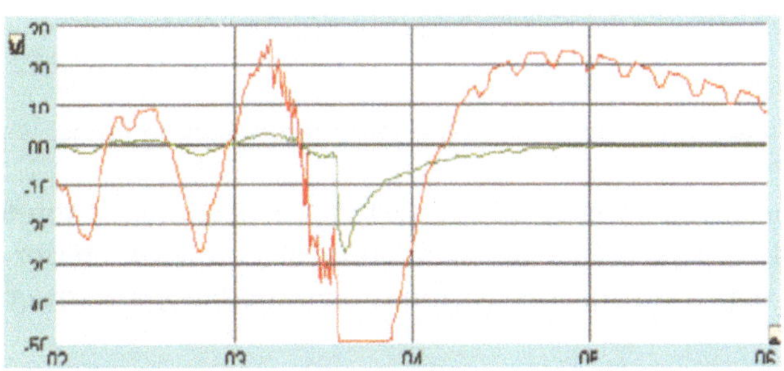

Fig. 5. Two outputs on one mini amplifier.

By designing multi element and multi layer sensors or actuators, a broad variation in sensitivity to different mechanical forces can be applied. Figure 6 shows some actuator functions, which are always sensor functions as well, such as bending detectors or strain gauges and torque detectors.

6 PVDF- Sensors in Automotive Applications

Due to their enormous dynamic range, PVDF Film Sensors are very comfortable in use for any crash detecting systems, such as detecting the pressure forces at a movable crash barrier plate, or the bending angle of thin metal structures. The sensor systems are always lightweight constructions.

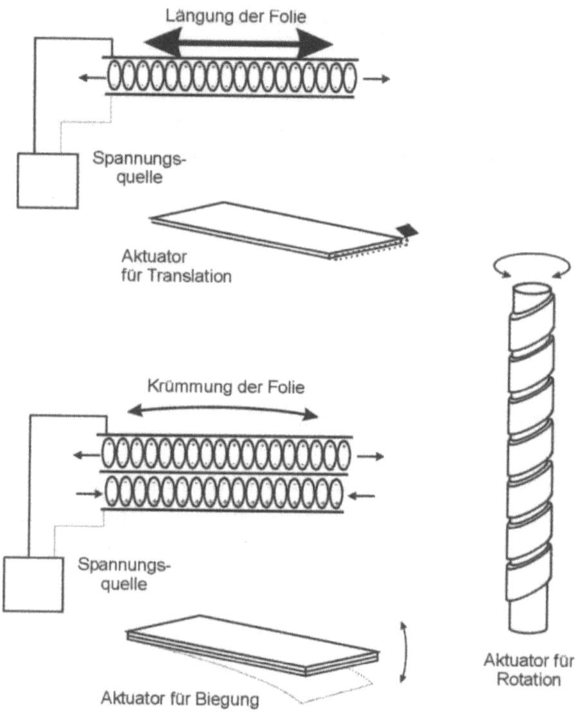

Fig. 6. Sensor/actuator assemblies for different motion forms.

A, patents pending, bumper sensor system for detection of the forces as well as the bending of the bumper, consisting of four twin sensor systems are mounted at the colour marked positions onto the bumpers structure.

A patent applied multi-element sensor system records forces and bending moments on a bumper simultaneously and has been created in a demo version by Mirow Systemtechnik GmbH. In case of a crash it allows a early online calculation of the dynamic forces and their angle of attack in order to estimate the vehicle reactions and thereby may be used for a optimised control of vehicle and passenger safety systems. For example, instead of activating the airbags, it might be better to stabilize the vehicle by ESP following a collision. The bumper sensors will give very early and fast information, even after the vehicle structure was displaced by a high degree.

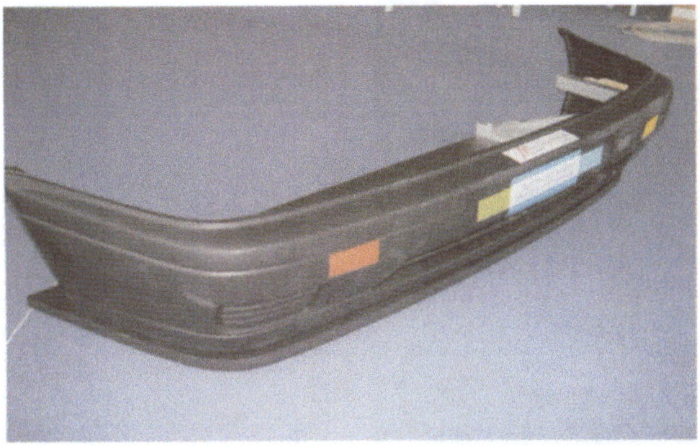

Fig. 7. Piezo film sensors applied behind the marked areas to a bumper demonstrator construction.

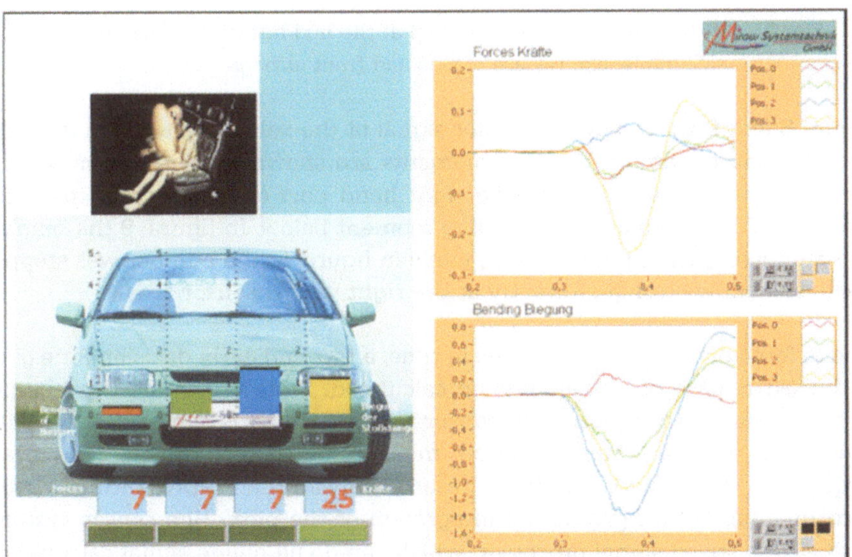

Fig. 8. PVDF- Bumper Sensor, pushed frontal in the right area, e.g. only front airbag release.

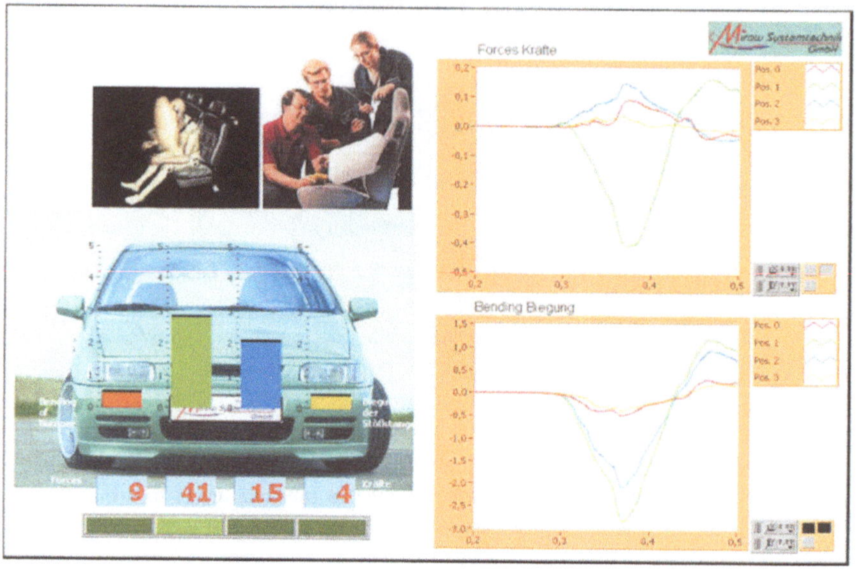

Fig. 9. PVDF- Bumper Sensor, pushed at the mid left green area with 50° side angle, e.g. release of side and front airbags.

Figure 8 and 9 show a demonstrator signal of the sensor system. At the vehicle front the maximum bending moments are shown as bars, the maximum forces are noted as numbers. The right hand part of the pictures gives the force reaction on top and the bending moment below. In figure. 9 the bumper was stepped upon in the left third, while in figure. 8, the bumper was stepped upon, somehow stronger, directly at the right yellow sensor area.

Due to force and bending reactions in time, a fast and early decision for e.g. an optimised release of airbags can be calculated. State of the art is a bumper demonstrator with a piezo film sensor system which is sampled and digital processed with an Laptop based measurement system (PCMCIA Data Acquisition Card and LabView based analysing and presenting virtual Instruments). Future goal is the integration of the piezo film sensor systems with a high speed signal processor, which can do all online signal calculations and generate several digital outputs for CAN- bus communication with the on-board computer systems in order to release airbag systems or take influence on e.g. vehicle stabilising systems like ESP.

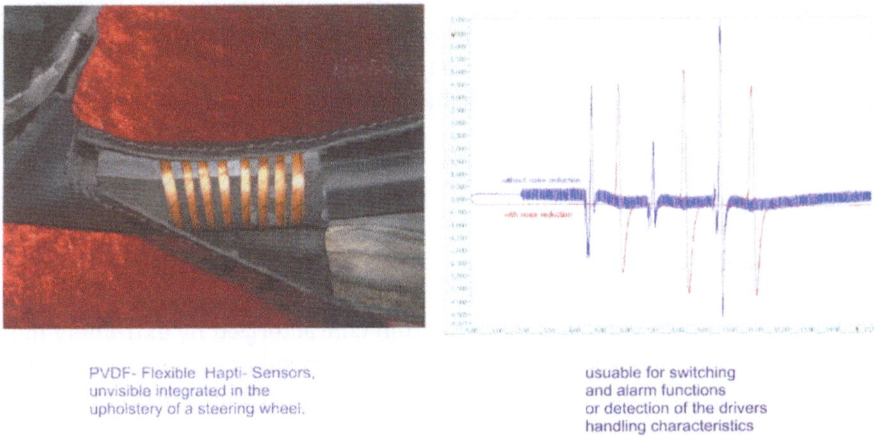

Fig. 10. Steering wheel sensor, optimized for switching functions under soft surfaces, reducing large vibration noise.

Figure 10 shows a soft- touch- sensor system and signal samples from a steering wheel. The sensor system is designed to minimize vibration and strong temperature disturbances.

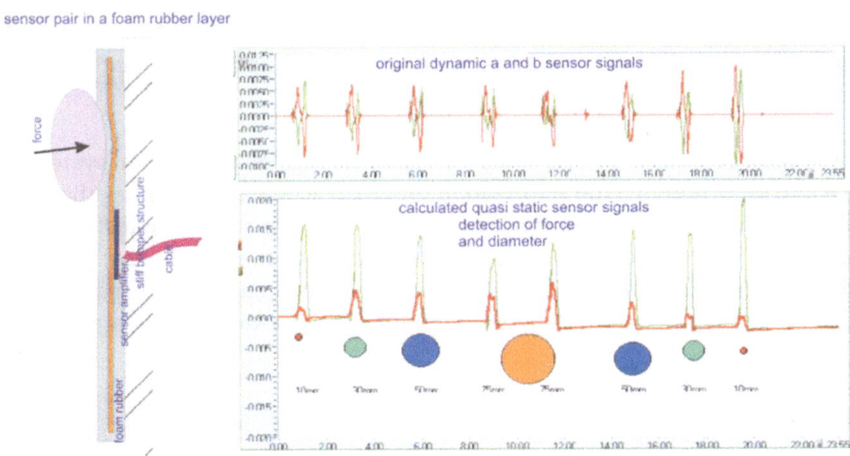

Fig. 11. PVDF- Foam- Rubber Twin Sensor, pushed with cylinders of different masses and diameters, at top are the sensor signals, at bottom the corrected quasi static force and bending curves.

Figure 11 shows the sensor signals of a sensor pair which is mounted within a foam rubber on the front of a bumper. The signals are produced by several hits to the rubber sensors with different cylinder diameters. As to be seen in the corrected lower signals a significant Force/Bending relation corresponds to the used diameters. This type of sensor can serve Airbags for the clearing for pedestrians, since during the collision can be calculated, if a leg-similar body is met. The sensor systems can be produced in very thin layer constructions, so it's possible to create a nearly invisible design in harmony with the bumper construction.

7 Conclusions

Mechanical events in vehicle safety are often characterized by extremely high amplitudes (forces and deformations in case of a crash) at comparably low frequencies, and the amplitudes normally decrease significantly with increasing frequencies. Due to its vast dynamic range of 286dB and high flexibility, the piezo film material offers many measuring solutions for strong forces or deformations, which can be sampled and calculated with a high resolution by a, application specific, analogue/digital amplifying system.

Piezo film sensor systems may be designed complementary with a high resolution at high frequencies and decreasing sensitivity at lower frequencies. By means of a numeric correction of the non-linear amplifier characteristics the dynamic range for slow processes of the digitised signal may be enhanced by some order. Thereby very small as well as extraordinary forces and deformations may be processed by the same system.

This high resolution on one hand enables the use of the sensor system for force and deformation characterization for crash scenarios as well as the detection of small forces, as for example in a slow collision with a pedestrian. Also when using their actuator properties piezo foils, by means of their flexibility and mono-axial orientation, may lead to a number of solutions. Areas of interest are mainly the excitation and dampening of light sheet metal structures.

Paul Mirow
Mirow Systemtechnik GmbH
Kaiserin Augusta Allee 10
10553 Berlin
Germany
paul.mirow@mirow.de

Keywords: PVDF - film sensor systems, extreme dynamic range, analogue/digital signal amplification, force, strain, vibration and bending measurements, noise reduction sensor systems

Solutions for Tire Pressure Monitoring Systems

T. Bever, M. Kandler, Infineon Technologies AG

Abstract

The demand for tire pressure monitoring systems is driven by increasing comfort and safety requirements. In order to make such systems available to a wide market, cost optimized solutions are necessary. Costs can be reduced by high integration, low power consumption, small size of the module and the reduction of components by monolithically integrated and intelligent on-chip algorithms.

To meet these requirements Infineon is developing highly integrated tire pressure sensor chips fabricated in a 0,5µm BiCMOS technology. The pressure sensor cells are realized in a surface micromachining process which allows the integration of complex electronic circuits as well as memory to store calibration data, configuration data and an identification number. A unique wake-up algorithm is implemented to eliminate the need for an additional wake-up switch or acceleration sensor.

1 Introduction

Interest in tire pressure monitoring has considerably grown in the past years [1], primarily due to safety issues. Tire pressure monitoring systems (TPMS) will help to avoid accidents caused by under inflated or defect tires. As a further benefit properly conditioned tires reduce fuel consumption and increase tire lifetime. Pushed by US legislation TPMS will be rapidly introduced into new car platforms.

According to recent market analysis the total market of TPMS will rise to approximately 35 Mio units in 2008. More than half of this number will be based on direct systems, i.e. these systems include sensors that directly measure the pressure in the tire cavity. With 4 to 5 sensors per car this comprises a considerably big market which has to be served by cost effective solutions.

In this contribution we present a new product platform which is based on advanced microsystem technologies. By combining sensors, control and memory on a single chip it offers the flexibility needed by an emerging and fast growing market.

One of the major challenges in TPMS are long lifetime requirements combined with the limited power of the battery. Minimizing the power needed for the measurement itself is just one aspect. The current saving capacitive pressure sensor is of big advantage here. Analyzing the measured data and adapting the sensor activity and thereby the power management to the requirements of the actual situation (vehicle parked or in motion) is the key strategy to reach long lifetimes. From technology point of view this requires the possibility of digital processing in a very early phase of the signal change. Surface Micromachining combined with a powerful IC technology is used to realize microsystems that are best suited to meet these requirements.

2 System Aspects

Each direct TPMS consists of 4-5 pressure sensing tire modules, one in each tire, and a central unit. The tire modules measure pressure and temperature in the tire cavity. They are battery driven and transmit the sensor data via RF to the central receiver. For power saving reasons the interval between transmissions needs to be adapted to the actual situation.

The module thus has the following three basic functions: sense – control – transmit.

These functions can be realized with various system approaches and different solutions for the partitioning between sensor, controller and transmitter. The new Infineon transmitter PMA5100 for example will have microcontroller functionality and ideally matches with KP505 one of the tire pressure sensors of the KP500 series. Alternatively more controller functions can be implemented on the sensor. KP510 will be such a smart sensor and can directly be connected to a standard UHF-transmitter chip like the TDA5100. This transmitter can be run in ASK or FSK mode at 434 or 869MHz.

With the development of the KP500 series a platform of sensors becomes available which can flexibly be adapted to varying system specifications.

3 KP500-Series: A Sensor Platform For TPMS

The Infineon smart Tire Pressure Sensor KP500 offers all functions that are required for the sensor front-end of a tire module, without using further external sensor elements. It measures the relevant physical parameters pressure, temperature and battery voltage.

A schematic block diagram is shown in figure 1. The sensing elements are monolithically integrated together with an A/D converter, signal preprocessing, calibration data memory, power management and a SSI communication interface for easy and flexible system integration.

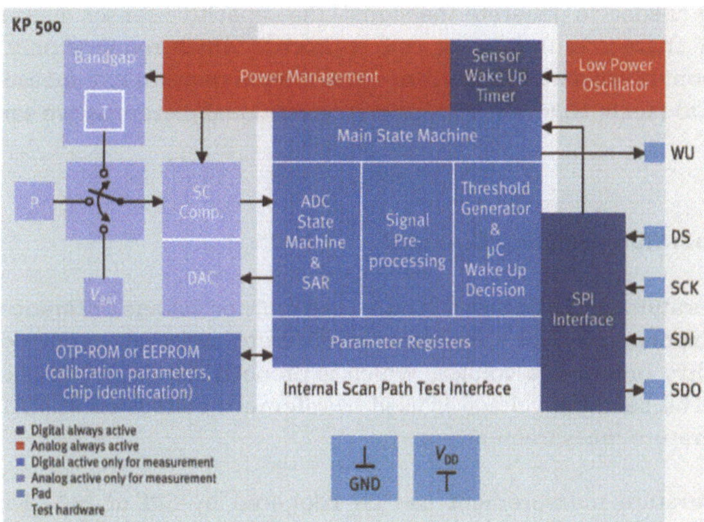

Fig. 1. Block diagram of KP500.

A sophisticated power management is mandatory to reach long lifetimes. The KP500 sensors are designed to reach lifetimes of more than 10 years with a single battery including the power consumption of a low power microcontroller and a transmitter.

Most of the circuitry has to be active only during the short time interval of the measurement. The state machine controls the measurement cycle by commanding the power management. A measurement cycle can be initiated either by an external command or by the internal wake up algorithm. Samples are taken typically once or twice per second. This frequency is sufficient because the pressure changes that have to be detected are slow.

4 Pressure Sensor

The pressure sensor consists of surface micromachined capacitive sensor cells. The device is designed for a pressure range up to 650kPa (6.5bar). Details of the technology are described below. Two sensor arrays and two reference arrays form a capacitive bridge circuit. Applied pressure changes capacitance of the sensor arrays and thereby unbalance the brigde. The resulting bridge voltage is fed to a fully differential A/D converter. Conversion is done using successive approximation register method (SAR).

The capacitive sensor cell is ideal for this application because of its low power consumption. In contrast to piezoresistive pressure sensors where a current is inherently needed to generate the signal, the capacitive sensor measures the amount of charge on the sensor cell, a process which requires only a very small amount of current. The current consumption per pressure measurement can be up to factor 10 lower than using conventional piezoresistive sensors.

5 Temperature Sensing Circuitry

The temperature sensor is based on the base emitter voltage difference of two identical bipolar transistors operated at different emitter current densities. This circuitry produces a voltage, which is proportional to the absolute temperature. The bandgap voltage is used as reference voltage for the ADC during temperature measurement.

The temperature measurement can be triggered by SPI or by an internal counter. The counter is clocked by the pressure measurement rate so that in this case the temperature measurement rate is proportional to the pressure measurement rate

6 Battery Voltage Sensing Circuitry

The battery voltage is divided to fit the range of the ADC. The bandgap voltage is used as reference voltage for the conversion of the battery voltage. The measurement can be triggered by SPI or by an internal counter, which is clocked by the pressure measurement rate.

7 Intelligent Wake-Up

Accelerometer or mechanical switches are commonly used in TPMS to detect whether the car is in motion or not. When the vehicle is parked the measurement frequency can be low [2]. This saves energy, however the motion switch increases the system costs considerably.

The KP500 sensors in contrast utilize a more cost-effective strategy for power saving. The measured pressure data themselves provide a very relevant criteria for the determination of the transmission interval. The KP500 sensors determine the optimum frequency of transmission from the change of pressure in time. The challenge here is to minimize the power consumption for the measurement as well as for computing.

The extreme low power consumption of the capacitive pressure sensor enables continuous pressure monitoring. A careful analysis of the pressure data is performed on-chip resulting in an intelligent wake-up. This algorithm calculates the appropriate time for sending data. If no change in pressure is observed no signal needs to be transmitted and valuable energy can be saved. On the other hand a rapid change in pressure requires frequent transmission of data ('Alert Mode'). Further signal processing in the central unit will decide whether the driver has to be warned or not.

Figure 2 illustrates the principle. The top curve shows an exemplified pressure lapse. While the sensor measures in fixed time intervals, the transmission of values depends on the pressure gradient. At constant pressure the frequency of transmission is low. Frequent data transmission will only occur in phases of rapid pressure changes.

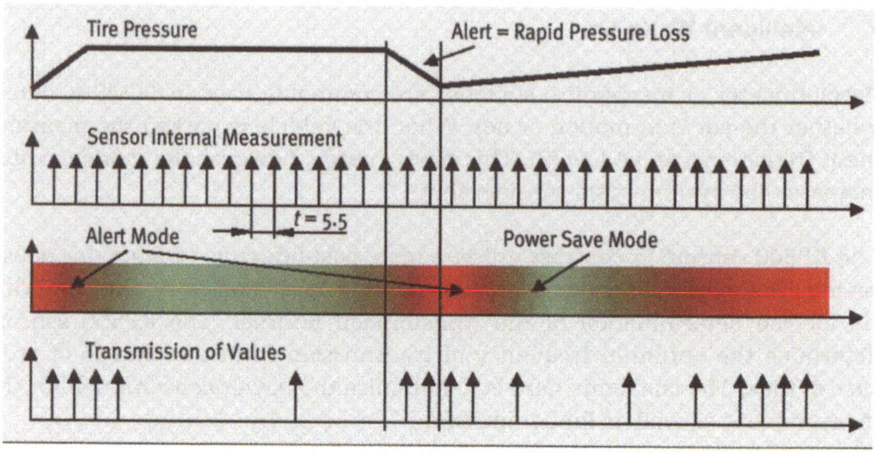

Fig. 2. Illustration of Intelligent Wake-Up.

Tuning the number of transmissions intelligently rather than following rigid clock schemes contributes to the overall power savings approach of the KP500 tire pressure sensor series.

8 EEPROM and Chip ID

The pressure sensor is calibrated in the IFX production line and the calibration coefficients are stored in an internal PROM memory. No calibration is needed for the temperature sensor. Additionally the EEPROM contains a unique 32bit serial number, which is traceable through the whole fabrication process at IFX.

The chip ID can be read out via SSI and can be used for wheel identification. This is important for the assignment of the received sensor data in the central unit. Also interference with unwanted data from e.g. other cars is avoided.

9 Interface and Programming

The synchronous serial interface (SSI) is a commonly used interface for the communication between a microcontroller and peripheral devices. In the KP505 a basic version of this protocol is implemented, which allows the sensor to be operated as SSI slave with standard microcontrollers. The KP505 always operates as a slave, whereas the microcontroller is the master. Therefore a write-operation means data sent from the microcontroller to the sensor and a read-operation means that the microcontroller requests data from the sensor.

The interface uses four lines: serial clock, serial data out, serial data in and data synchronization (DS). By setting the general configuration register the DS line can be configured to serve as a wakup-trigger line for the sensor. In this case a "low" pulse on DS without any data transfer triggers the wake up. After the processing time, the new values can be directly read out without further communication.

Besides the actual measurement value, the sensor provides additional information like average values and noise level. The settings for the averaging levels and the parameters for the sensor wake up as well as the transmitter wake up can be programmed via SSI. Thus the sensor can be adapted to different types of tires and vehicles.

10 Surface Micromachining (SMM)

The sensor technology is based on the meanwhile well established and high-volume proven surface micromaching (SMM) technology. Infineons SMM integrates a pressure sensor cell into a BiCMOS process. Since BiCMOS offers both bipolar and CMOS devices this technology represents the most advanced microsystem technology available to date.

In 1998 Infineon started mass production with this novel micromachining technique for pressure sensors. SMM sensors with both analog and digital output are in high-volume production now. Since only the front side of the wafer is processed together with the integrated circuitry this technique is called Surface Micromachining. In SMM the sensors are implemented using standard semiconductor processes. Currently a 0.8µm BiCMOS process is in SMM-production. A shrink down to 0.5µm minimum line width is in development and will further boost the potential of SMM technology.

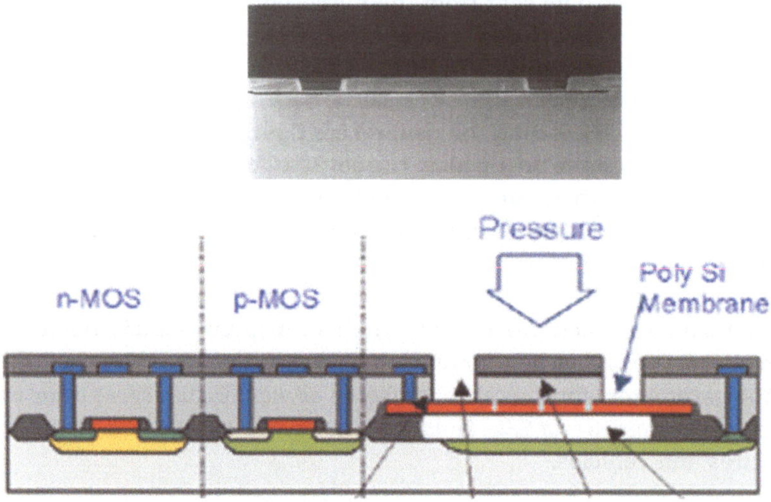

Fig. 3. SEM picture of a pressure sensor device (top) and schematic drawing of the sensor cell.

The pressure sensor itself is based on a capacitive sensor cell consisting of a thin flexible membrane, the silicon substrate and a sealed cavity. Membrane and substrate form the two plates of a capacitor. The membrane bends under pressure leading to a change in capacitance of the sensor cell.

The membrane is made of Polysilicon. This layer is part of the BiCMOS process and serves at the same time as resistor layer and as part of a linear capacitor. The thermal oxide which is used as the **isolation** layer within the BiCMOS process serves as sacrificial oxide in the micromachining part. In one of the few extra processes the sacrificial oxide is removed underneath the polysilicon membrane by wet-chemical etching which has a very high selectivity to the Polysilicon of the membrane. The sacrificial layer etch is performed through 1µm holes in the sensor membrane. The lateral isotropic etch lengths determines the cavity dimensions

This unique technology offers cost-effective solutions especially for those applications where immediate digital processing of the pressure signals optimizes the overall system. At the same time the integration in a standard BiCMOS process comprises a cost effective way to realise highly reliable smart sensors for automotive applications. For TPMS with its high requirements for low overall power consumption and high accuracy SMM is ideally suited.

11 Summary and Outlook

The development of advanced microsystem technologies enables the creation of smart sensors. As described in this contribution these sensors are able to preprocess the collected data and control e.g. the frequency of data transmission. In TPMS this strategy helps to save valuable battery power and at the same time reduces system costs.

Battery driven systems will dominate TPMS for quite a long time thanks to its relatively simple and robust design. Nonetheless batteryless systems are in development and the integration of the sensor directly into the tire is in discussion [3]. This will further challenge sensors and electronics so that microsystem development has by far not reached its end.

References

[1] J. Becker; "Tire Pressure Monitoring Systems- the new MEMS based safety issue" Advanced Microsystems for Automotive Applications 2002, Springer Verlag, ISBN 3-540-43232-9, pp 243-250.

[2] M.L.Shaw; "Motion Sensing Techniques and Analysis for Direct Tire Pressure Monitoring" SAE-technical Paper 2003-01-0202, SAE2003 World Congress, Detroit Michigan.

[3] R. Benedictm, P. Shepler, M.Fischer and D. Wagner; "Talking Tires – A Basis for Tire Diagnosis" SAE-technical Paper 2003-01-1279, SAE2003 World Congress, Detroit Michigan.

Thomas Bever, Michael Kandler
Infineon Technologies AG
Balanstrasse 73
81541 Munich
Germany
thomas.bever@infineon.com

Keywords: tire pressure monitoring system (TPMS), lifetime requirement, pressure sensor, synchronous serial interface (SSI), surface micromachining (SMM), BiCMOS

Subsystem Integration for Low-Cost Tire Pressure Monitoring Systems

A. van der Wiel, Melexis NV

Abstract

The scope of this paper is to analyse the trade-offs for system integration for TPMS: Tire Pressure Monitoring Systems. System integration is necessary to meet the cost requirements for systems in the automotive industry.

Two ways of integration are discussed: SOC: System On Chip and SIP: System In a Package. The key issue for SOC products is process complexity in combination with testability, whereas success of SIP products is mainly based on packaging technology. The analysis will show that the key factors for system integration are cost, packaging and test.

The SOC and SIP trade-offs for TPMS are presented based on Melexis' expertise in manufacturing all required components that are made by IC technology and low cost packaging. These elements are a pressure sensor, a motion sensor, a microcontroller and a RF transmitter.

1 Introduction

Tire Pressure Monitoring Systems (TPMS) will become mandatory safety devices. Therefore, they have to be produced in a very cost effective way. However, TPMS are complex systems which have to survive the harsh environment inside a car tire for 10 years. Today these systems are too expensive to fulfil the cost requirements for all new cars. The reason is that nowadays these systems are still too complex.

The TPMS systems that are available today consist of many components, typically more than twenty. Each functional subsystem in the module is represented by a dedicated circuit.

Figure 1 shows the basic silicon components: the microcontroller for signal conditioning, the pressure sensor, the motion detector and the RF / LF transmitters. Other essential components are the antenna, some passive components and the battery.

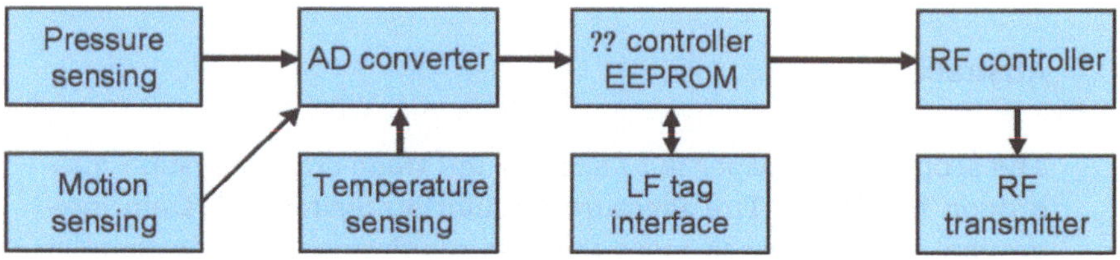

Fig. 1. TPMS functional blocks to be integrated in silicon.

The key to build cost effective electronic systems for automotive applications is combining certain system functions to reduce the number of components at final assembly. The utmost integration consists of realising the complete system on one single silicon chip. Such a system is called a SOC: System On Chip. However, complex systems are often fabricated as hybrid solutions as they are either more cost effective or do not require specialised IC technology. A hybrid solution where several silicon components are placed in one package is called a MCM: Multi Chip Module. When such a MCM fulfils a complete function, one can call it a SIP: System In a Package. Figure 2 depicts a SIP for TPMS containing a pressure sensor and a controller chip.

This paper first discusses the advantages of SOC and SIP in respect to cost, packaging and testing effort. Then the system requirements for TPMS are outlined. The main part of the paper discusses the choices for system integration for TPMS, leading to different SOC and SIP integrations. Also integration possibilities are discussed which are not cost effective.

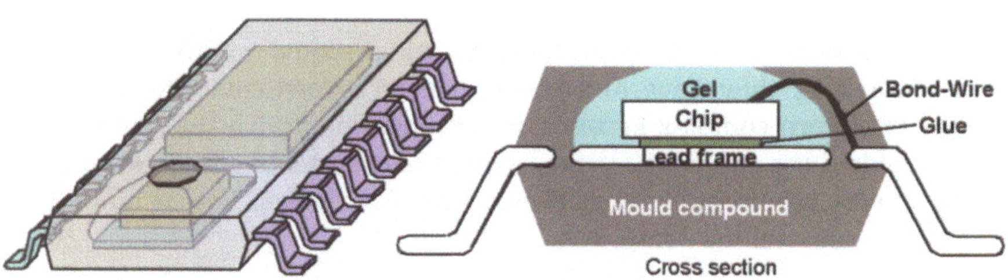

Fig. 2. Concept of a SIP containing a pressure sensor and a signal controller forming a calibrated pressure and temperature sensing system. The pressure is applied to the sensor via a gel.

2 System On Chip versus System in a Package

In general one cannot favour SOC or SIP as is demonstrated by the success of the accelerometers of Analog Devices [1] and Motorola [2]. The accelerometer of Analog Devices is a SOC where the accelerometer sensing elements are integrated on top of the read-out electronics during the wafer fabrication. The accelerometer of Motorola is a SIP where the sensing device is fabricated with a dedicated MEMS process and the controller with a low-cost CMOS process. Both products are currently produced in large quantities with more or less the same performance.

System On Chip

SOC technology focuses on IC technology to integrate all system functions onto one chip. This normally increases the complexity of the IC process. The complexity gives rise to higher costs per unit die area and it also becomes more difficult to maintain high yields. When integrating a pressure sensing membrane into a CMOS process, one cannot put CMOS circuitry on the membrane. Membrane bending would cause improper functioning of the circuitry and the circuitry on the membrane would introduce offset drift to the sensor over time. So, one ends up not only with a higher price per unit area, but also with a larger die and lower yield. Analog Devices avoided this problem by building the acceleration sensor on top of the CMOS circuitry. The packaging is straight forward for SOCs, as one deals with just one chip. Even the packaging of bad die is only of slight importance, as the package is cheap in comparison with the cost of the complex chip. Hence, one does not have to test exhaustively before packaging. However, the testing after packaging can become expensive and cumbersome for complex SOCs as sometimes different testers are required to test all system characteristics.

System in a Package

SIP and MCM technology focuses on advanced packaging technology to integrate several dies into one package. This normally only slightly increases the complexity of the packaging process, but requires a more carefully designed package. Especially stress within the package can be difficult to control [3]. The main advantage of MCM technology is that one does not have to compromise on IC technology. Each chip can be fabricated with its own dedicated process. In contrast with SOC packaging, one has to avoid using bad die at assembly, as one has to scrap the entire MCM if only one die is bad. Hence, SIP and MCM packaging requires exhaustive testing of the dies before assembly

(Known Good Die concept). As a consequence, the test after packaging only has to focus on system performance and less on die functionality.

3 TPMS Requirements

The block diagram of figure 1 shows all the functions that have to be realised with a TPMS system. The passive components such as battery, crystal, antennas, capacitors, coils and resistors are not taken into account.

The pressure sensor and the temperature sensor are the main sensors to provide the basic information one expects from TPMS. Note, that the pressure sensor requires a dedicated package for the harsh tire environment. The sensor is not only exposed to moisture and organic fumes at temperatures ranging from -40 to 125°C, but also exposed to accelerations higher than 1000g at speeds around 250km/h.

The motion sensor is normally used to detect movement of the wheel. With this information the system is set in a sleep mode when the car is not moving to save battery life. In this mode the system only measures pressure and temperature at long intervals and does not transmit the values. Data transmission takes the major part of the energy consumption. Note that battery free systems do not require a motion sensor, as the system is activated as soon as the car is moving by LF-TAG transmitters near the wheel.

Fig. 3. Pressure sensing unit with gel coating in a SOIC16 package.

The functionality of the system is provided by a microcontroller in combination with ROM, RAM and EEPROM. These memories all have specific functions. The ROM contains the instructions for basic functioning, the RAM allows proper functioning of the microprocessor and the EEPROM is not only used to store the part serial number and calibration data, but also contains the embedded software to define specific system functionality.

Nowadays a RF transmitter is used to send the TPMS pressure information to the Remote Key Entry receiver. As there is no standard for TPMS transmission, the RF controller has to be capable to work at several frequencies. The LF TAG interface allows bidirectional communication with the system after fabrication of the final product. It can be used to activate the system at the moment it is mounted in a wheel or, for example, it can be used to adapt alarm levels for low pressures when wheels are changed. The LF interface should be configured in such a way that it also allows transmission of the TPMS information in systems that do not rely on transmission to the Remote Key Entry receiver [7].

4 System Implementation of TPMS

On-chip Integration of the Pressure and Motion Sensor

For battery powered TPMS the first evident on-chip integration is the combination of the motion sensor with the pressure sensor [4]. Both pressure and motion can be detected by piezo-resistors on respectively a membrane and a cantilever. Therefore, one can define a dedicated MEMS process that provides both sensors on one chip. By simultaneously fabrication of the sensing resistors, one can simplify the calibration. Both sensors will have very similar temperature drift of the sensitivity.

On-chip Integration of a Programmable VCO with RF Transmitter

The circuitry needed for RF transmission is normally realised in a BICMOS process. For TPMS the RF transmitter needs to be adapted for different applications and countries. Europe, the USA and Japan do not allow the same frequencies for Remote Key Entry systems that are used to receive the TPMS signals. So, the RF transmitter needs a frequency controller for different frequencies. Other needs are tuning of the output power and allowing the use of frequency modulation (FSK) or amplitude modulation (ASK). Implementing these options in one chip reduces the passive components on the final assembly board and reduces also the amount of different parts for different applications. Melexis developed such a transmitter that can be used for all TPMS applications [5].

MCM Integration of the Controller with the Sensor

Melexis combines the sensor chip and a programmable signal conditioning chip in one MCM as is outlined in figure 2. The temperature sensing is realised on the CMOS chip as it offers resistors with better temperature characteristics then the sensor. Within the package, the temperature of the sensor and the CMOS chip can be considered the same. The calibration of the two-chip package results in a digital pressure signal which is compensated for temperature. Note that calibration can be carried out before final assembly of the complete TPMS module. At final assembly the system with EEPROM and microprocessor allows a lot of flexibility for programming embedded system functionality such as measurement intervals, identification number, transmission protocols, self test etc.

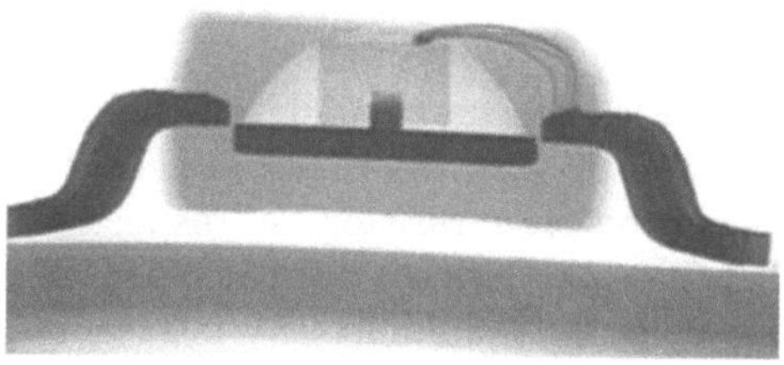

Fig. 4. X-ray graph of the gel feed-through of a pressure sensor packaged based on a standard SOIC8 process.

The key for this MCM package is the transfer of the outside pressure to the sensor. Therefore, a dedicated package technology has been developed to combine low-cost assembly with the severe TPMS requirements. A standard packaging process for SOIC packages was extended to realise a pressure connection by using a soft gel [6]. Figure 3 shows pressure sensors realised with this process. The connection hole with gel is located in the middle of the package. This gel provides the transfer of the pressure from the package surface to the sensor surface. Figure 4 shows an X-ray graph of a pressure sensor in a SOIC8 package. The dark areas show the lead frame and the very light area the gel around the sensor. The position of the bond-wires on the right needs further optimisation. Note that this package is very thin in comparison with sensors that are built with pre-moulded packages. This results in a very low sensitivity for acceleration.

Integrated System Overview

The three integration steps described above lead to the TPMS system implementation shown in figure 5. The sensor chip and the controller chip are packaged in one SOIC or SSOP package and the RF dedicated parts are realised in a MLP package. The pressure sensor and motion sensor are integrated without circuitry on one chip. The TPMS controller is realised in a standard CMOS process and includes a 16bit microprocessor, temperature sensor, analog-digital converter, ROM, RAM, EEPROM and a LF send-receive module at 125kHz. Note that the sensors and controller are packaged as a SIP, providing a complete calibrated measurement system. The RF transmitter is realised with a BiCMOS process and put in an MLP package. Due to integration of the Voltage Controlled Oscillator and a tuneable gain, a SOC is almost obtained. It only needs a crystal and some additional passive components to form a complete system.

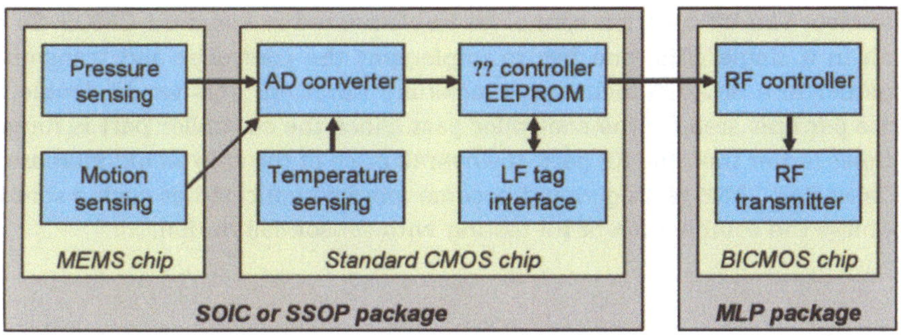

Fig. 5. Melexis solution for TPMS system integration.

Discussion Further Integration

Integration of the sensors onto the controller chip is not a cost effective solution. The three dimensional structuring of the sensors requires several additional masks to the IC process. Furthermore, the motion sensor needs a cap to assure free movement of the seismic mass. It is difficult to imagine the implementation of these sensors on top of CMOS circuitry. Hence, the silicon area of the combined controller and sensor functions will probably not be smaller than the areas of the separate dies. However, the cost per unit area of the separate dies is considerably lower than for an all integrated system.

A market requirement also favours the proposed SIP solution. Battery free systems do not require a motion sensor. For a SIP, it is relatively easy to replace the combined motion-pressure sensor with just a pressure sensor and qualify a product without a motion switch. Probably the same package with the same pin connections can be maintained. However, for an all integrated device one not only needs to qualify a new product, but also a new extended IC process.

Integration of the RF chip into the MCM package is not considered to be cost effective. The presence of the RF chip in the MCM would increase the test effort enormously as RF products need dedicated high frequency testers.

The market requirements for battery free systems also favour a separate package for the RF transmitter. Battery free systems do not rely on RF communication but on bidirectional LF communication at much lower frequencies such as 125kHz [7]. The MCM solution as it is proposed can serve both kinds of applications as a bidirectional LF communication is available on the controller chip. This communication circuit is relatively small.

Integration of the RF chip onto the controller chip is not considered to be cost effective. The RF function cannot be implemented in standard CMOS. To get both in a single chip, one has to implement the controller and transmitter together in a BiCMOS technology including EEPROM. This would double the price per unit area for the controller part. Since the controller part is roughly 5 times larger than the RF part, the overall price of the chip would increase by at least 84%. Also testing would become more complicated as such a solution requires the combination of RF testing with sensor calibration.

5 Conclusions

The main trade-off considerations for SOC and SIP are presented and made applicable in a general way. It is shown that the choice for a SOC and SIP depends on a comparison between IC technology, packaging technology and test.

These trade-off considerations are applied to define a low-cost solution for TPMS and lead to the integration of all functions in 2 packages. Therefore Melexis proposes for TPMS a MLP package containing a highly integrated RF transmitter and a SOIC based package containing a MEMS chip with a pressure sensor and motion sensor and a CMOS chip with the system control electronics, EEPROM, LF transmitter and temperature sensor.

References

[1] "ADXL50-monolithic accelerometer with signal conditioning" Analog Devices, norwood, MA, data sheet, 1993.

[2] "MMAS40G10D-micromachined accelerometer" Motorola, Phoenix, AZ, data sheet, 1997.

[3] Gary Li and Ampere A. Tseng; "Low Stress Packaging of a Micromachined Accelerometer" IEEE transactions on electronics packaging manufacturing, VOL. 24, NO. 1, January 2001.

[4] "Tire pressure sensor SP12" Sensonor, web-page on www.sensonor.com.

[5] "TH72005 Transmitter - 315MHz, FSK/ASK (10L MLP 3x3) " Melexis, web-page on www.melexis.com.

[6] Appo van der Wiel, Jian Chen and Henri Vanderhenst; "Low-Cost System-In-Package for TPMS" A3.4, Proceedings Sensor2003, Neurenberg, Germany, 2003.

[7] Robert Benedict, Peter Shepler, Martin Fischer and Dieter Wagner; "Talking Tires – A Basis for Tire Diagnostics" SAE International, 2003-01-1279, 2003.

Appo van der Wiel
Melexis Tessenderlo
Transportstraat 1
3980 Tessenderlo
Belgium
awl@melexis.com

Keywords: TPMS, system integration, system in a package, MCM, pressure sensing, motion sensing.

BioVolume: The Seat Integrated Human Based System to Meet FMVSS208 Automatic Suppression Concerns

M. Pajon, L. Chabert, H. Dirand, L. Jaillet, Faurecia

Abstract

Motor vehicle Occupant related indications such as morphology class and dynamic position are important information to be taken into account by future passive safety systems in order to increase protection of occupants. Since 1998, the National Highway Traffic Safety Administration has initiated the first step that will require introduction of occupant sensing means. Occupant information will have to be considered as prior in airbag deployment decision in the event of a frontal crash. A first rule that amends the occupant crash protection standard will require application of improvements in order to reduce risk of severe or fatal airbag induced injuries to occupants, particularly young children and small adults [3, 4]. This paper presents the Faurecia BioVolume sensing system which has been developed for the purpose of occupant monitoring. This system which establish relevant information on occupant morphology and dynamic position can be used for Static or Dynamic Automatic Suppression Option required by the new FMVSS208.

1 Introduction

1.1 Background

Airbag as one part of a vehicle's occupant protection system is reported to save many lives in case of frontal accidents. At the beginning of year 2003, the US National Highway Traffic Safety Administration reported that frontal airbags has help to save more than 10000 lives in the front passenger seats. However, deployments of airbags occur with power that can induce injuries to occupants. In the NHTSA summary document, 221 fatalities attributable to airbag deployment were reported [5]. This recent data confirm the need for application of the new US requirements that are designed to reach the goal of improvement of the protection for occupant of all sizes, belted or unbelted, from moderate to high speed crash, and minimise the risk posed by airbags to occupants, especially at low speed accidents. The use of occupant sensing means are to be introduced in vehicles in order to take into account the type of occupant and his position prior an airbag deployment decision.

The BioVolume system has been designed by Faurecia to fulfil the airbag Automatic Suppression Options requested by the FMVSS208 in order to improve safety of both drivers and front outboard passengers. The system which consists in data processing of measurements from conducting electrodes inserted in the passenger seat has been developed following different axes: Safety performance, Smart sensor technology and Seat integration. This paper describes each of these axes and also the interaction between.

1.2 Occupant Sensing Methods

A number of sensor technologies are available in the field of occupant sensing in motor vehicle. This sensing principles include pressure distribution, strain gages, electric fields, active infrared, ultrasonic or video. Sensor performances, as well as related strengths and weaknesses can be found in Bevan [1]. It appears that sensors can be classified in two groups according to the performance they are designed for. On the first hand, several sensors aim at establishing a calculation of the morphology of the occupant. These sensing principles are based on the measure of an effort, pressure or load revealing the interaction of the occupant towards its supports in the vehicle [2]. For these devices, the seat is a very relevant location for sensor implementation. On the other hand, the sensors dedicated to occupant position calculation are based on a non mechanical and non contact principle and use information from a fixed point in the vehicle, far from the occupant.

1.3 BioVolume System

The technology we present is based on capacitive sensors which consists is conducting electrodes placed on the "A" surface of the foam pads. The following paragraphs describe how data from seat sensors can give information related to occupant type, morphology and position.

2 BioVolume System Description

BioVolume system is aimed at giving information about morphology or position of the occupant in the seat. This system is based on capacitive sensors integrated under the foam pads (figure 1). A sensor is made of at least two electrodes and the measurement is based on the values of capacitors generated by each sensor. The measures applied on sensors are split into two successive sequences: field application followed by a charge counting: The number of charges kept on each electrode is directly linked to the occupant morphology and distance to the sensor.

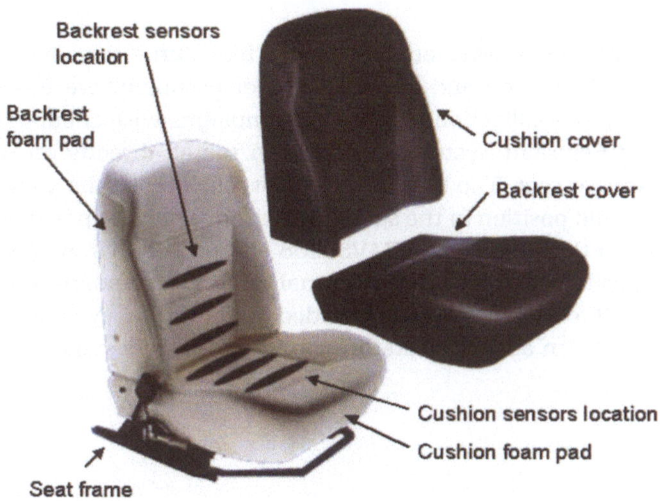

Fig. 1. Description of sensor area location at the "A" surface of the seat foam pads.

This system can be seen as a capacitance measurement system which is described in a Lucas document [6]. The different capacitors that are involved in the measurement as well as methods to separate distance to sensor and body surface covering sensors are also described in this document.

3 Occupant Classification

A study was conducted in order to evaluate the capability of the system to assess information related to type and morphology of individuals. Different performance criteria are to be taken into account for this evaluation: first is the accuracy of the classification provided by the system and second is the stability of the result according to the various positions that an occupant can have or have in a seat. The result provided by an occupant classification system should not change when an occupant has moved from a position to another, as a result of individual comfort purpose or in the event of an external dynamical solicitation. A seat with integrated BioVolume sensors was used to record data from occupant of different morphologies in different positions. The front outboard passenger seat was considered in order to take into account the important diversity of occupants and positions.

3.1 Methods

Mass produced seats were equipped with BioVolume sensors at the cushion level. Series of children and adults volunteer occupants were selected for an experimental data collection phase. Test campaigns with children included the use of Child Restraint System and children seated directly on the front outboard passenger seat. The type of child seat, position on the passenger seat as well as the child position in the specific restraint or on the passenger seat were established so that it included FMVSS208 test procedures. Adults were seated in various positions other than the normal reference standard. The Stature and body mass of occupants were recorded in order to evaluate results with respect to children and small stature adults as listed in standard 208.

3.2 Results

The output signal resulting from the presence of an element (child seat, occupant,...) in the neighborhood of a sensor can be expressed as $Vs = f(\varepsilon,S,d)$ with:

ε: dielectric constant of the element

S: surface of the element in the filed generated by sensor probes.

d: distance between sensor and element

Child Restraint Systems

For situations with child restraint systems, because the dielectric constant of materials that child restraint systems are made of, are at very low values and because the child is far from sensors, the resulting signal given by each sensor is very low. This fact leads to a direct distinction between children in child seats and adults occupancy. Additional loading that can be set on the child seat with the seat belt cinch has very limited influence on the output, this is an interesting level of robustness for reaching in the simplest way the Static Automatic Suppression requirement for children in child restraint systems.

Classification of Children versus Adults

The S surface in formula $V = f(\varepsilon,S,d)$ expresses the amount of occupant which is in the neighborhood of each sensor. An analysis of the allocation of signal over the different sensors allows to estimate the category of the occupant. After analysis of experimental databank, an algorithm dedicated to a binary classification of occupant was elaborated. The limit between the two classes was chosen according to static automatic suppression requirements: 6-year old children (21 to 25.6kg and 114 to 124.5cm) and under must be classified in class 1 and adults of morphology greater than a 5th percentile (46.7 to 51.25kg and 139.7 to 150cm) must be classified in class 2. The aim of the classification module was to classify occupants in both classes for many positions corresponding to real life / real world events. We considered the FMVSS208 test procedure for the positioning of volunteers and we added other positions such as: transversal displacements to side of the seat, lateral rotation of pelvis and thighs, longitudinal displacement of pelvis, backrest inclinations, movements of the legs and feet and kneeling for all children. The graph below shows the result of the classification for all occupants in positions configurations.

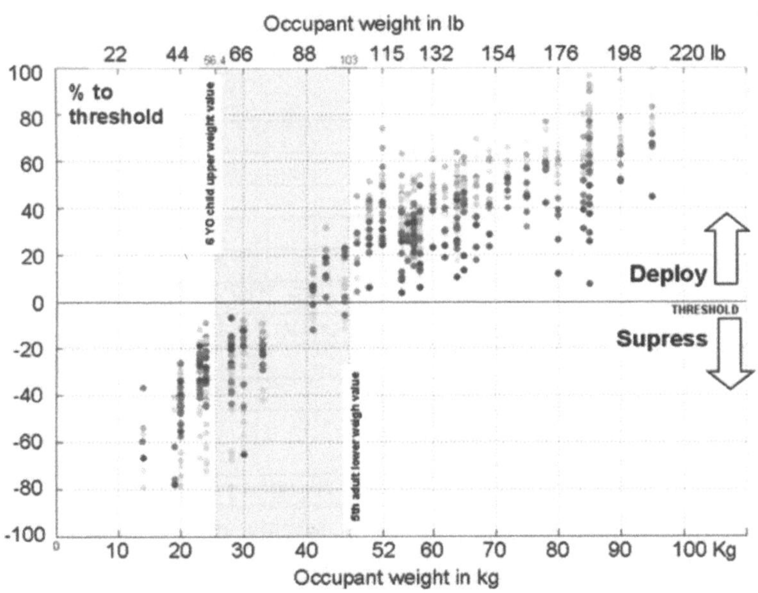

Fig. 2. Binary classification of occupant in various seated postures for a purpose of AirbBag suppression decision. A vertical set of points (at a given occupant mass indicated on the X axis) corresponds to the different positions of occupants.

The results of figure 2 shows the capability of the system to assess the proper classification in an important range of postures. The initial raw algorithm provides instantaneous output and does not take into account any history data processing. The use of BioVolume available information such as pelvis position will improve the accuracy of the result in an overall filtering algorithm.

Occupant Mass Assessment

In a complementary task to binary occupant classification, an evaluation of the capability of the system to estimate the mass of an individual was investigated. As described above, sensor response is linked to the surface of occupant covering the different probes. In a given seated configuration, the surface is directly an expression of the width of the occupant in the area of sensors locations. The mass estimation capability is dependant on the level of correlation that can be found between different widths of occupants pelvis/thighs region (impacting the cushion sensors) and their mass.

An algorithm using information of occupant pelvis position extracted from the seat cushion sensor allocation was used to build a continuous classification. The purpose of an optimization was to minimize estimated output versus measured occupant weight. Figure 3 shows the result of mass estimation for occupant in a reference position.

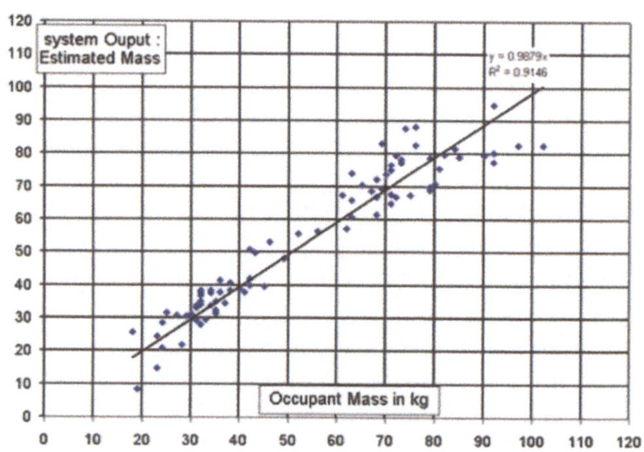

Fig. 3. Mass estimation result. Databank of adults and teenagers volunteers.

The graph above shows an interesting estimation of the weight of occupants. The input needed to assess acceptable accuracy is the position of the pelvis in order to use adequate weight estimation.

The results reported below were related to occupant classification. The possibility to take into account occupant position for weight estimation was described. The following section is completely dedicated to the occupant position calculation.

4 Occupant Position Monitoring

We conducted a study in order to establish a method to compute information of motor vehicle occupant position. The aim was to establish criteria of proximity of the occupant thorax/head region to the instrument panel where airbags systems are located.

The purpose of the process is to use the following information:

- Position of the seat in the vehicle.
- Position of the torso in the seat.
- Position of the head compared with the torso.

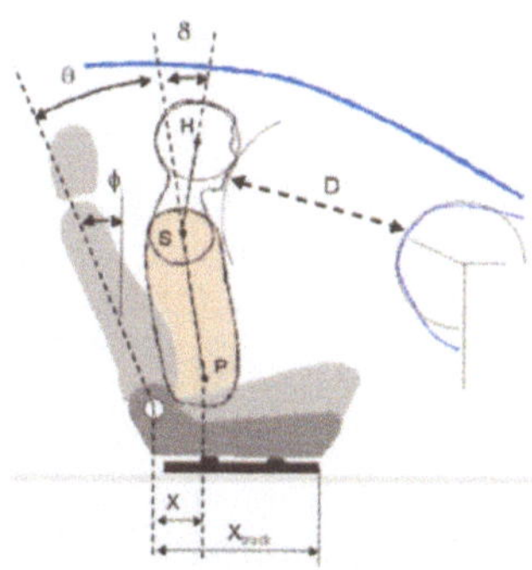

Fig. 4. List of parameters for estimation of the distance (D) between the occupant and the instrument panel. x_{track} and (Φ) are seat adjustment parameters, (θ) is the occupant trunk to seatback angle, (x) is the occupant pelvis position, PS is the length of the Pelvis to Shoulder segment, SH is the length of the shoulder to head segment, (δ) is the neck longitudinal inclination, (θ) and (x) are measured with the BioVolume system.

Figure 4 shows all the parameters which come into play in the classification of position. The position of the seat in the vehicle is provided by a displacement and angular sensors that output the longitudinal adjustment (x_{track}) of the seat and the angular inclination (Φ) of the backrest. The position of the body in the seat is computed by the BioVolume system which provides the torso inclination (θ) and the longitudinal pelvis position (x). The position of the head compared to the body is computed with the use of anthropometrical and ergonomical laws (PS, SH, δ).

The signal provided by a BioVolume sensor is given by the relation as follows: Vs=f(S,d). An other characteristic of the BioVolume is that the normalized signal Vsn=Vs(S,d)/Vs(S,0) is an independent function of (S). Vsn=g(d).

4.1 Occupant to Seat Position

A seat equipped with three BioVolume sensors in the backrest was used to determine the torso angle calculation module. Sensor are located in the upper, mid and lower part of the backrest. We have $Vsn_i = g(d_i) = g(x,\theta)$ where (i) is the index of the sensor (between 1 and 3). For a given pelvis position, the evolution of Vsn_i versus the angle (θ) were established during experimental tests. Thus, we can easily compute an estimated angle (θ_{est}) function of Vsn_i.

Figure 5 represents the estimated angle (θ_{est}) versus the real angle (θ).

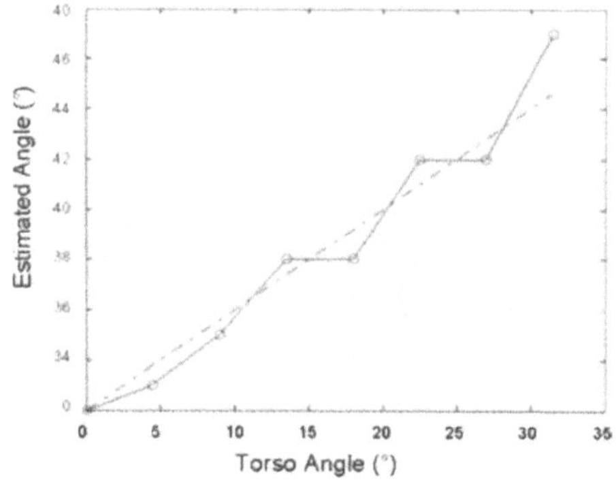

Fig. 5. Function for the estimation of the (q) torso angle.

To compute the (x) pelvis position information, the cushion sensors are used. The method is based on the checking of cushion signal repartition. Figure 6 shows the evolution of a combination of signal values of sensors positioned at the rear part of the cushion.

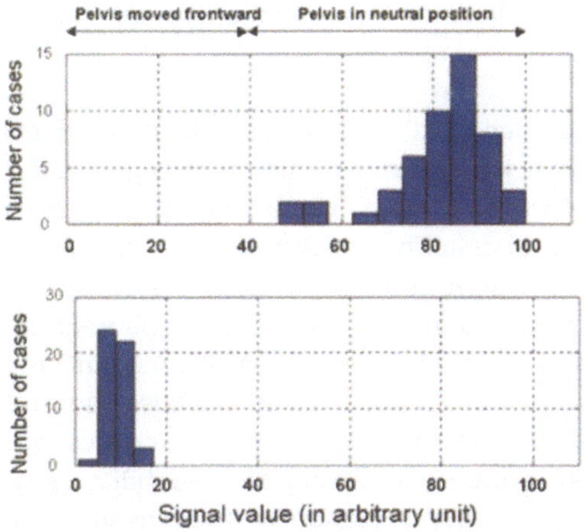

Fig. 6. Estimation of the x position. The upper part of the graph displays the allocation of the signal with 50 occupants in reference position. Results with pelvis moved forward are shown on the lower part of the graph.

The presence of sensors at the backrest and cushion level allows to establish the classification of the occupants in terms of torso inclination and pelvis longitudinal movements. Figure 7 shows a BioVolume demonstrator presented during several exhibitions.

The system, first presented in 2000, illustrated the capabilities of the technology to classify the occupant in position according to four levels of inclination for the torso and two longitudinal displacements for the pelvis. Lateral movements of the upper body were also computed with the help of sensors in each side of seat backrest bolsters.

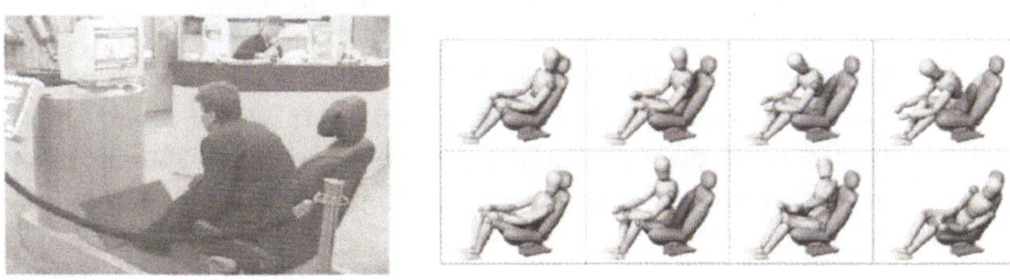

Fig. 7. Faurecia BioVolume exhibition Mock up. The Positions of occupant were computed in 8 class and displayed on the screen.

4.2 Estimation of the Distance Between the Occupant and the Instrument Panel

Following the estimation of (x) and (θ) that describe the position of the pelvis and torso in the seat, we use an estimation of (x_{slide}) and (Φ) to transfer data into the vehicle reference system. The information of the seat position is provided with the help of specific position sensors that are not described here. The measurement of the position of the head is not the purpose of BioVolume seat sensors. The measurement of PS , SH and the neck to torso angle (δ) are estimated with the help of statistical laws. We use anthropometrical descriptions for the estimation of PS and SH and we base the probability estimation of the neck inclination with the help of ergonomical description. With all these parameters, we can compute a density of probability of the distance (D) and consequently the probability of presence of the head and torso area in a specific zone.

For example, we can divide up the space of the occupant compartment in 2 zones. A first zone, close to the airbag, to regard as danger zone, and a second one, far from the airbag, in which the airbag can be deployed without risk (figure 8).

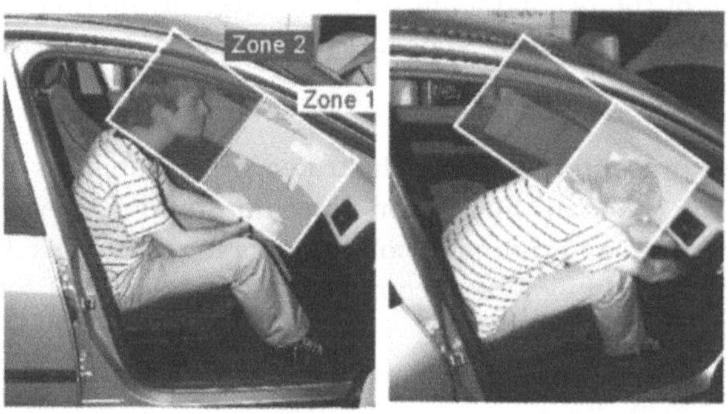

Fig. 8. Example of space division. Volunteer n°12 (178cm / 74Kg) in position corresponding to D=500mm on the left part of figure and D=100mm in the right figure. In this study, we used a 350mm distance from the instrument panel to define the limit between zone 1 and zone 2.

Two probability of presence are computed: p1 is the probability of presence of the occupant in the first zone and p2 is the probability corresponding to a presence in zone 2. An information of occupant classification is then established: zone 1 if $p1>p2$ or zone 2 if $p1<p2$. In the case of figure 8 left part, we have $p1=1\%$ and $p2=99\%$, in the case of figure 8 right, we obtain $p1=85\%$ and $p2=15\%$.

We conducted series of validation on 50 people. Table 1 gives characteristics of the data bank (morphological characteristic of humans and seat adjustment parameters).

	Min	Average	Max
Slide adjustment: x_g	27 mm	128 mm	240 mm
Backrest adjustment (Φ)	20°	27°	60°
Weight of the occupant	53 Kg	71 Kg	105 Kg
Size	151 cm	174 cm	190 cm

Tab. 1. Occupant and seat parameters.

For each person, three positions of seat adjustments were tested and for each adjustment, the occupant was placed in six different positions related to six different head-dashboard distances (100mm to 600mm). Figure 9 shows the results of the computation and displays the percentage of estimation in the proper zone.

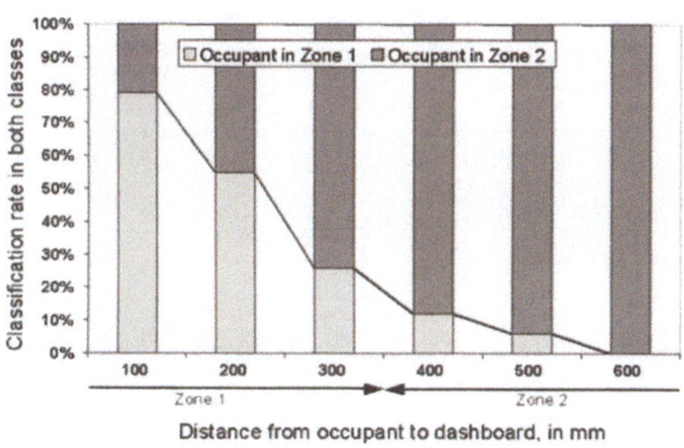

Fig. 9. Allocation of decision with 50 occupants. Vertical bars indicate the percentage of decisions according to two classes of positions: percentage of occupants detected in zone 1 at the bottom of the bar and percentage of occupants in zone 2 in the upper part. X axis indicates the physical distance between the occupant and the instrument panel. Occupants are in zone 1 for x lower than 350mm.

Results indicate that 100% of cases in far position were classified accordingly (50 occupants in normal postures with 3 different adjustments of seat). The distribution is about 80% of proper classification for occupant in the most inclined position. The fact that occupants seated far from the dashboard are classified in a proper way is due to the better accuracy of the technology for situations related to proximity between sensors and occupant.

5 Road Condition Tests

In order to gain insight into the evaluation of the capabilities of the system to perform occupant classification, series of tests with a vehicle embedded mock up were performed. Road test with 10 adult volunteers occupants placed in a normal comfortable position on the front outboard passenger seat were performed. The test consisted in the acquisition of raw signal as well as processed occupant morphology and position data in the moving vehicle, riding on specific road tracks (figure 10). The vehicle speed was set to 20km/h in order to reach important vertical motion of the vehicle due to road profiles.

Fig. 10. View of the tracks for road tests. One lap consisted in a 300m of resonance track (lower part of the figure) and a 300m of Belgian block part (top of the figure). The resonance track is a waved longitudinal profile characterised by an low amplitude with a low period and the paved road part has a profile producing important amplitude at a low period.

During the tests, the registrations of raw signal from each sensors were considered. The following figure displays the output for the 7 sensors that were integrated into the seat. Sensors 1 to 4 are located in the seat cushion and sensors 5 to 7 are integrated in the backrest. The test started with an empty seat configuration, all sensors output a zero value. At t = 10 seconds, the occupant sits in a normal reference position while the vehicle starts to ride on the tracks. The resonance track is crossed during the 75 to 125 seconds time interval and the belgian block paved part is crossed at time 160 to 200 seconds.

The display of the raw data in figure 11 shows a limited noise produced by the movements of the occupant induced by the severe profile of the road. The occupant in reference position is classified accordingly and the computed confidence index remains at an important value during all the test. The computation of the morphology of the occupant, which takes into account the instantaneous raw signal output and the class of position with the corresponding confidence index shows an important stability. The use of a filtering process does not improve the classification.

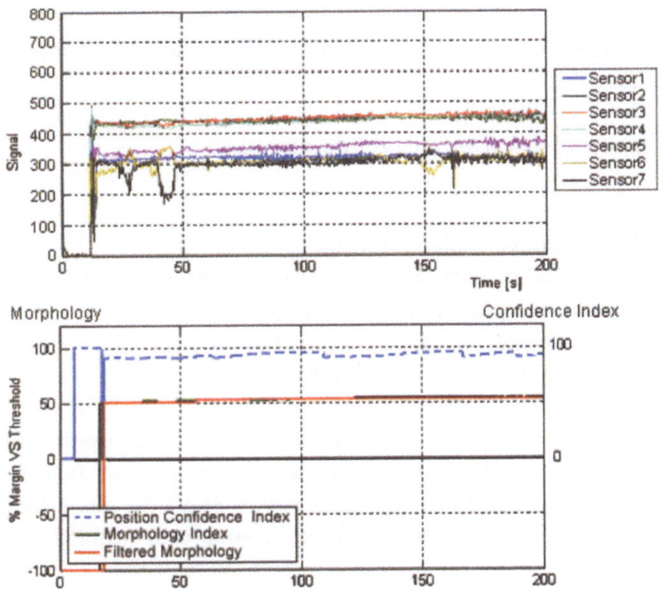

Fig. 11. Results of data acquisition and computation during a 200 second onboard riding test with a 62kg and 166cm occupant. The raw sensor output are displayed in the upper part of the graph ; the morphology and position processed data are visualised in the bottom part of the figure.

T = 0 to t = 10 seconds corresponds to an empty seat condition. Occupant in reference position from t = 10 to t = 200 seconds. The system outputs a reference position with an important confidence level index (85 to 98%). The occupant is detected in the Adult morphology class (The Morphology Index reaches values between 0 to 100% value).

This preliminary road test campaign showed relevancy of the use of a the BioVolume electric field technology to reach a stability output during motion of the vehicle.

6 System Integration

Faurecia, as a Tier-1 seat supplier, is expert in integrating systems into seats. Thus, integration was a major requirement from the very beginning of the development.

Many aspects have been taken into account:

- Comfort.
- Compatibility with seat design.
- Compatibility with other systems integrated into the seat.
- Compatibility with trimming processes.
- Suitability for JIT plant current operation.
- Durability in seat environment.

The integration of a BioVolume sensor mat can be compared with the integration of a heating mat. Both mats are located on the A-surface of the seat, between the foam pad and the seat cover. Faurecia has a long experience in heating mat integration, and this experience was utilized for the BioVolume integration.

6.1 Comfort

One of the goals was to have no impact on the seat comfort. For that reason, two main solutions were studied: printed polyester mats and wire. Antenna size and shape have been optimized to made possible the printed polyester mats solution.

A study have been run by Faurecia experts on the integration of different type of sensors, and the results of the blind test show (figure 12) the possibility to reach a good level of integration in term of comfort, comparable with an heating mat integration.

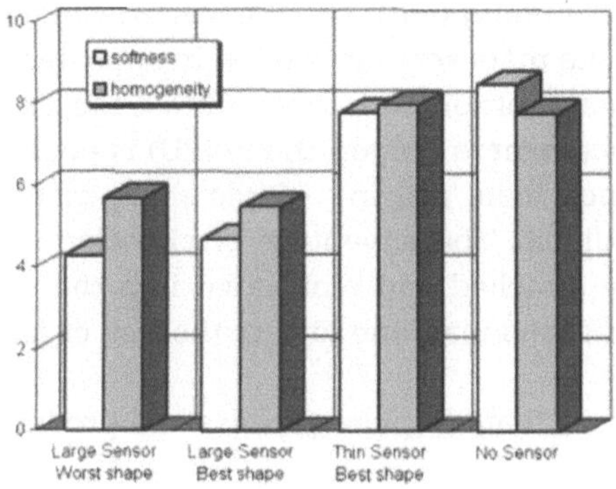

Fig. 12. Subjective blind test quotation of different sensors in different foam pads. A higher value corresponds to a better appreciation. Criteria are homogeneity and softness.

6.2 Compatibility with Seat Design

A BioVolume mat is constituted of several individual sensors that can be located in different region of the A-surface. Location of the sensors must be compliant with a performance topography, but can be tuned to match the seat style: in order to follow seat lines, sensors can be straight or curved, and the sensor length can be adapted. Sensors shall be placed on seat flat surfaces.

6.3 Compatibility with Other Systems Integrated into the Seat

There is no particular compatibility issue with the other systems. The only system that can be a concern is the heating mat. By driving the heating mat by the BioVolume system – and placing the BioVolume mat above the heating mat –, the BioVolume system is not affected during data acquisition.

A restriction has to be added: seat cover shall not be made of conductive fabric which is not a critical issue as those are not a common use in seating materials.

6.4 Compatibility with Trimming Processes

The goal was to be adaptable to various seat-trimming technologies offered by the seating market.

- For traditional trimming (with mechanical attachments between cover and foam pad), the mats are very similar in appearance to heating mats. Layers of soft materials on both sides protect the mat (figure 13).
- For foam in place, where polyurethane (PU) is poured inside the cover to obtain trimmed foam pad in a single shot), mats, tools and specific process are available. The advantage of choosing such process is that the mat is fully attached and embedded into the foam. This provides good protection of the mat, and lowers the risk of damage during trimming.
- For bonded covers, foam in place process and product can be adapted.

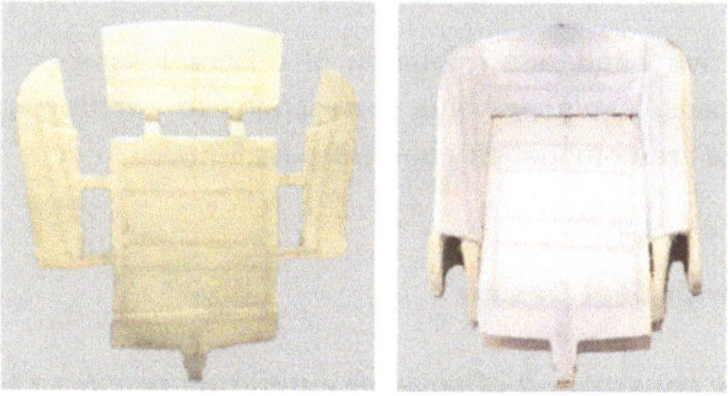

Fig. 13. Example of BioVolume sensor mat positioned on the cushion foam pad, prior to the assembly with the cover.

6.5 Suitability for JIT Plant Current Operation

The main advantage of BioVolume in JIT is that there is no need of calibration, because there is no mechanical measurement. The only need is to perform a unique and short functional End of Line test. This test will check each sensor, the complete system, and will be able in case of damage to locate the problem.

Otherwise, BioVolume mat handling is very similar to heating mat, except the safety class of the BioVolume system.

6.6 Durability

For both traditional and foam in place trimming processes, durability tests have been performed. Faurecia experience has shown that entry-exit and combination of entry-exit + driving conditions are the most critical tests. Printed polyester mats have proven their resistance to this type of test. The critical areas like along tie down or marking lines can be reinforced with no impact on comfort.

By a diagnostic made on each sensor, the system is able to warn in case of damage and ensures his function under failure mode.

7 Conclusion

The development of the BioVolume system has proven the possibility of this electric field sensing technology for occupant classification performances regarding morphology computation, position monitoring as well as seat integration.

One of the main advantages of this non-mechanical system is the simple distinction between children in child restraint system (CRS) and adults, due to the low signal provided by the system in case of CRS, whatever a potential seat belt pre-load on the child seat.

The use of sensors located in different areas provides information that clearly distinguishes a child from an adult in any position. In term of occupant classification, results show that the system is able to perform with the same technology the morphology and the position of the occupant.

The accuracy of the detection of occupants in far mode from the airbag reaches important values with BioVolume sensors located in the seat. Introduction of sensing systems designed to provide information for minimizing risk of occupants in out of position must be done with a lot of care and considerations. A suppression decision resulting from an erroneous classification will result in the loss of the protection benefit provided by the airbag system. This situation is to be considered with the same concern level than the primary goal which is the reduction of the risk of severe or fatal injuries to occupant closed to the airbag at time of deployment. The accuracy of detection in far zone was found interesting when BioVolume technology was evaluated.

In the industrialisation and integration point of view, BioVolume technology fits many constraints.

Some remaining items related to management of electric field technology difficulties are underway. These intrinsic troubles correspond to elements that can affect the surface, or distance of the object to detect, can modify the dielectric characteristic of element to detect or can add dielectric elements in the neighborhood of the probes. The separation of the distance and surface information in order to provide proper classification of occupants seated on different obstacles has already been performed.

The presence of water in the neighborhood of sensor produces increase of dielectric element that is read in a raw capacitance measurement. Enlargement of operating zones in high humidity conditions as well as separation of signal from liquid on seat with those of occupant are currently items that are under development.

The results presented in this article present BioVolume as an answer to FMVSS208 for both passenger and driver places. The choice of BioVolume by Faurecia as seat leader and safety expert and Hitachi as electronic leader is, in that philosophy, an insurance for the right technology with the right seat integration, for the best final user protection.

References

[1] Bevan M., "Introduction to occupant sensors." Proceedings SAE Seminars N° 99008. Apr 19, 2002. Troy, MI.

[2] Jones R.C., "Technologies for Static Airbag Suppression Systems." Paper number V11, Airbag 2002 6th International Symposium on Sophisticated Car Occupant Safety Systems. Karlsruhe, Germany, December 2-4, 2002.

[3] Federal Motor Vehicle Safety Standards; Occupant Crash Protection. Department of Transportation. National Highway Traffic Safety Administration. Federal Register / Vol. 65, No. 93, May 12, 2000.

[4] Federal Motor Vehicle Safety Standards; Occupant Crash Protection. Department of Transportation. National Highway Traffic Safety Administration. Federal Register / Vol. 66, No. 243, December 18, 2001.

[5] "Air Bag Fatality Summary Report & Tables". Special Crash Investigation Reports, NHTSA web site. www.nhtsa.dot.gov.

[6] Lucas J., Batis C., Holé S., Ditchi T., Launay C., Da Silva J. "Distance corrected capacitive sensors for advanced air bag applications.", paper number 2003-01-0508, SAE 2003 World congress.

Marc Pajon, Laurent Chabert, Hervé Dirand, Laurent Jaillet

Faurecia

Nanterre

France

mpajon@brieres.faurecia.com

Capacitive Angular-Position Sensor

V. Ferrari, A. Ghisla, D. Marioli, A. Taroni, Università degli Studi di Brescia
P. Fasolato, F. Maggi, S.Pelloso, V. Rossetti, Finmek S.p.A.

Abstract

This paper describes the design, realization and experimental characterization of a capacitive angular position sensor. The sensor is composed of two parts, namely the sensing element and the dedicated signal conditioning electronic circuit. The sensor is based on a low-cost Printed Circuit Board technology for the electrodes, and the use of a thick conductive rotor that is electrically floating and does not require any electrical contact to the electronic circuit. The electronic circuit is based on a relaxation oscillator and the output is a DC voltage proportional to the angular position of the rotor. The sensor and the electronic circuit are electrically equivalent to a voltage-excited resistive potentiometer. The fabricated device provides redundancy by including two identical and independent sensors and electronic circuits in the same housing that work as separate measuring sections using the same rotor. Measurement results on the prototype over a measurement range of $\pm 11°$ gave a best-fit line non-linearity within $\pm 1\%$ of voltage supply.

1 Introduction

Angular position sensors find many applications in the automotive field, such as in the measurement of pedal position (drive by wire), throttle valve, mirror adjustment, seat adjustment, light adjustment to name a few. Harsh environmental conditions, including wide temperature range, vibrations, moisture and dust, combined with requirements of good measurement performances, high reliability and low cost make the design of angular position sensors a particularly demanding task.

Capacitive sensors are in general attractive for their low power consumption, simple geometry and easy construction. Moreover, they can be arranged in contactless structures that avoid friction between electrically active parts and therefore improve the long-term reliability of the sensors.

On the other hand, capacitive transducers tend to be critical devices because of their low-level signals and the necessity of adequate electronic circuits, sensitivity to humidity and dust, and possible susceptibility to electromagnetic interference (EMI).

So far, these critical aspects have limited the use of capacitive sensors in the automotive field. Examples of angular-position sensors and rotary encoders which use the capacitive principle are reported in the literature [1-7], with some of them explicitly devoted to automotive applications. Some designs use open electrode configurations [1, 2] with possible lack of sufficient shielding to EMI, some others tend to be rather complicated [3, 5]. In most cases, except for instance in [6] and [7], the rotating element requires an unpractical and failure-prone electric contact to the electronic circuit.

This work proposes a solution that aims at maximizing the performance/cost ratio by means of thorough study of the application and a co-design of the sensing element together with the dedicated conditioning electronic circuit. The application of the sensor regards the measurement of angular position of a pedal shaft (for example an accelerator). The device is intended to be electrically and mechanically compatible with a standard resistive potentiometric sensor, while improving reliability.

2 Design and Realization of the Prototype

The capacitive sensor is composed of two parts: the capacitive sensing structure and the readout electronic circuit. They will be described in the following.

2.1 The Sensing Structure

Figure 1 shows the layout of the capacitive sensing structure. The capacitive structure is composed of three main parts: the two electrode planes (stator and counter-stator) realized on two Printed Circuit Boards (PCB) which are mounted face to face in a parallel plate configuration, and the conductive rotor which is placed in between and is electrically floating without any contact required to the electronic circuit.

The central part of the rotor is connected to a perpendicular shaft that transfers the angular displacement to the rotor with respect to the fixed stator and counter-stator planes.

The sensing structure provides redundancy by including two identical and completely separate sensing elements, namely the main element (M) and the redundant element (R). Each element is connected to its dedicated electronic circuit, thereby forming two separate and independent measuring sections within the same device.

Each of the M and R sensing elements is formed by a number of segmented electrodes on the stator and the counter-stator. In particular, for each element there are four interleaved electrode sectors on the stator (Ma, Mb, Mc, Md, and Ra, Rb, Rc, Rd), and two interleaved electrode sectors respectively opposed on the counter-stator (Me, Mf, and Re, Rf). Selected electrode sectors within both sensing elements are shorted together on the appropriate PCBs, namely Ma and Mc, Mb and Md, Me and Mf for the M section, and Ra and Rc, Rb and Rd, Re and Rf for the R section.

The floating rotor is suitably thick and has four conductive blade sectors, two for the M and two for the R sections respectively.

In this way, each of the sensing sections M and R is actually composed by two capacitances C_1 and C_2 that have a common contact on the counter-stator and two separate contacts on the stator. The capacitances C_1 and C_2 are sensitive to the angular position of the rotor by varying in a push-pull fashion, i.e. when C_1 increases, C_2 decreases by the same amount.

The interleaved configuration of the electrode sectors provide a spatial averaging effect which compensates for possible mounting misalignment and tilt of the rotor. To reduce electric field bending, each electrode sector on the stator is surrounded by guard strips which are all connected together at the G contact. The structure is enclosed in the shield formed by the metallized outside layers of the PCBs connected to ground to reduce the EMI susceptibility.

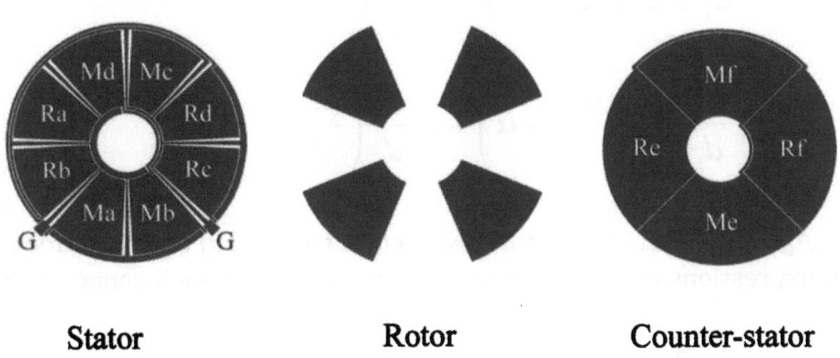

Fig. 1. Layout of the sensor structure including the main (M) and redundant (R) sections.

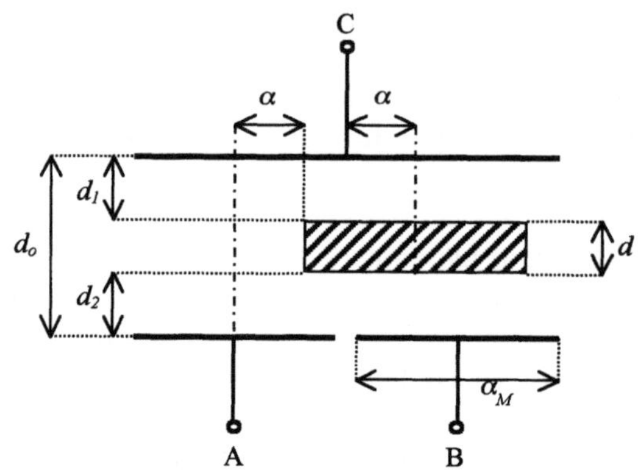

Fig. 2. Side view of the sensing element.

Given that the M and R sensing elements are exactly equal, the description of operation can be made with reference to one of them, say the M element.

Figure 2 shows the side view of a simplified equivalent scheme of the M sensing element, where all the electrode sectors have been represented by their combinations that globally result in three electrodes. The electrode A on the stator (Ma-Mc), together with the common electrode C (Me-Mf) on the counter-stator form the capacitance C_1, the electrode B (Mb-Md) and common electrode C form the capacitance C_2. It can be observed that the thickness of the rotor d is a significant fraction of the spacing between stator and counter-stator do.

Both the capacitance C_1 and C2 depend on the angular position a of the rotor, where $|\alpha| < \alpha_M/2$:

$$C_1 = \varepsilon_o \frac{A_o}{d_1 + d_2}\left(\frac{\alpha_M}{2} - \alpha\right) + \varepsilon_o \frac{A_o}{d_o}\left(\frac{\alpha_M}{2} + \alpha\right) \quad (1)$$

$$C_2 = \varepsilon_o \frac{A_o}{d_1 + d_2}\left(\frac{\alpha_M}{2} + \alpha\right) + \varepsilon_o \frac{A_o}{d_o}\left(\frac{\alpha_M}{2} - \alpha\right) \quad (2)$$

where A_0 is the total area of the two sectors composing each electrode A and B. The expressions of the minimum and maximum for each capacitance are:

$$C_{MIN} = \varepsilon_o \frac{A_o}{d_o}\alpha_M \quad (3)$$

$$C_{MAX} = \varepsilon_o \frac{A_o}{d_1 + d_2} \alpha_M \tag{4}$$

When the rotor is in the middle position between the electrode A and B, i.e. $\alpha = 0$, the capacitances are equal and their expression is the following:

$$C_1\big|_{\alpha=0} = C_2\big|_{\alpha=0} = C_o = \varepsilon_o \frac{A_o}{d_1 + d_2} \frac{\alpha_M}{2} + \varepsilon_o \frac{A_o}{d_o} \frac{\alpha_M}{2} \tag{5}$$

From (1) and (2), the sensitivity of each capacitance with respect to the angular position of the rotor results:

$$\left|\frac{dC}{d\alpha}\right| = \varepsilon_o A_o \left(\frac{1}{d_1 + d_2} - \frac{1}{d_o} \right) \tag{6}$$

To increase the sensitivity, the spacings d_1 and d_2 should be made as small as mechanical tolerances allow, and do should be made large up to a tradeoff value between the increase of sensitivity and the decrease of the minimum value of the capacitances (3), that should be greater than the parasitic capacitances.

By arranging the previous equations, the relation between the capacitance values and the angular position can be rewritten as:

$$C_1 = C_o - \frac{C_{MAX} - C_{MIN}}{\alpha_M} \alpha = C_o (1 - kx) \tag{7}$$

$$C_2 = C_o + \frac{C_{MAX} - C_{MIN}}{\alpha_M} \alpha = C_o (1 + kx) \tag{8}$$

where:

$$k = \frac{C_{MAX} - C_{MIN}}{C_{MAX} + C_{MIN}} = \frac{d_o - (d_1 + d_2)}{d_o + d_1 + d_2} = \frac{1}{\frac{2d_o}{d} - 1} \tag{9}$$

and:

$$x = \frac{2\alpha}{\alpha_M} \tag{10}$$

The value of capacitance varies linearly with the normalized angular displacement x. In our case, given the dimension of the sectors, the angular measurement range is about ±20°.

The factor k is a numerical constant only dependent on a ratio of dimensions and, as such, it is mostly unaffected by influence factors. In addition, being the rotor a conductor, the sensitivity neither depends nor relies on the dielectric permittivity of any material, except air, which fact provides the sensor with an intrinsic stability and robustness against drift.

The vertical position of the rotor within the gap between the stator and counter-stator has no influence on k, hence mounting tolerances are not an issue.

Humidity also is expected to have minimal effect, because a thin moisture layer over a conductive surface does not change its electric properties, as opposed to what happens for a dielectric surface.

By examining equations (7) to (10), it can be observed that the angle information can be conveniently and accurately obtained by measuring the following dimensionless quantity:

$$\frac{C_1 - C_2}{C_1 + C_2} = kx \tag{11}$$

2.2 Signal Conditioning Electronic Circuit

The small values of capacitance involved and the requirements for suitable sensitivity and reduced circuit complexity make the design of the readout electronics a very important step in the design of the sensor.

A dedicated circuit has been designed and assembled in surface mounting technology (SMT) on the same PCB of the stator close to the sensing element outputs to minimize the parasitic capacitances. Two identical and separate circuits have been included for the M and R sections, respectively.

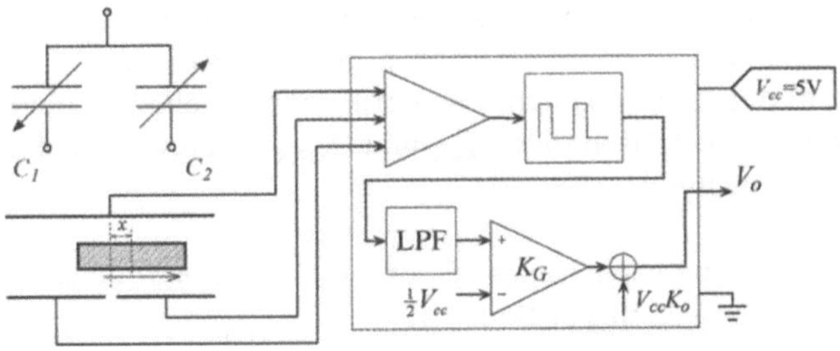

Fig. 3. Block diagram of one section of the sensor and the conditioning circuit.

Figure 3 shows the block diagram of one section of the device. The circuit allows to measure the ratiometric quantity $(C_1\text{-}C_2)/(C_1+C_2)$ introduced in (11) and converts it into a DC voltage V_0. The core of the circuit is a capacitive-controlled relaxation oscillator, where the high time fraction of the period of a rectangular wave is proportional to the capacitance C_1 and the low time fraction is proportional to the capacitance C_2. Therefore, the duty cycle of the oscillator output varies around the value of 0.5 proportionally to $(C_1\text{-}C_2)/(C_1+C_2)$. Then the rectangular wave is filtered by a low-pass filter and the output is amplified and level shifted. As a consequence, the unipolar DC output voltage Vo is given by:

$$V_o = V_{cc}\left(\frac{1}{2} + K_G\frac{C_1 - C_2}{C_1 + C_2} + K_O\right) \tag{12}$$

where Ko and KG are suitable calibration constants that can be adjusted for the desired output characteristic. It can be observed in (12) that V_0 is proportional to V_{cc}, therefore the output varies ratiometrically with respect to the supply voltage, as desired to emulate a resistive potentiometer.

3 Experimental Results

3.1 Preliminary Measurements

A capacitive structure like that described in figure 1 has been built and the capacitance variations versus the angle have been tested. The distance do between the stator and counter-stator plates has been made equal to 2.5mm and the thickness of the rotor d equal to 1.6mm, therefore the spacings d_1 and d_2 have resulted of 0.45mm.

Such dimensions, with an electrode external diameter of 37.6mm, were found to be an adequate tradeoff between a good sensitivity and a value of C_{MIN} greater than parasitic capacitances, moreover they allow to have small electric field bending. The capacitance C_{MIN} and C_{MAX} amounted to about 0.76pF and 2.11pF respectively.

The variation of the capacitances C_1 and C_2 have been measured subsequently by means of the circuit of figure 4.

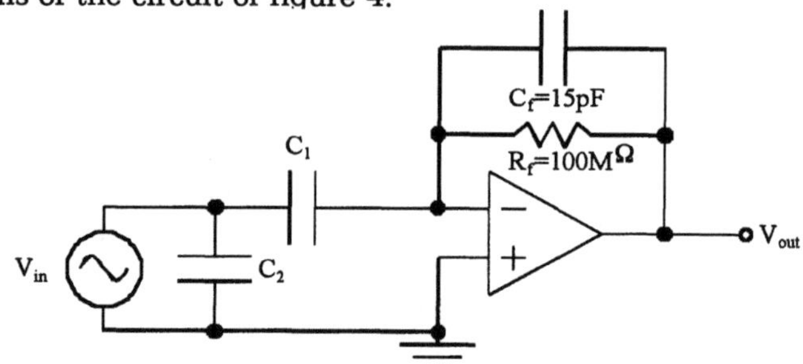

Fig. 4. Circuit for the measurement of C_1/C_f ratio.

For the situation in the figure, given a fixed input amplitude, at frequencies greater than $1/(2\pi R_f C_f)$ the output amplitude only depends on the ratio C_1/C_f in the following way:

$$V_{out} = -\frac{C_1}{C_f} V_{in} \tag{13}$$

Therefore, the behavior of C_1 versus the rotor angle could be derived. The capacitance C_2 and the parasitic elements have no effect, and because all of the stator electrodes are tied to ground or to virtual ground, there is a guard effect provided by the sectors of C_2 to those of C_1.

The same procedure has been done for C_2, by swapping the connections of the capacitances to the circuit accordingly.

The obtained results over a range of $\pm 20°$ are shown in figure 5. As expected, the capacitance variations are opposite in slope and linear in an interval around the middle position, with some deviations towards the end points of the range.

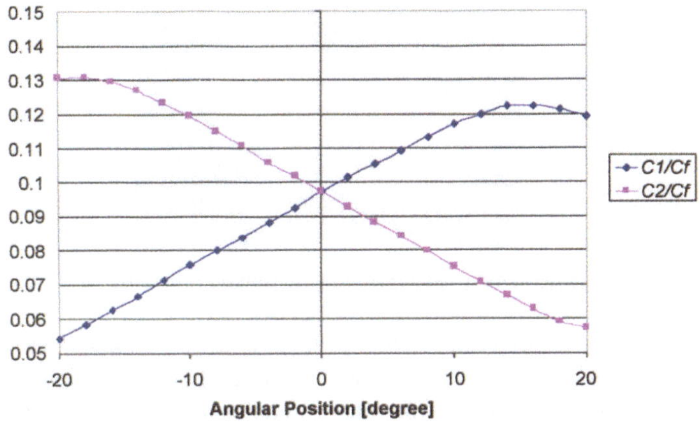

Fig. 5. Measured capacitance variations versus the angular position.

3.2 Measurement Results on the Complete Sensor

The capacitive angular-position sensor and the electronic conditioning circuit described in the previous section have been built and characterized. figure 6 shows the components of the device.

It can be observed that the electronic circuit has been assembled on the stator PCB with SMT components to reduce the size and improve the symmetry of the circuit with respect to the output of the sensing element. The rotor blades are made of brass, while the shaft is plastic. The complete device has been assembled in a molded plastic housing, shown in figure 7, mechanically compatible to a previous resistive potentiometric sensor, and mounted on a rigid fixture.

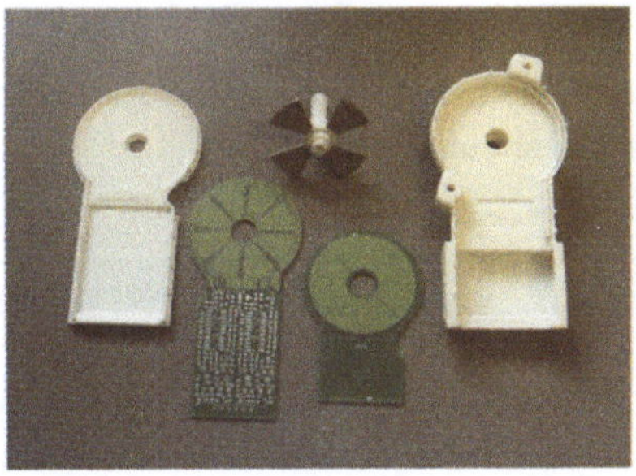

Fig. 6. Components of the sensor.

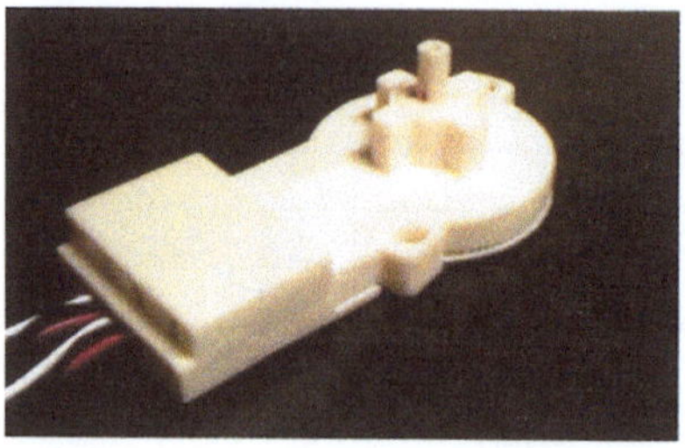

Fig. 7. Complete sensor assembled in its plastic housing.

The rotor shaft has been connected to a goniometer to allow for the testing and calibration of the sensor. For each section, the output voltage V_0 of the respective conditioning circuit has been measured versus the rotor angle over the range $\pm 11°$. Figure 8 shows the measured output and best-fit nonlinearity for the M section of the device. Similar results have been obtained for the R section. It can be observed a linear behavior as expected, with a nonlinearity error within $\pm 1\%$ of the voltage supply.

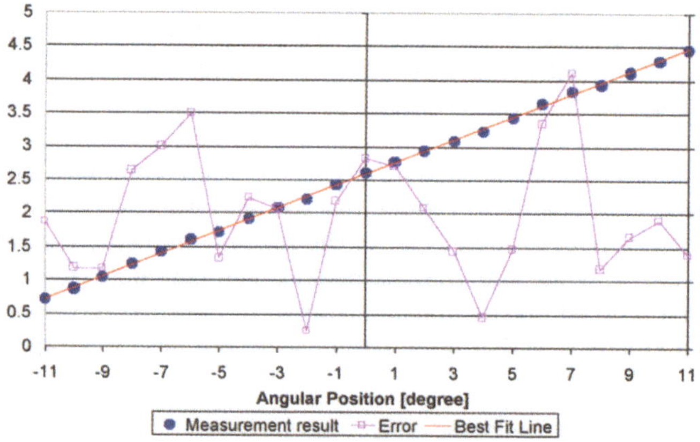

Fig. 8. Output voltage of one section and nonlinearity with respect to the best fit line versus angular position.

Parameter	Request value	Obtained value
Power Supply	5 V	5 V
Electrical consumption per section	< 10 mA	8 mA
Measurement Range	±11°	±11°
Nonlinearity Error	< ± 3%	< ± 1%

Tab. 1. Required specifications and obtained results for the sensor.

Table 1 reports a comparison of required specifications and obtained results for the sensor.

3.3 Influencing Factors

The output of the device is based on a ratiometric measurement principle, as shown in equations (9)-(12). Therefore, to first approximation, temperature variations should reflect to the output in proportion to the thickness ratio do/d, and, as such, they can be made virtually negligible by a suitable choice of materials.

Experimental tests on the complete sensor have shown a significant thermal drift which, however, was subsequently found to be due to the electronics. The preliminary results of a second version of the circuit, currently under test, have shown a substantial improvement, confirming the intrinsic advantages of the sensing principle. The effect of the environmental humidity is also currently being investigated.

Preliminary EMI tests of the sensor, carried out at about 20V/m at 460MHz, have shown an unacceptable degree of susceptibility for the present version, but a substantial immunity for the second version of the electronics.

4 Conclusions

The paper has reported on the design, realization and characterization of a capacitive angular-position sensor. The contactless principle and the redundant structure improve the reliability of the sensor. The use of a thick conductive rotor electrically floating without any contact to the circuit improves the performance and stability of the sensor while reducing its mechanical complexity [8].

The sensor and of the electronic readout circuit have been co-designed to improve the performance/cost ratio. Measurement results on the first prototype have shown a nonlinearity within $\pm 1\%$ of the supply voltage over a range of $\pm 11°$ for both the main and redundant sections. Tests at different temperatures have shown a significant thermal drift that was ascribed to the electronic circuit.

Further work is currently under way on a second version of the electronic circuit in order to reduce the sensitivity to temperature and improve the EMI immunity. The preliminary experimental results are extremely encouraging regarding both the above aspects.

References

[1] A.H.Falkner, "The Use of Capacitance in the Measurement of Angular and Linear Displacement", IEEE Trans. Instrum. Meas., vol. 43, no.6, pp 939-942, 1994.

[2] S.P. Cermak, G.Brasseur, P.L. Fulmek, "A Planar Capacitive Sensor for Angular Measurement", Proc. of IMTC 2001, Budapest, May 21-23, 2001, pp. 1396-1396.

[3] G.Brasseur, "A Capacitive 4 Turn Angular Position Sensor", IEEE Trans. Instrum. Meas., vol. 47, no.1, pp. 275-279, 1998.

[4] G.W. de Jong, G.C.M. Meijer, K. van der Lingen, J.W. Spronck, A.M.M. Aalsma, D.A.J.M. Bertels, "A Smart Capacitive Absolute Angular –Position Sensor", Sensors and Actuators A, 41-42, pp. 212-216, 1994.

[5] X. Li, G.C.M. Meijer, G.W. de Jong, J.W. Spronck, "An Accurate Low-Cost Capacitive Absolute Angular-Position Sensor with a Full-Circle Range", IEEE Trans. Instrum. Meas., vol. 45, no.2, pp.516-520, 1996.

[6] G.Brasseur, "Virtual Rotor Grounding of Capacitive Angular Position Sensors", IEEE Trans. Instrum. Meas., vol. 49, no.5, pp 1108-1111, 2000.

[7] M.Gasulla, X. Li, G.C.M. Meijer, L. van der Ham, J.W.Spronck, "A Contactless Capacitive Angular-Position Sensor", Proc. of IEEE Sensors 2002, Orlando, FL, USA, June 11-14, 2002, pp.880-884.

[8] "Trasduttore capacitivo per rilevare la posizione angolare di un organo rotante", Italian patent application UD2002A000144, Assignee: Finmek S.p.A., Via Lisbona 28, 35127 Padova, Italy.

V. Ferrari, A. Ghisla, D. Marioli, A. Taroni
Dip. Elettronica per l'Automazione
Università degli Studi di Brescia
Via Branze 38
25123 Brescia
Italy
vittorio.ferrari@unibs.it

P. Fasolato, F. Maggi, S.Pelloso, V. Rossetti
Finmek S.p.A.
Via Lisbona 28
35127 Padova
Italy

Keywords: angular-position sensors, capacitive sensors

Functions and Applications

Powertrain

Multi-Function Microsensor for Oil Condition Monitoring Systems

D. Wüllner, H. Müller, O. Lüdtke, HELLA KG Hueck & Co
H. Dobrinski, T. Eggers, Campus Micro Technologies GmbH

Abstract

Today's passenger cars provide a very long service interval. Because of this, the monitoring of the oil condition is becoming more important. It is necessary to monitor the oil condition to prevent engine failures as a result of extreme driving conditions, such as when one makes many short trips over a long period. This task can only be reliably realized by means of a sensor located in the difficult environment of the engine oil. As a significant supplier of oil level sensors, Hella developed, in cooperation with its partners, a low cost microsensor (Fig.1b). Based on simulation results, a sensor design was developed and later fabricated by employing standard CMOS-compatible materials and processes. The microsensor allows for the measurement of various electrical and physical parameters. This sensor can provide a significant contribution to the reduction of the running costs of the car. It can also protect the environment through extended oil change intervals by considering driving behaviour. Additionally, damage caused by oil deterioration in sophisticated engines can be prevented [1].

1 Significance of the Paper

The microsystem technology allows for the expansion of oil sensor functionality to provide oil condition monitoring. The focus has been the integration of the packaged sensor into a highly reliable oil level sensor as a stand-alone module with consideration for assembly concerns. A successfully integrated multifunction microsensor in an existing oil level sensor is a revolution in the field of oil applications (Fig.1a). Compared to existing or conventional oil condition sensor principles [2], this microsensor demonstrates the capability for working with multiple measurement principles such as capacitive read-out or impedance spectroscopy. This allows the system to measure the dielectric constant and the viscosity with the same microsensor.

A very important factor has been the successful development and integration of the microsystem in the existing automotive sensor design.

Fig. 1. Fully functional prototype of the oil condition, level and temperature sensor (a); Photo of the microsensor in an open-cavity SOIC-8 package in comparison to a one euro coin (b).

2 Sensor Design, Modeling & Fabrication

2.1 Sensor Design

The capacitive microsensor used for the measurement of the dielectric constant er is composed of a combination of CMOS-compatible materials (see Fig. 2): A thin metallic interdigitated electrode structure (IDS) fabricated on a dielectric substrate is coated by an ultra thin dielectric layer for resistance to the environment and for electrical insulation.

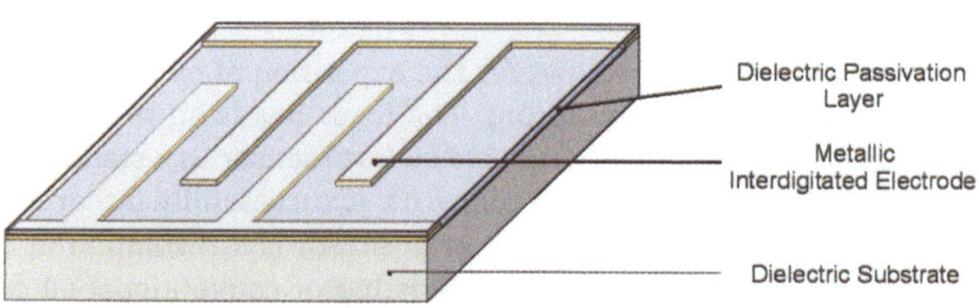

Fig. 2. Detailed view of the microsensor.

2.2 Modeling Aspects

For evaluation of the influences of the geometrical sensor parameters numerical simulations based on a 2-dimensional model of a unity cell of the IDS were done (see Fig. 3a). Thereby, the optimization was focused on the sensitivity and the sensor capacitance without oil loading, herein after referred to as "offset capacitance". The formulation of the optimal sensor design was distinguished by the following primary aspects:

- Maximization of sensitivity or relative capacitive swing due to changes in oil permittivity, resp.
- Matching of sensor capacitance and working range of signal pre-processing circuit.
- Adaptation of sensor total geometry to a surface mountable package with the JEDEC SOIC-8 outline.

The sensitivity is the most important parameter and was optimized through the simulation of the penetration depth of the electrical field into oil over a broad range of design variations and material combinations (see also figure 3b). In consideration of listed requirements the following sensor geometry was finally chosen:

- Electrode width: 10µm
- Electrode spacing: µm
- Overall dimensions of IDS: 2.1mm · 15mm

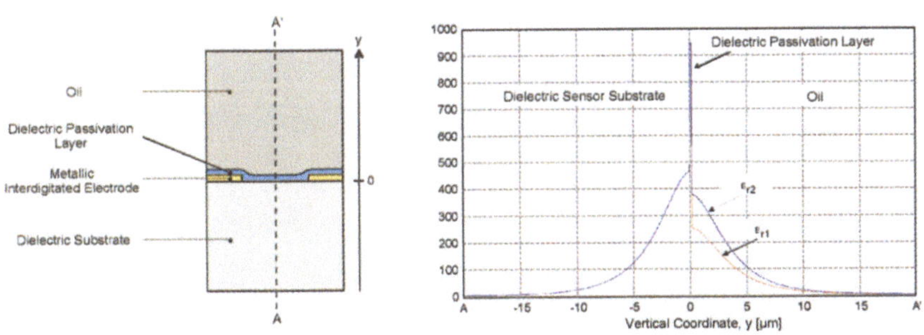

Fig. 3. Simulated changes in density of electrical field energy along path A-A′ due to different dielectric constants of oils (applied dielectric constants in (b): ε_r(substrate) = 3.75; ε_r(passivation layer) = 7.5; ε_r(oil 1) = 2.0; ε_r(oil 2) = 3.0).

2.3 Fabrication

The interdigitated electrodes were fabricated by rf sputter deposition of a tungsten titanium alloy (90% W, 10% Ti) on to amorphous quartz glass. After photolithography the finger structures were patterned by a special reactive ion etching process using a CF_4/O_2 gas mixture and coated by a thin silicon nitride film (SiN) in a low pressure chemical vapour deposition process (LPCVD). Afterwards for bondpad creation the SiN layer was locally opened by dry etching using a photoresist mask. Next aluminium was deposited and structured by wet etching.

After electrical tests at the wafer level the individual microchip was mounted onto a standard lead-frame with a JEDEC SOIC-8 outline. It was then encapsulated by a special plastic, open-cavity moulding technique, patented by Eurasem B.V. [3], whereas the sensitive electrode structure was protected against particle contamination during moulding. This package concept ensures a high resistance to the environment and is suitable for automotive temperature ranges and durability.

In the context of process total tolerance analysis, the technological variations of all single process steps were investigated regarding their relevance and weighting for sensitivity deviation and offset capacitance (see Fig. 4a). (For further consideration, the sensitivity was defined by the relative capacitive swing from ε_r = 2.0 to 2.1.) Due to selective optimization and observations of individual fabrication steps this work was designed to realize a calibration free, highly accurate structure.

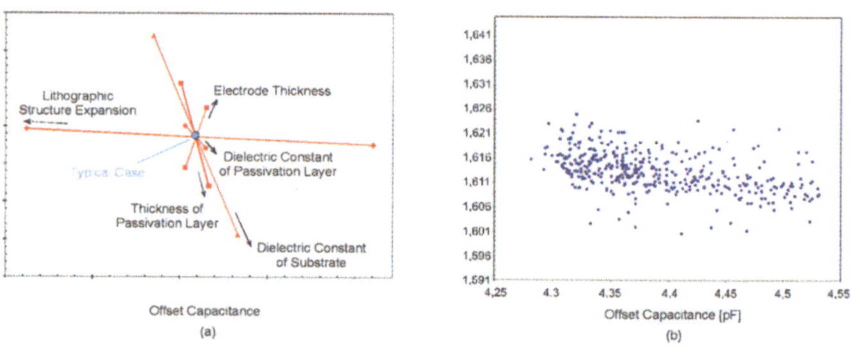

Fig. 4. Influence and significance of process tolerances on offset capacitance and sensitivity; (a) Simulated total tolerance analysis; (b) Measured offset capacitances and sensitivities.

As shown in Fig. 4a the lithographic structure expansion represents the most significant influence regarding deviation of the offset capacitance. By changing the ratio between width and spacing of the interdigitated electrodes the highest capacitive variation is induced while the sensitivity is still barely affected. In reference to this, a special attention is turned to the permittivity of the quartz substrate. This parameter must be monitored by either the material supplier or by in-coming inspection controls at the chip manufacturer's site using well-known permittivity characterization methods, e.g. by the so-called two fluid method [4].

After measurement of the offset capacitance of air (ε_r = 1.00054, 20°C, [3]) each single sensor element is loaded with a small amount of cyclohexan (p.a.) for sensitivity testing. This volatile unpolar fluid was chosen concerning its permittivity of ε_r =ε to oil in the range of ε_r = 2...6. As depicted in figure 3 the measured deviation of sensitivity and offset capacitance fits well to the simulation results.

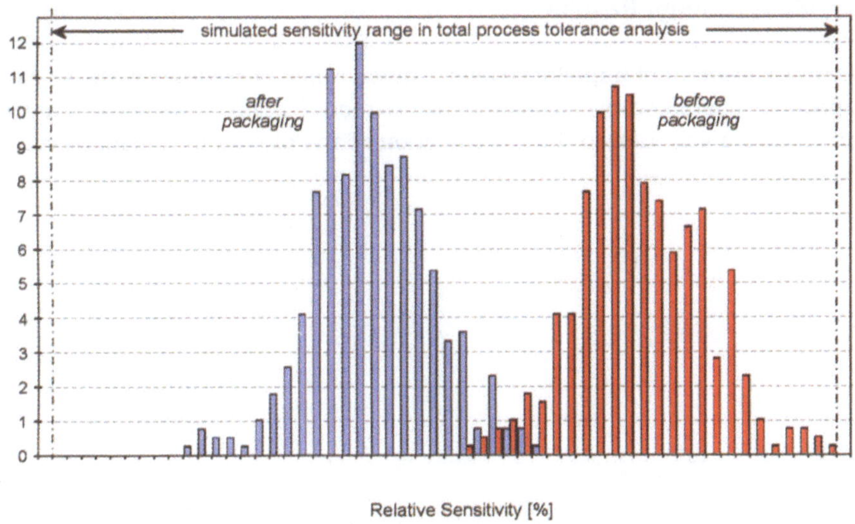

Fig. 5. Bar charts of relative sensitivity in comparison to simulated range of total process tolerance analysis.

The final packaging and moulding procedure did not lead to a measurable increase in deviation of sensitivity (see figure 4) because the covering of the sensitive area by mould compound particles was smaller then 1%. While before packaging an overall sensitivity variation of 0.77% – based on average – was detected, this value was 0.76 % after packaging. The small decrease of mean sensitivity from 1.63% to 1.61% , shown in Fig. 5, is directly related to parasitics caused by the capacitance of the interconnections inside the package.

A reduction in the deviation of offset capacitance could be achieved by further process optimization at the manufacturing site or by transition to integrated solutions at a system level, e.g. by a hybrid integration of sensor chip and signal processing circuit in a single multi-chip module (MCM).

3 Measurement Results

The measurement of impedance over a given range of frequency, the so called impedance spectroscopy, allows for the characterization of dielectric materials and creates a variation to the classification of the dipoles in this material.

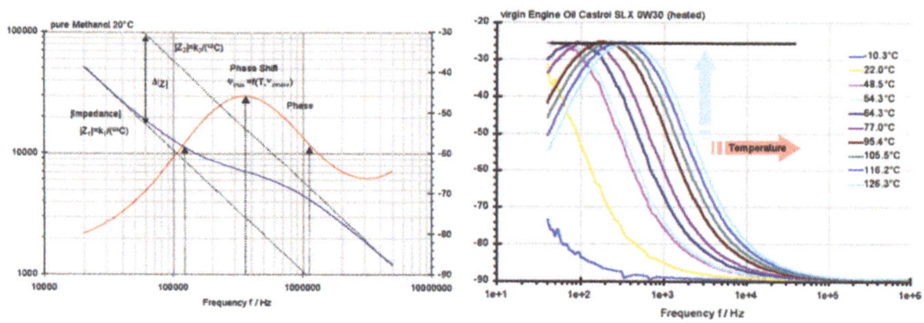

Fig. 6. Impedance spectra of methanol (a, left); Temperature dependency of phase shift (b, right).

Figure 6a shows the absolute value of impedance |Z| and the phase measured with methanol (20°C). The phase curve shows a maximum at a certain frequency f_{max}. For liquids of higher viscosities, e.g. engine oil we found the phase maximum at lower frequencies. Furthermore, the frequency at phase maximum depends on the temperature which can be seen in figure 6b. Apart from the temperature dependency figure 7a shows the dependency of the phase on the petrol dilution in oil. By increasing the content of petrol the phase decreases to lower values at a certain frequency. However, it has a beneficial effect to determine the frequency at a certain phase ($\varphi = 60°$) with the secondary condition $|Z| = \min$. The previous results lead to the assumption that the resonant frequency depends on the viscosity. By determining the frequency at a phase of $\varphi = 60°$ we found a dependency on the content of petrol which is shown in figure 7b. When comparing this data with the viscosity values, a connection between the viscosity and the frequency becomes very obvious in figure 7b. As it becomes apparent, among other things also from [6], the impedance spectroscopy is a suitable method, in order to determine the change of oil viscosity. The frequency at maximum phase can be interpreted as the resonant oscillation of the existing dipole collective in the dielectric material.

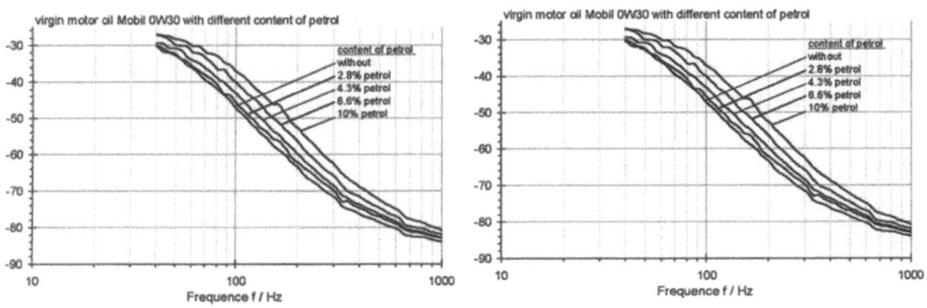

Fig. 7. Phase shift dependency from petrol dilution (a, left); Kinetic viscosity vs. content of petrol and frequency at $\varphi = 60°$ vs. petrol dilution (b, right).

4 Conclusions and outlook

A highly reliable microsensor has been developed. The packaging concept allows easy sensor integration at the board level as an SMT component. This microsystem will be a stand-alone working oil condition sensor providing permittivity and viscosity measurements and is capable of being integrated into various designs of automotive oil sensors. The main objective of this work is the successful integration of an oil condition monitoring system into an existing oil level sensor (figure 1a). The result is an upgraded automotive sensor without the requirement for a new mechanical interface to the oil pan.

References

[1] W. Warnecke et al., Belastungsgerechte Ölwartung mit ASSYST, MTZ Motortechnische Zeitschrift 59 (1998) 7/8, pp. 414-421

[2] B. Jacoby et al., A Multifunctional Oil Condition Sensor, AMAA 2001 Yearbook, pp. 157ff

[3] I. Van Dommelen, Plastic Packaging for Various Sensor Applications in the Automotive Industry, AMAA 2002 Yearbook, pp. 289-296.

[4] The Institute for Interconnecting and Packaging Electronic Circuits, Permittivity (Dielectric Constant) and Loss Tangent(Dissipation Factor) of Materials (Two Fluid Cell Method), IPC-TM-650 Test Methods Manual, 1987.

[5] D. R. Lide (Ed.), CRC Handbook of Chemistry and Physics, 81rd Edition, 2000-2001.

[6] Hellwig, Guenther; Normann, Norbert; UhL, Guenther "A sensor on dielectric basis for the on-line characterization of engine oils (Alkalinity, Viscosity)"

D. Wüllner, H. Müller, O. Lüdtke
Hella KG Hueck & Co
Beckumer Str. 130
59552 Lippstadt
Germany
dirk.wuellner@hella.com
info@campus-micro-technologies.de

H. Dobrinski, T. Eggers
Campus Micro Technologies GmbH
Universitätsallee 27-29
28359 Bremen
Germany

Keywords: automotive MEMS, microsystem technology, oil condition, oil level sensor, impedance spectroscopy

Determination of Fuel Quantities with a Micromachined Injection Rate Sensor

U. Schmid, EADS Deutschland GmbH

Abstract

To detect the injected fuel quantities "on board" of passenger cars at each cylinder, a novel micromachined mass flow sensor on an LTCC (Low Temperature Cofired Ceramic)-substrate is presented for high-pressure automotive applications. With the thermal injection rate sensor, fuel quantity measurements are performed at fuel pressures of 25MPa to 135MPa, whereas the drive pulses vary between 0.3ms and 1.5ms. Important injection parameters, as the injection begin/end and the injection rate, are discussed as a function of fuel pressure and drive pulse width. The sensor characteristics are compared to signals, gained from an injection amount indicator (EMI), which is part of the high-pressure hydraulic test bench. For the present application long-term measurements up to 380h of bare and a-SiC:H passivated thin film micro-heaters in oil atmosphere at atmospheric pressure yield the need and the benefit of a proper coating technology for the hot film anemometers.

1 Introduction

An increasing awareness of the public for environmental problems as well as strict governmental regulations on the exhaust emission for internal combustion engines force the automotive industry to further optimize the direct injection system of diesel engines [1, 2]. An "on board" mass flow sensor, to determine the injection rate as well as injection begin and end from cycle to cycle, could contribute to a smooth combustion process with lower NO_x and CO_y emission values due a closed-loop control of the valve especially in the pre- and post injection range. Therefore, a new type of micromachined mass flow sensor, using the thermo-resistive measurement principle, is developed and integrated into injection nozzles of a Common-Rail (CR) injection system. In a previous study, molybdenum (Mo) thin film sensors demonstrated their performance at a high-pressure hydraulic test bench up to injection pressures of 60MPa at different drive pulses widths t_0 for the magnetic-driven valve [3]. But despite good signal-to-noises ratios even down to t_0=0.2ms, further improvements had to be done to exploit the full potential of these novel hot film anemometers. The Mo heaters (d_{Mo}~100nm), which exhibit good adhesion to the LTCC-substrate, but a high resistivity $\rho_{Mo}=8.2 \cdot 10^{-7}$ Ωm in combination with a low temperature coefficient of resistance (TCR) $\alpha_{Mo}=9.3 \cdot 10^{-4}K^{-1}$ between room temperature and 473K are replaced by a new metallization system of titanium/platinum (Ti/Pt) [4]. The Ti layer serve as a "glue" between the dielectric substrate and the noble metal Pt, to achieve sufficient adhesion to the glass-ceramic substrate for a reliable operation of the thin film element. This new metallization reduces the resistivity by a factor of 4 and increases the TCR by a factor of 2 respectively. The evaluation of this type of thermal injection rate sensor at drive pulses between $t_0=0.3$ms and 1.5ms is done at a hydraulic test bench up to fuel pressures of 135MPa. Long-term measurements of bare Ti/Pt micro-heaters in oil atmosphere at atmospheric pressure yield the need for a proper passivation technology, to stabilize the sensor signals in the working point. In a previous study, diamond and diamond-like coatings, which promise excellent thermal material properties [5], were deposited with an arc discharge plasma jet CVD (Chemical Vapour Deposition) technology [6]. Investigations of the passivated glass-ceramics surface exhibit a poor coverage due to high surface temperatures of about 650°C during deposition. With a PECVD (Plasma-Enhanced Chemical Vapor Deposition) technology, hydrogenated amorphous silicon carbide (a-SiC:H) thin films can be homogenously deposited on the LTC-ceramics at substrate temperatures below 300°C [7]. With this passivation layer, drift effects of the metallic thin film micro-heaters in oil can be reduced up to more than one order of magnitude within an operation time of 380h.

2 Fabrication Process

In this section, a brief description of the fabrication process for bare and a-SiC:H coated thin film resistors is presented. The detailed fabrication process of the bare hot film anemometers is described in previous works [3,8]. For this high-pressure automotive application, the flow sensitive thin film element has to be placed on a mechanically stable and electrically insulating substrate. An LTC-ceramics is used, which allows the integration of electrical feedthroughs made of Au, Ag or PdAg thick film pastes with a low resistivity of 2-20Ω/sq [9]. To enable thin film technology with heights of the metallic structures of about 100nm, the surface of the glass-ceramics with a maximum initial roughness of R_{max}~1700nm (average roughness R_a~350nm) has to be reduced with a polishing procedure. A bi-layer of titanium/platinum (d_{Ti}~10nm/d_{Pt}~100nm) is e-beam evaporated onto the substrate at a chamber pressure below $3 \cdot 10^{-6}$mbar and the micro-heaters are patterned in aqua regia. For the reliability investigations in oil atmosphere at atmospheric pressure, a 1μm a-SiC:H layer was deposited on the glass-ceramic plate by a commercialized PECVD-process using a liquid source ($C_6H_{18}Si_2$) [7]. For all samples, the deposition temperature Td was chosen to 300°C. The sensor chips are cut out of the glass ceramic plate with an ultrasonic milling process operating at 20kHz. Figure 1 shows the sensor chip with a complete coverage of the front side with an a-SiC:H thin film layer. Finally, 80μm thick Pt-wires are micro-welded on the bond pads, which have a diameter of about 200μm.

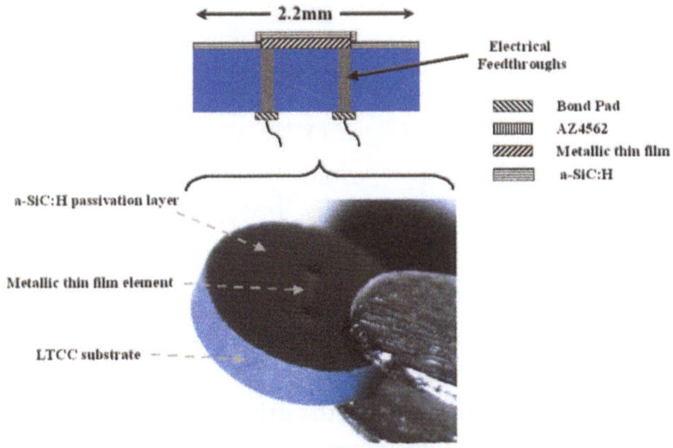

Fig. 1. Schematic picture and optical micrograph of a thermal injection rate sensor on a LTCC substrate, which is covered with a 1μm thick a-SiC:H passivation layer.

3 Results and Discussion

3.1 Injection Rate Measurements

Injection rate measurements up to fuel pressures of 135MPa, done with the nozzle integrated hot film anemometer, are presented in this section. The sensor resistance at room temperature is about 7.8Ω. For these investigations the electrical heating power is P=110mW, corresponding for the present sensor design (L=550μm, w=120μm) to a thin film overheat temperature of about 70K based on an ambient temperature of 298K. The hydraulic investigations are performed at a high-pressure test bed, described in detail elsewhere [8], where the injection amount indicator (EMI) allows to determine the injected fuel volume as a function of time by the displacement of a piston in a magnetic coil. The change in voltage drop across the thin film heater due to the convective heat transfer to the fluid is read out by a Wheatstone bridge followed by an amplification and filtering unit [4]. The latter effectively reduces in the electronic readout the electronic noise arising mainly from the high-pressure pump unit. At a rail pressure of p = 135MPa Figure 2 shows the drive pulse to open the magnetic valve. Due to the fuel mass flow, the sensor resistance changes, which is, as the EMI signal, recorded as a function of time.

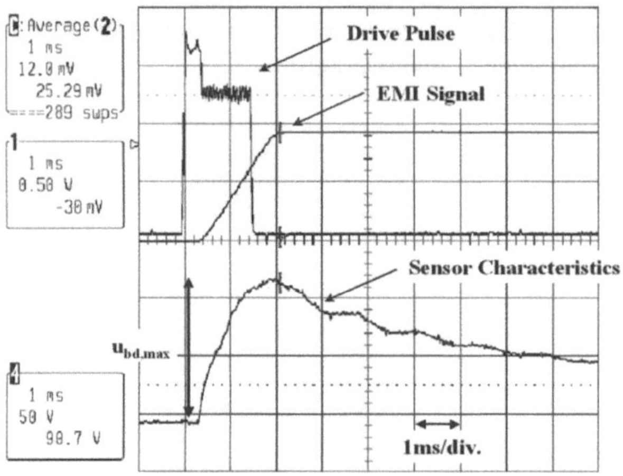

Fig. 2. Simultaneous recording of the drive pulse, the EMI signal, and the injection rate, determined with the micromachined hot film anemometer, at a fuel pressure of 135MPa and a drive pulse width of 1.5ms.

Pressure waves, occurring after the injection pulse, are measurable and arise from the fast deceleration of the fuel column at the injection end as well as by the movement of the needle. Especially these information about the flow condition, concerning the time after the nozzle is closed, are experimentally only gained with a nozzle-integrated flow sensor, as different measurement principles, as the injection amount indicator, provide only a signal of the injection rate, when a mass flow occurs out of the nozzle. Due to the fast change in sensor signal height, one flow sensitive element is sufficient to detect the width of the injection pulse, which is verified by the EMI characteristics. As shown in Figure 2, the maximum diagonal bridge voltage $u_{bd,max}$, which is related to the maximum value of the injection rate during operation, corresponds to that time, where the nozzle is closed, and represents a characteristic parameter for the injected fuel quantity. Figure 3 yields for different drive pulse widths a linear relationship between the rail pressure p and the injection volume V_{EMI}, recorded with the EMI set-up.

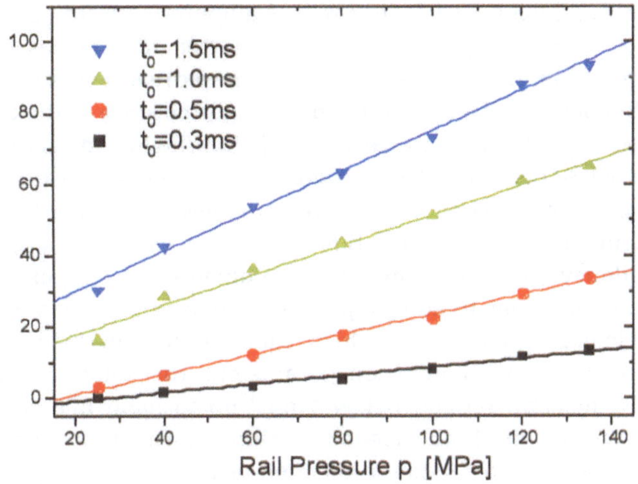

Fig. 3. Injection volume V_{EMI} vs. injection pressure p for different values of the parameter t_0.

According to the injection parameters in Figure 3, the maximum diagonal bridge voltage ubd,max is determined from the sensor characteristics and depicted in Figure 4.

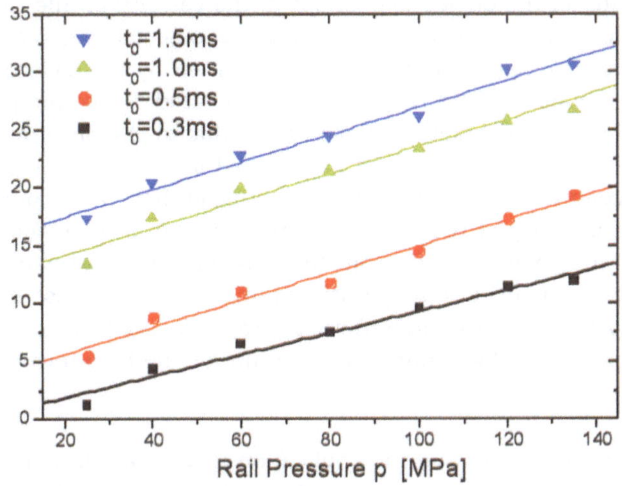

Fig. 4. The maximum diagonal bridge voltage ub,max as a function of injection pressure p and injection pulse width t_0.

As expected, for a given value of t_0, the amplitude $u_{bd,max}$ rises with increasing p, as a higher fuel velocity leads to a more effective cooling of the hot film anemometer and hence, to a higher sensor amplitude. On the other hand, for a given value for p, $u_{bd,max}$ decreases with shorter drive pulses. In this injection pressure range, the injected volume is up to drive pulses of about 1.0ms strongly influenced by the opening and closing procedure of the nozzle due to an incomplete needle lift, as verified by EMI/IMAD measurements [10]. Therefore, for $t_0 \leq 1$ms, the velocity profile of the fuel can be regarded as not fully developed for the applied pressure level, leading to a reduced value for ubd,max. Comparing figure 3 and figure 4, $u_{bd,max}$ as well as V_{EMI} exhibit the same dependency on the fuel pressure at a given value of t_0 except of different slopes for the linear relationships. This is due to the transfer function of the thermal injection rate sensor, which is incorporated in the data of figure 4. Due to the wide range of injection parameters, fuel quantities are covered from the pre-injection regime below 1mm^3 up to main injection quantities (~90mm^3), which are both needed during operation at diesel engines, equipped with a modern CR direct injection technology [10].

3.2 Long-term Stability of Bare and a-SiC:H Coated Micro-Heaters in Oil Atmosphere

The long-term stability of the temperature-dependant resistance of hot film anemometers is decisive for a proper interpretation of their electronic signals. The sensitivity of the device to changes in the velocity profile, determined by the temperature reduction due to a liquid flow over the micro heater, increases with increasing operating temperature, limited by the boiling temperature T_b of the surrounding fluid [11]. However, elevated device temperatures give rise to accelerated degradation of the device due to chemical processes at the heater surface, especially when it is operated in chemically aggressive environmental conditions, resulting in a pronounced drift behaviour of the resistivity and sensitivity of the device. To measure the drift in the micro-heater resistance, Mo as well as Ti/Pt thin film resistors are operated under atmospheric pressure at different operating temperatures in test oil (Shell1434, $T_b = 207$°C) which has the same physical properties as diesel oil. In figure 5, a schematic picture of the experimental set-up for the long-term investigations is shown.

The data acquisition unit and the application of the electrical heating power is computer controlled. To get a certain device temperature, the cold resistance of the thin film resistor R_a is determined at room temperature T_a (=23°C). The temperature of the probe T_s is according to Eq. 1

$$R_s - R_a = R_a \cdot \alpha \cdot (T_s - T_a) \tag{1}$$

linearly related to the actual resistance R_s, whereas a is the temperature coefficient of resistivity (TCR) of the metallic thin film [11]. The output power of the Keithley 2400 is increased gradually, until the computed value of $R_{s,o}$ is reached, which corresponds to the sensor temperature $T_{s,o}$ at the beginning of the long-term investigation. The heating voltage is kept constant and the heating current is measured as a function of time t. The temporal change in thin film resistance during the investigations is related to the original value $R_{s,o}$ and defined as follows:

$$R_{rel} = \left(\frac{R_s}{R_{s,o}} - 1\right) \cdot 100 \tag{2}$$

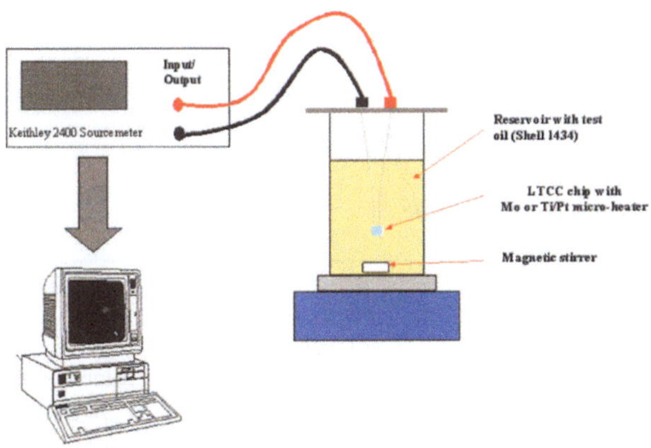

Fig. 5. Schematic picture of the set-up for the long-term measurements.

In figure 6, the value of Rrel of uncoated Mo and Ti/Pt thin film sensors in oil atmosphere is shown at different sensor temperatures as a function of time. At $T_{s,o} = 160°C$, the resistance of the molybdenum heater rises about 6% of its original value within a measurement period of 170h. In comparison, the Ti/Pt heater elements yield an increase in resistivity of about 75% and 200%, depending on the initial sensor temperature, which was 60°C and 160°C respectively. The most significant change in the heater resistance of 75% ($T_{s,o} = 60°C$) and 160% ($T_{s,o} = 160°C$) is observed within the first 30-40h of operation. Due the local increase of thermal loading, the oil is cracked on the probe surface and a subsequent polymerization, supported by the catalytic influence of the platinum on the hydrocarbon molecules, causes a coating on the hot metallic thin film surface, as reported in [12]. As the thickness of the thin resin layer rises with time, but with a lower rapidity at the heated non-noble molybdenum thin film, the heat transfer of the sensitive element to the fluid, moved by a magnet stirrer, is suppresses more effectively. The thin film resistance increases, as both the Mo and the Ti/Pt bi-layer have a positive coefficient of resistivity between room temperature and 200°C [4]. Sharp decreases in resistivity occur only in the Ti/Pt characteristics especially at higher temperatures of the probe, as uncontrollable delamination of part or almost the whole coating occurs, which leads to a decrease in Ts to almost the initial value. After 175h of operation, the test cycles are stopped, as no saturation of any of the output characteristics can be observed.

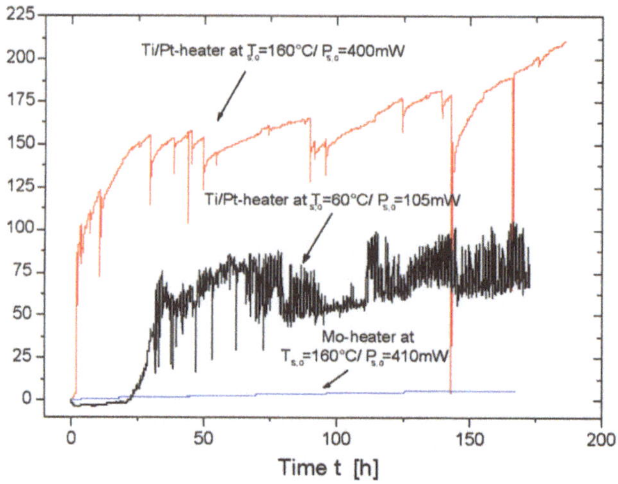

Fig. 6. Change in resistance of Mo and Ti/Pt micro-heaters with an ambient resistance of $R_{a,Mo}=104W$ and $R_{a,Pt}=21.2W$ as a function of time at different temperatures $T_{s,o}$ in oil atmosphere. The applied heating power at the beginning of the long-term measurements $P_{s,o}$ is given in brackets in the picture.

These measurements demonstrate, that as far as the thermal stability of the device is considered, non-noble metallizations as molybdenum are more preferable in oil atmosphere at comparable thin film temperatures than Ti/Pt elements. With bare hot film anemometers, which have a thin film element, consisting of a noble metal as platinum, precise injection quantity measurements, as requested for the target application in an injection system, cannot be performed within the automotive life cycle due to the unpredictable nature of the coating and the dynamics of its formation. As the Ti/Pt metallization exhibits a twice as high TCR than Mo thin films at a thickness of about 100nm and is more preferred for thermal mass flow measurements due to increased output signals, Ti/Pt thin film sensors are coated with a thin passivation layer of amorphous silicon carbide (a-SiC:H) to reduce drift effects as described above. In earlier investigations [6], it was found, that an a-SiC:H coating with a thickness of 1µm is sufficient to prevent the direct contact between the hot film element and the oil atmosphere and a clear reduction in R_{rel} can be achieved, as shown in figure 7.

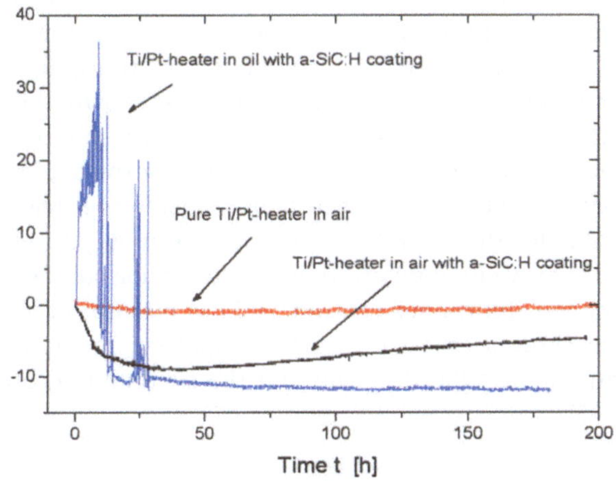

Fig. 7. Change in resistance of bare and a-SiC:H coated Ti/Pt micro-heaters as a function of time at $T_{s,o}$=160°C in oil as well as air atmosphere.

In accordance to figure 6, the most significant change in thin film resistance occurs during the first 30h of operation, followed by stable device characteristics with a reduction in resistance R_{rel}=-11%. In addition, long-term measurements of Ti/Pt thin film resistors with and without a-SiC:H coating in air are depicted in figure 7. For an initial device temperature of 160°C, the uncoated Ti/Pt thin film exhibits a maximum decrease of $R_{rel,max}$=-1.0% at t=33h and a R_{rel}=-0.48% at 180h respectively. The influence of the LTCC-surface, which has high roughness values in comparison to the thin film thickness and could promote "hot spots" due to an inhomogenous current density, as well as any diffusion or oxidation of the titanium adhesion layer, which starts to migrate along grain boundaries through the Pt thin film to the exposed surface at temperatures of 200°C [13], can be neglected. An enhanced drift behavior is observed for the a-SiC:H passivated thin film element on the LTC-ceramics with a maximum change in thin film resistance of $R_{rel,max}$=-8.9% at 35h and a subsequent decrease to R_{rel}=-4.6% after 180h of operation. Comparing the measurements in air, the higher drift of Rs at the protected Ti/Pt resistor can be attributed to thermal annealing of mechanical stress in the metallic thin film after the a-SiC:H coating. According to [14], thin metal films, which are deposited at moderate temperatures (room temperature to a few hundred °C), yield a tensile film stress of 102 to 103MPa for refractory metals. In contrast, the dielectric passivation layer has a compressive film stress with values of about -750MPa, when deposited at $T_d = 300$°C [7]. As the value of R_a rises about 15% of its initial value after the post-deposition process, mechanical deformation may take place in the thin Ti/Pt film, leading to an overcompensation of the original tensile state due to the a-SiC:H coating. With a thin film temperature of $T_{s,o} = 160$°C, structural rearrangements in the metallic layer are enhanced and lead to a thermally activated reduction in film stress and thin film resistivity. Independent of the surrounding atmosphere, a maximum value of Rrel occurs after 30h-35h of current load, as shown in figure 7. For the passivated micro-heater, which is operated in oil atmosphere, chemical reactions on the heated a-SiC:H surface with the test oil at the beginning of the test cycle, which have an influence on the heat transfer coefficient to the fluid, may be responsible for the high deviations of R_s from $R_{s,o}$ during the first 30h of operation. No indication is given, that this behavior can be attributed to a poor tightness of the amorphous coating, as SEM pictures of the coated glass-ceramics surface show a crack-free surface topology, providing no paths for contact between the metallic thin film and the surrounding atmosphere (figure 8).

Fig. 8. SEM picture of a 1μm thick a-SiC:H thin film on a LTCC-substrate.

As the stable device characteristics of the protected thin film element in figure 7 suggests a completed aging procedure in oil atmosphere, further reliability investigations are performed. To simulate the temperature cycles in the injection nozzle, the micro-heater is exposed to temperature pulses between 45°C and 160°C with a dwell time of 5 minutes at each temperature level. As shown in figure 7, within additional 200h of operation, the output waveform of the electrically heated device is square-shaped as the excitation by the heating power. R_s, normalized to $R_{s,o}$ at an initial sensor temperature of 160°C, increases about 3%, independent of the temperature level, and proves the stability of the metallic thin film as well as the effective protection by the 1µm a-SiC:H layer.

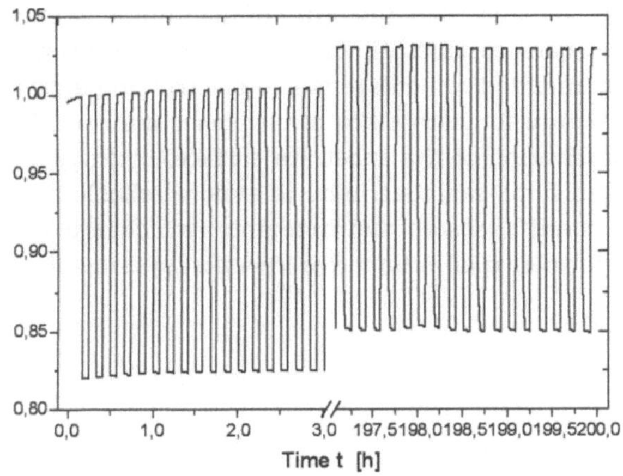

Fig. 9. Output characteristics of a passivated Ti/Pt micro-heater, normalized to the resistance at $T_{s,o} = 160°C$, to a square-shaped excitation by the heating power as a function of time in oil atmosphere.

4 Conclusion

A robust hot film anemometer, which consists of a Ti/Pt thin film element on a LTC substrate, was integrated into the nozzle body of a standard diesel injector, which is part of a Common-Rail (CR) injection system. With the integrated mass flow sensor, injection rate measurements are performed up to fuel pressures of 135MPa at different drive pulses, covering the whole range from pre-injection (~1mm^3) to main injection fuel quantities (~90mm^3). These investigations reveal a fast response to the injection pulses, verified with the EMI characteristics, and allow to determine the width of the injection pulse directly in the injection system. During operation, rising output levels of the thermal injection rate sensor are measured with increasing rail pressure due to higher fuel velocities. At a given fuel pressure level, however, the parameter $u_{bd,max}$ decreases with shorter drive pulses, as the needle is not completely lifted below t_0~1ms.

For metallic thin film micro-heaters on glass-ceramic substrates, a commercialized PECVD-process for amorphous silicon carbide (a-SiC:H) deposition is shown to offer a promising technology to minimize temporal drifts of the hot film probe. With the surface exposed to air the minimum drift in resistivity of Ti/Pt micro-heaters on the LTCC substrate is measured to 1% at an initial temperature $T_{s,o} = 160°C$. A comparison of bare and a-SiC:H coated Ti/Pt thin films in oil at $T_{s,o} = 160°C$ yield a reduction in the corresponding resistance for the passivated device by more than one order of magnitude to about 11% as well as a stable device characteristics. However, a pre-aging procedure of about 30h at 160°C is recommended to achieve stable electric properties of the investigated device structure for both oil or atmospheric ambient. Due to the tight and complete passivation of the substrate surface, hydrogenated amorphous silicon carbide thin films can contribute to a more reliable and stable characteristics of "on board" sensors and actuators, which are recommended to enhance the performance of automotive injection systems for future requirements.

References

[1] J. Schommers, F. Duvinage, M. Stotz, A. Peters, S. Ellwanger, K. Koyanagi and H. Gildein, „Potential of Common Rail Injection System for Passenger Car DI Diesel Engines", SAE Paper Number: 2000-02-0944, SAE World Congress, Detroit, 2000.

[2] W. Berg, „Political Requirements for the Diesel-Engine-A Power Source between Promotion and Critique", 20. Wiener Motorensymposium, Bd. 1, ISBN: 3-18-337612-1, pp. 81-126, 1999.

[3] U. Schmid, „A Robust Flow Sensor for High Pressure Automotive Applications", Sensors and Actuators A: Physical 97-98 (C), pp. 253-263, 2002.

[4] U. Schmid and G. Krötz, „A Robust Hot Film Anemometer For Injection Quantity Measurements", 6th Conference on Advanced Microsystems for Automotive Applications (AMAA), Springer Press ISBN: 3-540-43232-9, pp. 174-183, Berlin 2002.

[5] K.E. Goodson, „Impact of CVD Diamond Layers on the Thermal Engineering of Electronic Systems", Annual Review of Heat Transfer, Bd. 6, pp. 232-253, 1995.

[6] U. Schmid and G. Krötz, „Investigation On Different Passivation Technologies For Thin Film Sensors On Ceramic Substrates", Proc. of the 1st IEEE Sensor Conference, Orlando, pp. 617-622, 2002.

[7] A. Klumpp, U. Schaber, H.L. Offereins, K. Kühl and H. Sandmaier, „Amorphous silicon carbide and its application in silicon micromachining", Sensors and Materials, Vol. 6, No.6, pp.349-358, 1994.

[8] U. Schmid, G. Krötz and D. Schmitt-Landsiedel, „A Mass Flow Sensor For Fuel Injection Quantity Measurements in DI-Systems," 5th Conference on Advanced Microsystems for Automotive Applications (AMAA), Springer Press ISBN: 3-540-41809-1, pp. 139-155, Berlin 2001.

[9] R. Kulke, „LTCC-An Introduction And Overview" distributed by the IMST GmbH

under http://www.ltcc.de or http://www.imst.de.
[10] K. H. Hoffmann, K. Hummel, T. Maderstein and A. Peters „Das Common-Rail Einspritzsystem- Ein neues Kapitel der Dieseleinspritztechnik", MTZ, Vol. 58, No. 10, pp. 572-582, 1997.
[11] H.H. Bruun, „Hot wire anemometry-principles and signal analysis," Oxford University Press, ISBN: 0 19 856342 6, 1995.
[12] H. Eckelmann, „Hot wire and hot film measurements in oil", DISA Info, No. 13, pp. 16-22, 1972.
[13] M. DiBattista, and J.W. Schwank, „Determination of diffusion in polycrystalline platinum thin films", J. Appl. Phys., Vol. 86, No. 9, pp. 4902-4907, 1999.
[14] .I. Maissel and R. Glang, „Handbook of Thin Film Technology"; McGraw-Hill Book Company, 1970.

Ulrich Schmid
EADS Deutschland GmbH
Department SC/IRT/LG-ME
81663 Munich
Germany
Ulrich.schmid@eads.net

Keywords: injection rate sensor, LTCC substrate, long-term measurements, a-SiC:H, passivation technology.

Solder Alternative Electrically Conductive Adhesives with Stable Contact Resistance in Combination with Non-Noble Metallisations for Automotive Electronic Assembly

G. Dreezen, G. Luyckx, Emerson & Cuming

Abstract

Conductive Surface Mount Adhesives (CSMAs) are an alternative to traditional solders used in the electronics industry. CSMAs provide an environmentally friendly alternative to conventional Sn/Pb metal solders offering additional attractive technical advantages including low temperature processing, fine pitch capability and better resistance to thermal cycling. Traditionally CSMAs have been used in hybrid applications. These applications typically use noble metallisations such as gold and silver palladium on ceramic substrates. The major limitation of CSMAs has been their instability on common electronic metals such as copper and Sn/Pb solder. Recently, CSMAs have been re-evaluated in applications that were traditionally reserved for solders. The reason for this renewed interest can be attributed to the development of adhesives with stable contact resistance in combination with non-noble metals [1]. Data on the effect of eutectic tin lead and 100% tin terminations on the new technology CSMAs will be presented.

1 Background

Traditionially CSMAs have been used in hybrid applications typically using noble metallisations such as gold and silver palladium on both substrates and components. Benefits of CSMAs over traditional solder are the lower processing temperatures, fine pitch capability and improved thermo-mechanical performance [2,3]. Until recently applications with non-noble metallisations such as Sn/Pb and Sn have traditionially been assembled using solder. Sn/Pb eutectic solder has been the dominant interconnecting material in electronic assemblies. The drive towards leadfree electronics has increased strongly and legislative measures have been proposed to ban or limit the use of lead in Europe by 2006. The disadvantage of most of these leadfree solders is the higher melting point thus requiring higher reflow temperatures up to 250-260°C. This may damage temperature sensitive components and limit the lifespan of the product. This is an additional driver for using low temperature curing CSMAs.

The contact resistance instability of CSMAs in combination with non-noble metallisations is due to the occurrence of electrochemical corrosion at the interface between the adhesive and the metallisation. It has been shown during 85/85 temperature/humidity testing that the CSMA displays a stable bulk resistivity but that the contact resistance at the metal adhesive interface increases [4].

Fig. 1. Formation of an electrochemical cell at the interface between a CSMA and a non-noble metallisation.

Electrochemical corrosion occurs in the presence of two metals with dissimilar reduction potential and water (figure 1). When Ag, with a high reduction potential, is in contact with traditional electronic metals such as Cu, Sn and Pb or other low reduction potential metals, the addition of small amounts of water completes a galvanic cell and following reactions may occur:

The metal with the higher electrochemical potential (Ag/0.80V) acts as the cathode at which following reaction takes place:

$$2H_2O + O_2 + 4e^- \rightarrow 4\ OH^-$$

The metal with the lower electrochemical potential acts as the anode:

$$M - ne^- \rightarrow M^{n+}$$

Normal potentials for different metals are shown in table 1. For metals such as Ag, Au and Pt with similar or electrochemical potentials >0.40 V (potential of cathode reaction), corrosion is not taking place and the assembly is providing theoretically a stable contact resistance. For metals such as Cu, Sn, Pb and Ni with a potential below 0.40V, electrochemical corrosion will occur and the contact resistance will increase after exposure to humidity due to the formation of a non-conductive oxide at the interface.

Electrode Reaction	Normal Potential (Volt)
$Au - 3e^- = Au^3+$	1.50
$Pt - 2e^- = Pt^2+$	1.20
$Ag - 1e^- = Ag^+$	0.80
$H_2O + O_2 + 4e^- = 4OH^-$	0.40
$Cu - 1e^- = Cu^+$	0.52
$Cu - 2e^- = Cu^{2+}$	0.34
$Pb - 2e^- = Pb^{2+}$	-0.13
$Sn - 2e^- = Sn^{2+}$	-0.14
$Ni - 2e^- = Ni^{2+}$	-0.25

Tab. 1. Electrochemical potential of some electrode reactions.

Based on the above fundamental understanding Emerson & Cuming has been developing new and unique materials which exhibit exceptional contact resistance stability with Sn and Sn/Pb metallisations [5-7].

2 Adhesives

A select group of adhesives was chosen for comparison. One traditional adhesive (A1) was chosen as reference for comparison. Adhesive A1 has been commercially available for a number of years and is widely used in hybrid applications with noble metallisations. Adhesive B and C are two new formulations using specific corrosion inhibitors. Adhesive D and E are two new experimental adhesives using specific corrosion inhibitors together with low melting alloys.

3 Contact Resistance Test Device

A simple test device was constructed for ease of repeatability. The device consisted of a daisy chain silver palladium pattern on a ceramic substrate. The silver palladium pads were separated from one another by a 1.27mm gap suitable to place an 0805 component. Ten such separations were included in each daisy chain or loop. 0805 components with different metallisations have been evaluated including 63/37 Sn/Pb, 80/20 Sn/Pb and 100% Sn plating. It has already been reported that these metals are easily oxidised in the presence of silver and humidity due to the mismatch in their electrochemical potential. In addition, the oxides are non-conductive and create an insulating barrier at the adhesive metal interface. The test boards were completed by printing the adhesives onto the metallised pads (100µm stencil thickness) and by placing the components. The test boards were then cured for 1 hour 150°C.

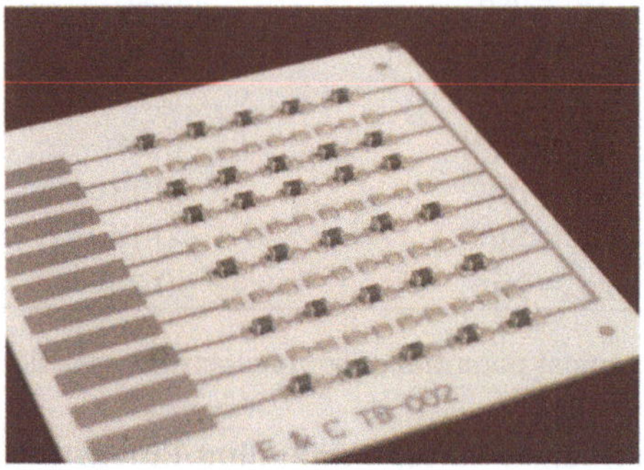

Fig. 2. Test devices with Sn terminated null-ohm resistors.

4 Results and Discussion

The contact resistance stability of the new conductive adhesives using specific anti-corrosion agents in combination with non-noble metallisations has been evaluated. The contact resistance stability under 85°C and 85% relative humidity (RH) with Sn and Sn/Pb will be discussed first. Secondly the performance of these materials with various component types is discussed followed by an evaluation of the contact resistance stability during thermoshock testing between -40°C and +150°C. Finally the contact resistance and adhesion stability of the new experimental adhesives combining anti-corrosion agents and low melting alloys are discussed.

4.1 Contact Resistance Stability Under 85°C / 85% RH Testing

A first test device with a traditional 63/37 SnPb surface metallisation as described above was used with 3 different conductive adhesives (A1, B, C). The low electrochemical potential of the Sn/Pb alloy makes it a potential problem. Contact resistance data up to 1000 hours at 85°C and 85% RH are shown in figure 3.

The control adhesive A1 shows a low initial joint resistance that is increasing rapidly under 85°C / 85% RH. This increase can be attributed to the electrochemical corrosion process forming an insulating oxide layer at the adhesive/solder interface. The new adhesives B and C containing anti-corrosion agents display a stable contact resistance up to 500 hours aging. After 500 hours, the contact resistance of adhesive B is increasing whereas adhesive C shows outstanding performance on Sn/Pb even after 1000 hours aging under the severe condition of 85°C and 85% relative humidity.

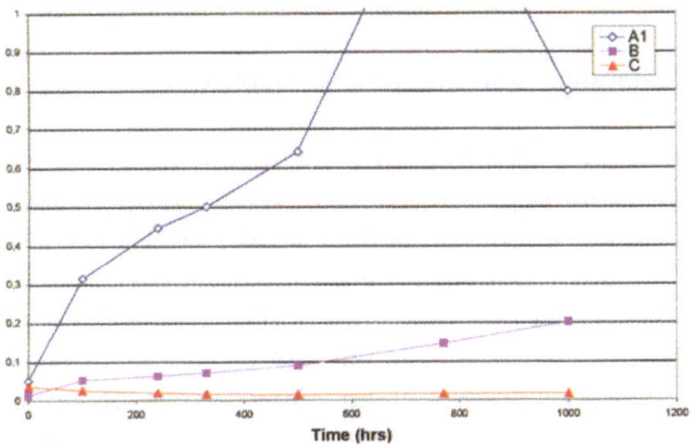

Fig. 3. Contact resistance versus time with SnPb under 85°C / 85% RH test conditions.

A second metal surface evaluated is 100% tin. This surface was selected for its potential to replace lead-containing terminations on electronic components. It has been well documented that high tin alloys will most certainly be used by the electronic industry in the future [8,9]. High tin alloys will pose similar corrosion issues due to its low electrochemical potential. In general, it has been more difficult to obtain acceptable and stable contact resistance on high tin surfaces (see figure 4).

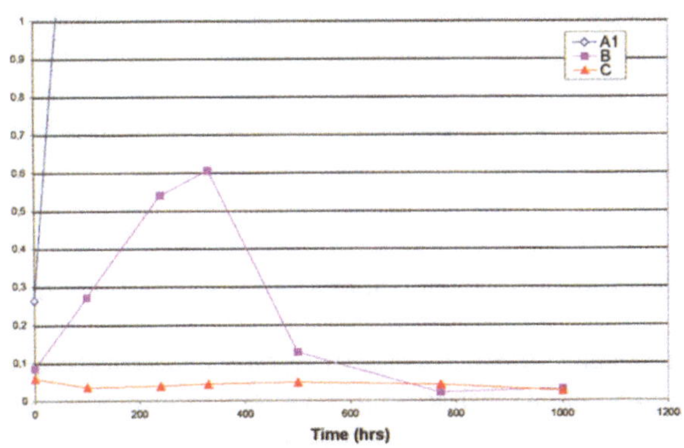

Fig. 4. Contact resistance versus time with tin under 85°C / 85% RH test conditions.

The contact resistance of the control adhesive A1 is increasing even more rapidly when tin is used as the contact metal. The performance of adhesive B shows a peculiar behaviour as it appears to detoriate rapidly but then briefly performs acceptable for a short time. Overall the performance of adhesive B with tin cannot be considered as being acceptable. Although the contact resistance is higher on Sn as compared to Sn/Pb, adhesive C shows an acceptable low initial joint resistance that maintains its low value through the aging study.

4.2 Contact Resistance with Different Components

The contact resistance of adhesive C has been evaluated with 0805 components from 4 different suppliers W, X, Y and Z. Components W and X contain a 80/20 Sn/Pb metallisation, component type Y contains a 100% Sn metallisation and the% Sn in component type Z was unknown (it was sold as a pure tin component). The contact resistance measured after cure for the different components is shown in figure 5. Components W and X show a very typical low single joint contact resistance of about 25mΩ. However for components Y and Z a much higher resistance has been measured; 780mΩ and 120mΩ respectively. The origin of this high initial resistance is suspected to be due to oxidation of the Sn metallisation before attaching the component to the test board.

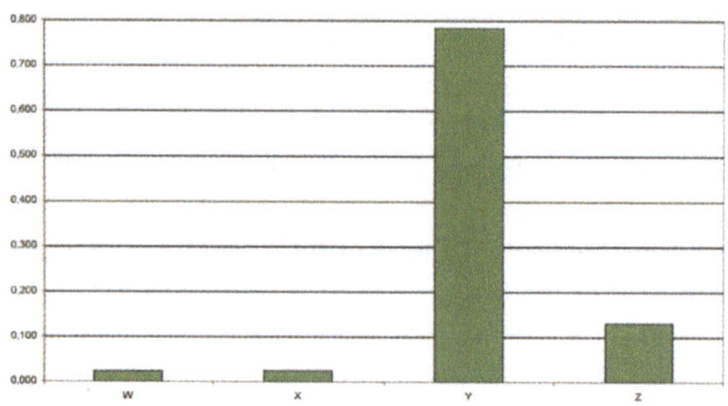

Fig. 5. Initial contact resistance of adhesive C with different component types.

4.3 Contact Resistance Stability Under Thermoshock Testing

Further work has been performed on the contact resistance stability after thermoshock testing between -40°C and +150°C. Figure 6 shows the resistance per joint after 850 thermoshocks for the 4 evaluated component types attached to the test board using adhesive C.

The high initial contact resistance with components Y (780mΩ) and Z (120mΩ) decreases significantly after 125 thermoshocks. Probably breakdown of the thin insulating layer on the component metallisation takes places due to mechanical stress caused by the differences in thermal expansion coefficient of the various materials. Components W and X also display a minor decrease in contact resistance after 125 thermoshocks due to post cure of the adhesive. The more detailed plot of figure 6 shows that for all component types a slight increase in contact resistance up to an average value of 30mΩ to 60mΩ occurs between 125 and 850 thermoshocks.

Although a very stable joint resistance has been established in 85/85 temperature humidity testing using specific anti-corrosion agents in the newly developed conductive adhesives some potential issues have been defined in a typical automotive reliability test such as -40°C/+150°C thermoshock, especially for the components with the highest Sn levels.

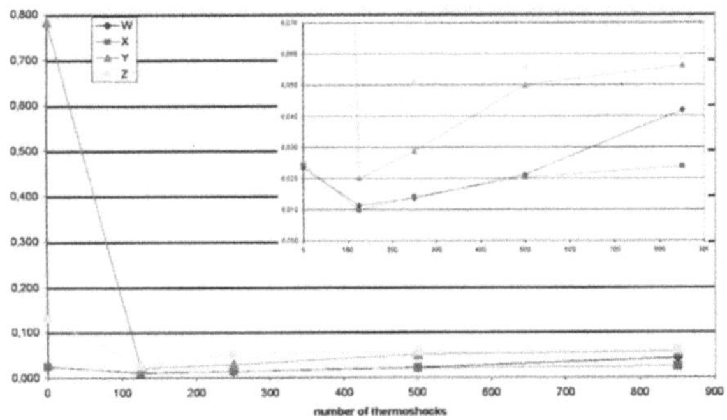

Fig. 6. Full scale and detailed plot of the contact resistance stability of adhesive C with different component types during thermoshock testing from -40°C to +150°C.

4.4 Improved Contact Resistance and Contact Resistance Stability Under Thermoshock Testing

The commercially available new conductive adhesive C shows excellent contact resistance stability with non-noble metallisations and was successfully implemented into various applications. However when studying the worst performing component types, an increased initial contact resistance together with a slight increase after thermoshock testing from -40°C up to +150°C has been reported. In order to lower the initial contact resistance and to improve the reliability of these adhesives towards the automotive electronics requirements new materials have been developed. These new experimental conductive adhesives are based on patented technology combining specific corrosion inhibitors with low melting alloys.

Figure 7 displays the initial contact resistance and contact resistance stability during thermoshock testing of adhesives C, D and E with component type Y. Component type Y was giving poor performance with adhesive C that only contains the anti-corrosion agents. Adhesives D and E are two new experimental adhesives containing both corrosion inhibitors and low melting alloys. As discussed before, the single joint contact resistance of adhesive C with component Y is very high directly after cure. The new experimental adhesives D and E display a significantly reduced initial contact resistance as compared to adhesive C. Adhesive E is the best performing material with an average single joint contact resistance of 15mΩ. During thermoshock testing up to 850 times the contact resistance of adhesive E remains low and stable at approximately 5mΩ to 15mΩ. The use of a low melting alloy results in the formation of an intermetallic layer during cure of the conductive adhesive providing such a low and stable contact resistance.

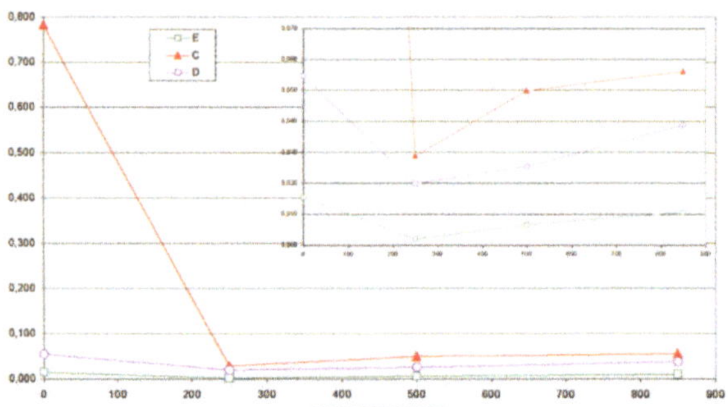

Fig. 7. Full scale and detailed plot of the contact resistance stability of adhesive C, D, E with component type Y during thermoshock testing from -40°C to +150°C.

4.5 Adhesion Stability after Thermoshock Testing

One of the other main requirements for the new conductive adhesives is adhesion and adhesion stability. Thermoshock testing has been defined as the most critical reliability test for components having non-noble metallisations. In this study the performance of the new adhesive E has been evaluated versus standard eutectic solder using component type Y. Figure 8 show the adhesion data before and after 850x thermoshock testing between -40°C up to +150°C. It is clear that adhesive E displays very good adhesion retention. Whereas the adhesion using eutectic solder drops about 25% from 4.2kg to 3.2kg; adhesive E only shows a decrease of 10% till 3.8kg. The improved performance of adhesive E might be explained from: a) the soft low melting alloy stopping the formation of microcracks at the interface and b) the low melting alloy can form a metallic layer at the interface preventing diffusion of Sn into the conductive adhesive which is suspected as the major driving force behind adhesion drop in contact with Sn terminated components (Kirkendale voiding). Microscopy work to obtain a better understanding of the different phenomena taking place is ongoing.

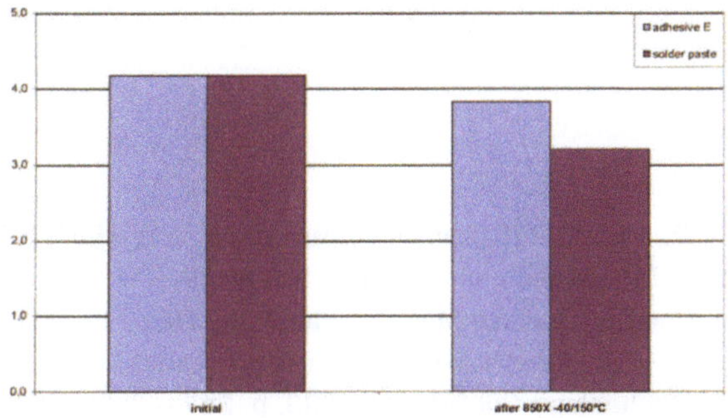

Fig. 8. Adhesion before and after thermoshock testing of adhesive E as compared to standard solder paste.

5 Conclusions

The purpose of this study was to evaluate the performance of commercially available and recently developed conductive adhesives in contact with non-noble metallisations such as Sn and SnPb. It was shown that adhesive C containing anti-corrosion agents displays excellent contact resistance stability with both Sn and SnPb metallisations under 85°C and 85% relative humidity conditions up to 1000 hours. During thermoshock testing between -40°C and +150°C a dependency on the component type used has been reported as well as a limited increase in contact resistance.

Recently developed formulations are showing an improved electrical and mechanical performance. When both anti-corrosion agents and low melting alloys are used in the conductive adhesives a very low contact resistance is observed even with the worst performing component types. Contact resistance stability and adhesion retention after thermoshock testing between -40 and +150°C was reported indicating that these new conductive adhesives can meet automotive reliability requirements when using non-noble metallisations. The excellent performance of these new materials is based on patented technology. The anti-corrosion agent is preventing electrochemical corrosion at the interface with the non-noble metal and the low melting alloy improves the contact resistance and mechanical strength from the formation of an intermetallic layer at the interface.

The development of these new adhesives compatible with non-noble metallisations on both substrates and components allows the replacement of expensive noble metallisations and does generate new routes for cost reduction into automotive electronics assembly.

References

[1] Lu D., Wong C.P., Tong Q. IEE Trans.Elec.Pack.Man., July 1999, 2-3, p. 223
[2] Gilleo K. Surface Mount Technology, 1995, 9.1, p. 39 – 45
[3] Gilleo K. Soldering and Surface Mount Technology, 1995, 19, p. 2 – 17
[4] Tong Q., Markley D., Fredrickson G., Kuder R. and Lu D. 1999 Electronic Components & Technology Conference, 1999, p. 347 – 352
[5] Dreezen G., Luyckx G., On Board Technology, September 2001, p. 46
[6] Dreezen G., Luyckx G., IMAPS 2001 Proc. International Symposium on Microelectronics, 2001, p. 19
[7] Dreezen G., Deckx E., „ G., IMAPS 2002, Proc. European Microelectronics Packaging & Interconncetion Symposium, 2002, p.164
[8] Fredrickson G., Cheng C.-M., Electronic Components and Technology Conference, 2001
[9] Grusd A., Circuits Assembly, August 1999 p.32

Gunther Dreezen, Geert Luyckx
Emerson & Cuming
ICI Belgium N.V.
Nijverheidsstraat 7
2260 Westerlo
Belgium
gunther.dreezen@nstarch.com
geert.luyckx@nstarch.com

Keywords: conductive adhesive, CSMA, non-noble metallisations, leadfree, surface mount, corrosion

Functions and Applications

Comfort and Human-Machine-Interface

A Low-g Accelerometer for Inertial Measurement Units

K. Kapser, M. Aikele, R. Gottinger, B. Hartmann, ContiTemic
R. Burghardt, Continental Teves
H. Seidel, Saarland University

Abstract

Inertial Measurement Units (IMU) for upcoming automotive chassis control systems require sensor clusters with multi-axial sensors for acceleration and angular rate. In this paper, we present a low-g accelerometer that fits the IMU requirements for advanced automotive applications and is capable of monolithic multiple-axes integration. For the target application a high offset stability and an overcritical damping of the sensing element are required. With simple design changes the squeezed-film damping of the sensor can be adjusted within a wide range (5Hz to 400Hz) using atmospheric pressure inside the cavity. Due to its symmetrical design and the differential readout principle, an offset drift of less than 50mg over the full temperature range (-40°C...120°C) was achieved. The sensor principle presented in this paper can be utilized to realize a sensitivity axis parallel to the wafer surface as well as perpendicular to the surface. Furthermore, monolithic integration into a two-axis (x/z) or a three-axis (x/y/z) sensing element is possible. The tri-axial element exhibits a four-mass design providing redundancy and thus an ongoing self-test capability.

1 Introduction

In future advanced automotive systems an IMU (Inertial Measurement Unit) is used to provide comprehensive vehicle inertial data consisting of angular rates and acceleration values. All six degrees of freedom are covered (i.e. yaw, roll, and pitch for angular rate as well as lateral, longitudinal and vertical acceleration) with some safety relevant functions (e.g. yaw rate) potentially being redundant (figure 1). Various vehicle systems can take advantage of this information for applications such as dynamic drive control or roll over detection for the airbag. The compact spatial setup within one relatively small IMU allowing improved alignment of the individual sensor elements helps to greatly reduce cross-axis sensitivities. Nearly optimal alignment can be achieved by the monolithic integration of several axes on one chip, as demonstrated in the three axial low-g sensor concept presented in this paper. Furthermore, a mathematical modeling concept of the known individual cross-axis sensitivities can be applied for further error reduction. The orientation of the IMU within the vehicle is not critical as the local coordinate system provided by the IMU can be transformed mathematically by signal processing (internal or external) to fit the desired vehicle coordinate system. Further integration with other sensors (e.g. wheel speed, GPS, etc.) is possible, thus enabling a dynamic reduction of the bias error, further increasing the precision of such systems.

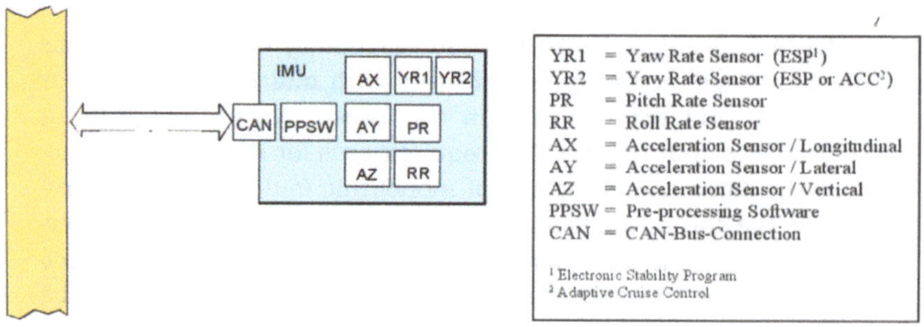

Fig. 1. Principle of an inertial measurement unit (IMU) containing different sensors and signal conditioning electronics.

One key component of such an IMU is a low-g acceleration sensor [1]. The required measurement range for this application is about ± 3g including DC stability, as absolute acceleration values have to be measured.

Different approaches were proposed for sensing elements which are capable of two-axes [2] and three-axes [3, 4] measurements. In this paper we focus on a silicon micromachined low-g acceleration sensor which can be used to realize one-, two- and three-axes sensing systems based on the same fabrication technology.

2 Sensor Concept

The acceleration sensor presented in this paper consists of a silicon seismic mass suspended from two torsional beams. An applied acceleration results in a torque and thus in an inclination of the mass. The change in distance with respect to a pair of electrodes in the cap wafer is detected with a differential capacitive principle, thus reducing cross-axis sensitivity and enhancing offset stability. All readout electrodes are on one side of the seismic mass thus increasing the symmetry of the design, and reducing the influence of packaging stress. The electrical connection to the sensing electrodes is provided by a special feedthrough technology resulting in a hermetically sealed sensor and in a nearly standard setup for the bonding pads. The fabrication is based on a bulk micromachining process including KOH wet chemical etching for the seismic mass, RIE etching for the torsional beams, and anodic bonding of glass wafers for wafer level packaging of the device.

By introducing slight modifications in the design a sensor with a sensitivity axis in wafer plane (x- or y-sensor) or perpendicular to the wafer surface (z-sensor) can be realized, respectively. Figure 2 shows the principle setup of such a sensor with a sensitivity axis parallel or perpendicular to the wafer surface.

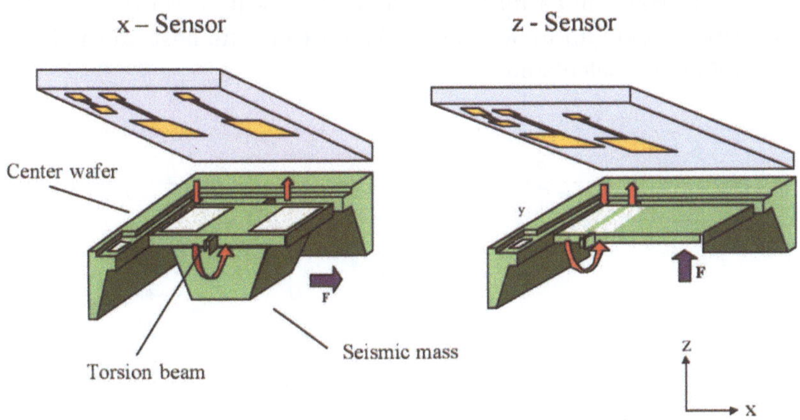

Fig. 2. Principle of the capacitive low-g sensor for a sensitivity axis in plane (x-sensor) and perpendicular to the wafer plane (z-sensor).

The gap between the capacitive readout electrodes was designed to be below 2µm yielding a capacity value of about 5.6pF. The geometry of the torsional beam, particularly its width, can be used as a parameter for adapting the sensitivity of the device. The following investigations were made with a beam width of 12µm resulting in a sensitivity of about 150fF/g (x-sensor). The z-axis sensor with its reduced seismic mass exhibited a sensitivity of approx. 75fF/g.

3 Measurement Results

For characterisation purposes the sensors were assembled on a ceramic substrate with a signal conditioning ASIC and packaged in a metal housing. Some applications require a premolded or overmolded plastic package due to cost reasons. This will necessitate additional investigations.

The ASIC output is proportional to the capacity difference C1-C2. Offset and sensitivity of the device could be programmed with the ASIC and were adjusted to 2.5V and 1V/g, respectively.

3.1 Linearity

Capacitive devices generally exhibit a non-linear behavior, as the change of capacitance is inversely proportional to the distance variation of the electrodes. A linear approximation is valid only for small distance changes (less than approx. 10%) with respect to the initial electrode gap. The following figure 3 shows measured linearity (left scale) and linearity error (right scale) for the acceleration range up to 3g. The deviation from the best linear fit to the measured value was calculated.

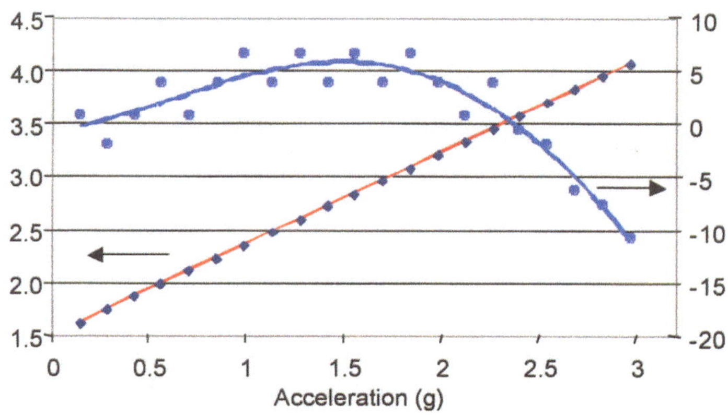

Fig. 3. Measured linearity error of the low-g sensor.

The linearity error is less than $\pm$ 10mg in the complete acceleration range which is well below the required specification of $\pm$20mg.

3.2 Frequency Response

For automotive applications the frequency response of the device is critical as there exists a wide vibration spectrum in the car which must not influence the sensor performance. Although electronic filtering is often applied to reduce this problem, it is generally preferable to use the sensor itself as a mechanical "pre-filter" with limited bandwidth. In this way an overload of the readout electronics due to vibrational noise can be avoided.

The bandwidth limitation of such a device is mainly due to "squeezed film" damping effects within the narrow gap between the relatively large capacitive electrodes. One way to adjust this damping is by adapting the cavity pressure inside the sensor. Reducing the pressure increases the bandwidth. This, however, can only be achieved by hermetically sealing the sensor on a wafer level with precisely controlled ambient pressure and requires a long term stability (>15 years) of this seal.

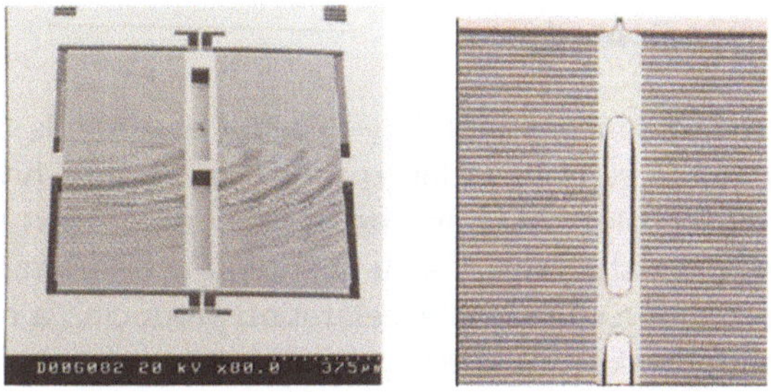

Fig. 4. Top view of the seismic mass including etched air flow channels (left), close-up view of the air flow channels (right).

A different approach was pursued for our sensor by reducing the squeezed-film damping through etched air-flow channels in the capacitor plate and maintaining atmospheric pressure within the cavity. This setup is shown in figure 4.

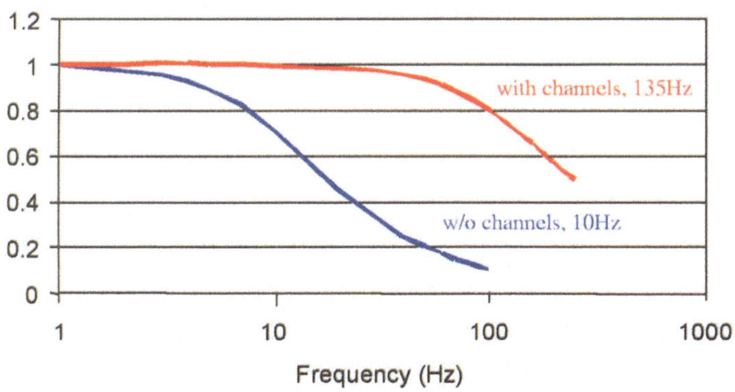

Fig. 5. Frequency response of the acceleration sensor with and without air flow channels.

The geometry and numbers of channels influences the damping behavior and allows to adjust the mechanical cut-off frequency of the device within wide limits by varying the design without the need of an additional lithography step. The experimental data indicate that cut-off frequencies in the range from 5Hz to 400Hz can be achieved. A further increase of the bandwidth could be achieved by enlarging the distance of the electrode gap.

In figure 5 the frequency response of two sensors with and without air-flow channels is shown.

3.3 Offset Temperature Dependence

One parameter of major concern in safety-relevant systems is the offset stability of the implemented sensors. A low-g sensor in an IMU has to be a real DC sensor measuring absolute acceleration values (in contrast to e.g. airbag accelerometers, detecting the impact as a changing acceleration). Any offset drift will be interpreted as a change of acceleration producing a reaction of the corresponding system (brakes in an ABS system, steering in a drive-by wire system). The relevant temperature range for automotive applications is typically from -40°C to 120°C.

Due to the differential readout principle and to the highly symmetric design, a high offset stability could be achieved in the presented device (Fig. 6).

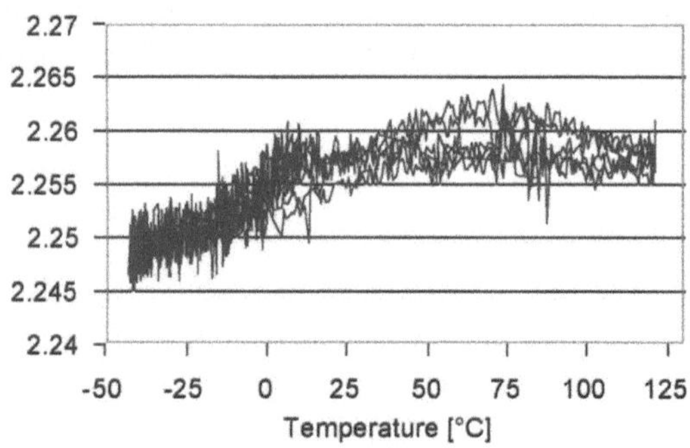

Fig. 6. Offset drift vs. temperature.

In Fig. 6 the offset-drift over temperature is displayed. Several temperature cycles were performed with a temperature gradient of 1.5K/min. The sensitivity of the measured device was 1V/g. The measured drift of less than ±10mg over the full temperature range was well within the specified value of ± 50mg.

4 Multi-axial Low-g Sensor

For future applications a major interest lies in the integration of several sensitive axes on one single chip in order to reduce size, cost, and alignment errors. The presented concept allows the integration of two or four sensing elements in a x/z or in a x/y/z configuration, respectively.

The basic idea as shown in [5] is a sensing element with a sensitivity axis which has an angle with respect to the wafer surface. This is realized by an asymmetric suspension of the torsional beams with respect to the center of mass. In this way, the sensing element has a sensitivity in x-direction (parallel to the wafer surface) and z-direction (perpendicular to the wafer surface) as well. A design with equal sensitivities in x- and z-directions results in a sensitivity direction of 45° with respect to the wafer surface. For special applications it is also possible to implement a sensing element with different specified sensitivities in the different directions (resulting in a "misalignment angle" other than 45°).

By using two sensing elements rotated by 180° with respect to each other two signals U1 and U2 are obtained from which the corresponding accelerations ax and az in x- and z-direction can be calculated (figure 7).

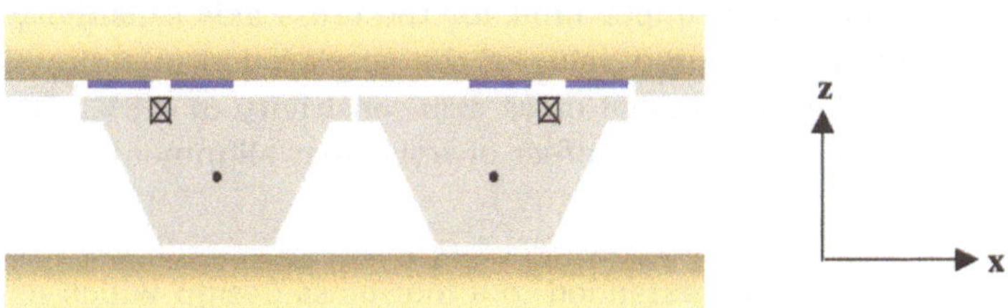

Fig. 7. Cross section of a dual-axis accelerometer.

In the case of a 45° angle of the sensitivity axis with respect to the wafer surface, this is simply done be the following equations (U1 and U2 are the output signals of the two elements, both sensing elements have identical sensitivity S):

$$a_x = \frac{1}{2S} \cdot (U1 - U2)$$

$$a_z = \frac{1}{2S} \cdot (U1 + U2)$$

Figure 8 shows measured results for the x/z dual axis acceleration sensor. The sensor is rotated in the field of gravity with z as its rotation axis. The signals

U1 and U2 from the two sensing elements are displayed as well as the calculated acceleration values a_x and a_z.

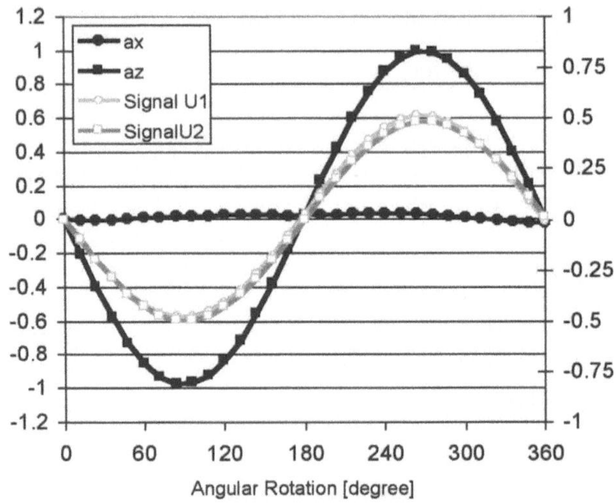

Fig. 8. Measured results for the x/z dual-axis acceleration sensor being rotated in the field of gravity around its z-axis.

From this measurement an upper limit for the cross axis sensitivity can be given. The maximum measured value of the acceleration in z-direction was less than 25mg giving a maximal cross axis sensitivity of 2.5%. This value, however, most likely includes an effect of imperfect alignment of the measurement setup.

The above concept is easily extended to a four-mass design which exhibits a pair of x/z and a pair of y/z sensing elements (Fig. 9). With this configuration the acceleration in all three directions x, y, z can be calculated with the shown principle. All sensors have identical layout and are monolithically integrated in plane on one chip.

A further advantage of this design is that the z-signal can be calculated in two redundant ways, using the x/z pair and the y/z pair, respectively. By comparing these two values, a built in ongoing self-test capability is available, covering all three axes of acceleration (not only the z-component). It returns an error signal as soon as one of the acceleration components has a wrong value.

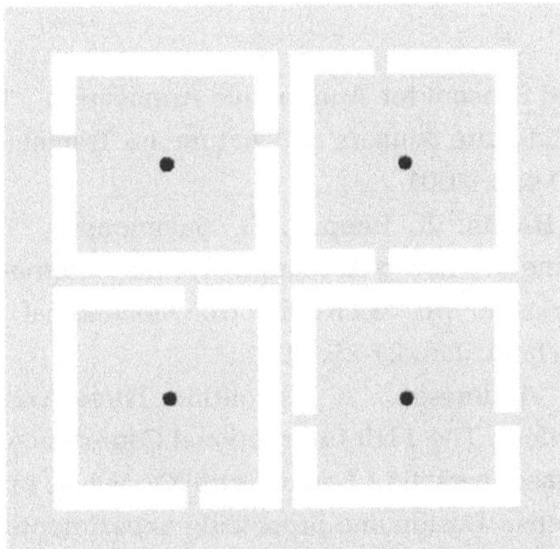

Fig. 9. Schematic top view of a tri-axis acceleration sensor with self-test capability consisting of one pair x/z and one pair y/z dual axis elements.

5 Conclusion

A new low-g acceleration sensor suitable for an inertial measurement unit was presented. The design allows a customization of the mechanical cut-off frequency between wide limits (5Hz to 400Hz). A small bandwidth is necessary for some vehicle application in order to suppress vibrational noise. The measured values for linearity error and offset drift over temperature are well within the specified limits of the target application. The sensor principle can be extended to realize planar dual axis (x/z and y/z) and tri-axis elements (x/y/z) based on the same fabrication technology. Due to the four mass design of the tri-axial sensor, an ongoing self-test capability is available.

References

[1] H.Kuisma, "Inertial Sensors for Automotive Application", The 11th International Conference on Solid-State Sensors and Actuators, Transducers '01 / Eurosensors XV, Vol. 1, pp. 430-433, 2001

[2] D. Lapadatu, S. Habibi, B. Reppen, G. Salomonsen, T. Kvister, „Dual-Axes Capacitive Inclinometer / Low-g Accelerometer for Automotive Application", IEEE International Workshop on Micro Electro Mechanical System (MEMS '01), Interlaken, Switzerland, Jan. 21-25, 2001

[3] H. Rödjegard, G. Andersson, "A Monolithic Three-Axis Accelerometer with Symmetric Properties", The 11th International Conference on Solid-State Sensors and Actuators, Transducers '01 / Eurosensors XV, Vol. 1, pp. 442-445, 2001

[4] R. Puers, S. Reyntjens, "Design and processing experiments of a new miniaturized capacitive triaxial accelerometer", Sensors and Actuators, A 68 (1998) 324 - 328

[5] M. Aikele, M. Rose, R. Gottinger, U. Prechtel, J. Schalk, T. Ohgke, M. Weinacht, H. Seidel, „A New Generation of Micromachined Accelerometers for Airbag Applications", in Advanced Microsystems for Automotive Applications 99, D. E. Ricken and W. Gessner, edts., Springer, Berlin 1999, pp. 251-260

Konrad Kapser, Matthias Aikele, R. Gottinger, B. Hartmann
ContiTemic
MEMS Development Center
Eugen-Sänger-Str.
85649 Brunnthal-Nord/München
Germany
konrad.kapser@temic.com

Roland Burghardt
Continental Teves
Guerickestr. 7
60488 Frankfurt/Main
Germany
roland.burghardt@contiteves.com

H. Seidel
Saarland University
Chair for Micromechanics
66123 Saarbrücken
Germany
seidel@lmm.uni-saarland.de

Keywords: accelerometer, low-g, multi-axial sensor, inertial measurement unit, bulk-micromachining

Reliable, High Quality Infrared Sensors Have Found Their Way Into Automotive Climate Control

Roger Diels, Melexis NV
Domenic Pompei, Melexis Inc

Abstract

Infrared temperature sensors, used for many years in process control and manufacturing, are now accepted for use in passenger vehicles. Production, reliability and cost issues had to be solved. In addition, complex signal conditioning and calibration issues had to be addressed. As a result, automotive engineers are becoming more familiar with the science and technology of IR sensing and its potential benefits to automotive systems.

1 Background on Infrared Temperature Sensors

Infrared temperature sensors have the unique ability to measure the temperature of objects without touching them. In addition, IR sensors have very fast response, making non-contact temperature measurement in less than one second.

Infrared sensors have been used extensively in manufacturing processes for over 30 years. They were adopted because they allowed direct temperature measurement of products that are moving during manufacture. Some examples of production products that are dominated by infrared temperature measurement are: the manufacture of paper, cloth, printed materials, and similar.

The important point is that IR sensing allows manufacturers to directly measure and tightly control product temperatures as they undergo important heating and cooling stages in manufacturing processes. The IR measurement is correct on those products regardless of the means used to heat or cool them, whether it is forced convection, radiation, or conduction through contact with rollers. Fast, accurate IR temperature measurement and control systems now allow minimum waste, and maximum throughput, quality, and manufacturing profits.

Prior to the use of IR, such manufacturing processes depended on indirect measurements: air temperature, air velocity, roller temperatures, and so on, in an attempt to try and control product temperatures as tightly as possible. Climate control of motor vehicle occupants is exactly the same.

2 Automotive Climate Control

2.1 Traditional Method – Indirect Measurement

To measure and control cabin temperature, automotive climate control manufacturers traditionally have used air temperature thermistors, and more recently, a combination of thermistors and solar sensors. Thermistors are almost universally mounted behind the dashboard. Solar sensors are mounted where they have a clear view of the sun and sky, located where the front windshield meets the dashboard.

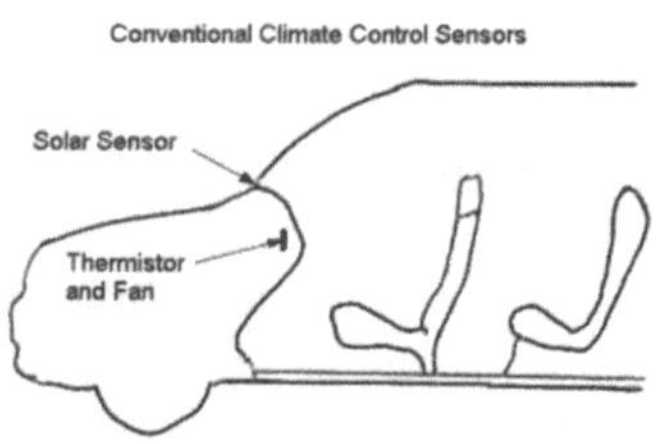

Fig. 1. Conventional Climate Control Sensors.

A thermistor, like any other conventional contact thermometer, can only measure its own temperature. On its own, this sensor is somewhat ineffective when mounted behind a dashboard. Thus an aspirating fan is installed to draw sample cabin airflow over the hidden thermistor.

Measurement accuracy of actual cabin air temperature conditions is less than ideal with this design, though, for three reasons: (1) the sample air is drawn from directly in front of the dashboard and can vary significantly from the air surrounding the occupant in the cabin; (2) the temperature effects from the neighbouring objects behind the dashboard can significantly affect the thermistor; and (3) the sensors are slow, often taking minutes for full response.

Also, because the air thermistor sensors are very slow to respond to the effects of radiant heat from the sun pouring through the glass surfaces, solar sensors are used. These sensors can measure sunlight intensity and direction to help predict the approximate radiant heat effects on the cabin and occupants.

To approximate the temperatures felt by the passengers, distinct vehicle types generally require a unique algorithm based on these local indirect measurements, amount of glass surfaces, and other cabin variables. Extensive testing is sometimes required to achieve reasonable predicted performance. Despite these efforts, there can be significant deviations between what the system delivers and the temperatures felt by the passengers.

2.2 Newer Method – Direct Measurement by IR Sensing

Infrared sensors allow a direct, faster, and more accurate measurement of passenger comfort. The principles are the same as those used with IR to control temperatures of moving products during manufacturing. The IR measurement is correct regardless of the means used to heat or cool the target, whether it is forced convection, radiation, or conduction.

IR sensors, aimed at the clothing of passengers in a vehicle, make a direct measurement of the surface temperature of the clothing (or any solid objects in the cabin). The surface of clothing is extremely sensitive to both airflow and solar radiation inside the cabin. Both effects are measured directly and combined into a single temperature output from the IR sensor that can be conveniently used by the climate control system.

Monitoring of the intensity and direction of the solar load are irrelevant when using IR sensing. For the target area that the IR sensor is monitoring, the solar load is automatically accounted for. In effect, the passenger clothing becomes a large combined, air and radiant temperature sensor representing the best possible direct measurement of passenger comfort.

The IR technique is also quite tolerant of the area sensed and its impact on the climate control system. As the sensor averages the temperature of the entire area it sees, seat surface and other interior surfaces, if seen by the sensor, also provide reliable temperature feedback information for the system. As such, a good approach is simply aiming the sensor in the direction of the occupant(s) where control is desired.

IR sensors also have numerous mounting options as can be seen in the following figure:

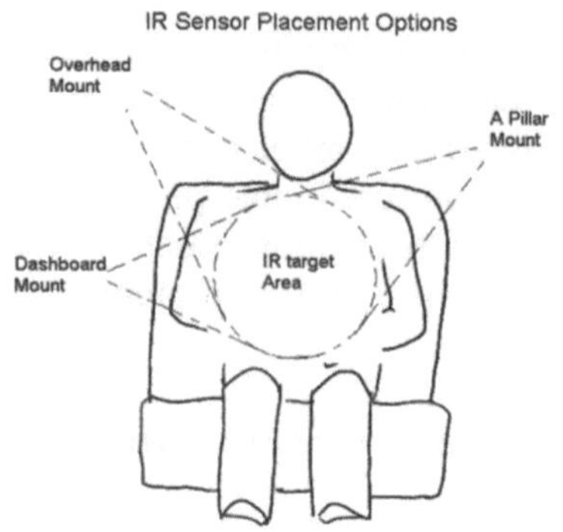

Fig. 2. IR Sensor Placement Options.

2.3 Some Common Questions About IR Sensing For Automotive Use

Q. Do clothing colors affect IR temperature measurements?

A. No. Clothing (and skin, and all common seating surfaces) are virtual black-bodies for the long wavelengths sensed by IR sensors for vehicles.

Q. What if the occupant is not wearing any clothing?

A. Then the sensor will see the temperature of the skin of the occupant and use it as a temperature sensor for control. It makes no difference. Clothing, skin, and seating surfaces can be used reliably for IR sensing.

2.4 Other Comfort Performance Improvements

With IR sensors, passenger comfort is measured and maintained regardless of vent airflow direction. Vent airflow can be aimed directly at a passenger or aimed away into the general cabin space. The effects of either airflow are automatically accounted for by the IR measurement. This feature is impossible with current air thermistor and solar sensor systems.

In very cold, or very hot, vehicle ambient starting conditions, IR sensing allows the climate control system to heat or cool an occupant with no uncomfortable temperature overshooting or undershooting. There is absolutely no time lag in the sensing system. System fans and supply air temperature can be regulated precisely to deliver maximum heating or cooling with no discomfort to the occupant.

When used in discrete zone control, IR sensing even accounts for situations when an occupant is perspiring or has wet clothing. Perspiration on clothing, or on skin, combined with airflow from vents causes evaporative cooling effects. IR sensors see and measure the evaporative effects directly. No control system algorithms are needed to accommodate this feature.

2.5 Comfort Control Algorithms

Complex comfort algorithms currently used in climate control systems to accommodate the limitations inherent in indirect measurements by the thermistor and solar sensor combinations can be reduced to simpler temperature control system instructions by using IR.

2.6 Zone Control

IR sensors allow true vehicle multi-zone climate control. Sensors can be placed anywhere in the cabin where a zone is required. Or, for lower installation costs, multiple zone sensors can be mounted in a single location. Using an overhead console mounting location allows a multizone module to reach all desired zones in most vehicles. This is especially useful for current popular SUV and Minivan vehicles. This approach drastically reduces the wiring cost, since the mounting location of the sensor is decoupled of the physical location of the object to measure.

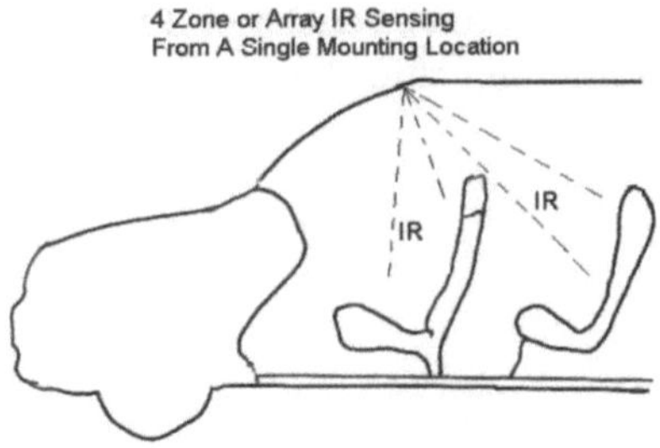

Fig. 3. Zone Sensing From a Single Location.

3 The IR Sensor Design

3.1 Background

Until recently, thermopile sensors were constructed primarily by vapour deposition on mylar (a form of polyethylene) substrate, and typically packaged in a TO can filled with Xenon gas. Such sensors have a very large mV output per deg C temperature measurement, and thus seemed attractive to use with existing inexpensive amplifiers and signal conditioning schemes for automotive use. Such detectors were used in the first mass production of IR sensing for automotive use. However, the attempt to use such IR sensors in vehicles led to unacceptable failure rates. The primary cause of failure was the mylar thermopile element. In summary, it was found that traditional manufacturing methods for IR sensors could not be scaled for the high volume, high reliability necessary for automotive use.

IR sensors made from silicon, however, can meet the reliability requirements for automotive use. Silicon thermopiles, when manufactured at costs low enough for automotive use, generate very low level output signals. As a result, signal amplification and calibration become difficult, if not impossible, with "off the shelf" generic low cost components.

3.2 IR System Solution

The proposed IR system solution, capable of coping with automotive mass production and reliability requirements, consist of 2 basic components: an IR sensor chip and a signal-conditioning chip. They are shown in a sample assembled form in figure 4.

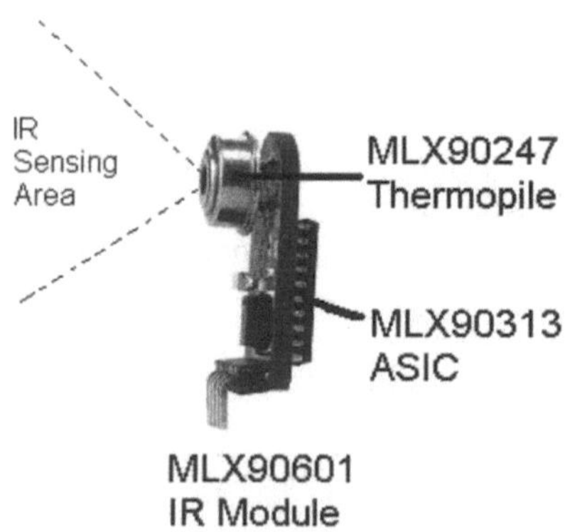

Fig. 4. MLX90601 Module.

Silicon thermopiles, when manufactured at costs low enough for automotive use, generate very low level output signals. As a result, signal amplification and calibration become difficult, if not impossible, with "off the shelf" generic low cost components.

Field-of-View Is Configurable

The optical field of view of the module can be configured to meet virtually any vehicle requirements with the use of apertures and lenses. The programmable amplifiers automatically accommodate added optical systems during factory calibration process. Circuit re-design costs and qualification costs are not necessary for adding such options.

3.3 The Thermopile

The thermopile die is manufactured in a full cmos compatible process, with KOH micromachined backside processing. . As such, Melexis is able to achieve high reliability with well-established CMOS processes that have been proven over time for automotive reliability.

Standard CMOS technology applies to the thermocouple junctions fabricated on the thin membrane as well as the n-well thermistor structure embedded in the bulk silicon.

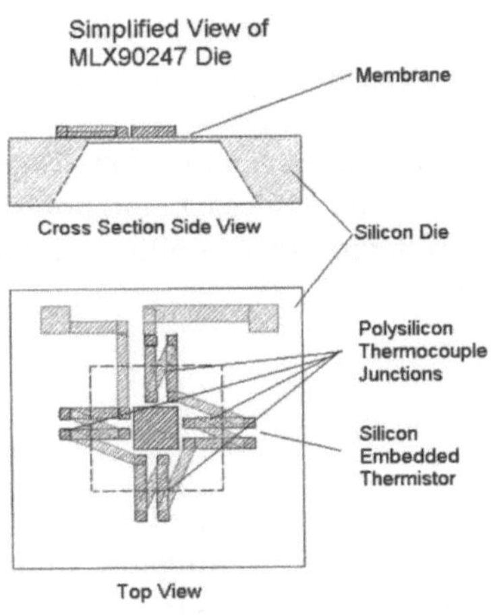

Fig. 5. Simplified View of MLX90247 Die.

Here is a brief explanation of how the silicon thermopile is constructed and operates:

After manufacturing the appropriate structures onto standard wafers, the backside of the wafer is selectively etched to remove silicon. The removal of silicon leaves a thin membrane in the centre of the silicon die. Please see figure 5.

The thin membrane structure absorbs heat radiation as it is aimed at remote objects. As such, the membrane temperature changes rapidly relative to the bulk silicon edges. On the membrane are a series of 40 thermocouple "hot junctions", made of two different types of polysilicon. The "cold junctions" of the polysilicon are located on the bulk silicon die edges. The temperature difference between the membrane and the bulk silicon die generates microvolt DC signals. That microvolt signal represents a temperature difference between the sensor and whatever target it is aimed at. Typically, the sensors generates 40µV signal for a 1 °C difference between the object-to-measure and the sensor itself (at room temperature).

However, as in standard thermocouples, a thermopile only measures temperature differences. Thus an absolute temperature reference sensor must be added. For highest accuracy and reliability, an absolute temperature sensor right onto the die, next to the "cold junctions" of the thermopile has been added.

By design, this technique allows to use high reliability standard wire bonds to connect both the thermopile and thermistor inside any package of choice.

This is in contrast to traditional thermopile construction where separate thermistor devices are installed. The devices, being non-silicon, are difficult to connect with standard silicon device wire bonding equipment, and thus interconnect reliability is unproven.

3.4 The Signal Conditioning Chip

A signal conditioning chip has developed to interface with Melexis thermopiles as well as thermopiles and thermistors from other manufacturers. (It can also be used for other types of sensors requiring signal processing, calibration, and linearization of two non-linear low-level signals.)

The thermopile output is in microvolts, and is self-generating. The on-chip thermistor is a resistance change device that requires a very stable current supply to generate an output voltage. In addition, the current supply must be low enough to prevent self heating of the thermistor and the silicon die.

Manufacturing sensitivity tolerances of the output of thermopiles and thermistors can easily go to ±40%. To cope with such variations the interface chip uses two separate programmable chopper-stabilized amplifier chains. The programmable amplifiers allow both sensor signals to be brought into a ±6.5% tolerance before entering the digital signal processing area.

After amplifier calibration, the voltages go into a 12 bit A/D converter for digital processing.

Final calibration and linearization are performed with 12 bit processing to assur high accuracy and tight output signal interchangeability of IR modules.

For more technical details about the signal conditioning functions, we refer to the appendix.

3.5 Outputs Options

After the in-line calibration, the IR module will act as an ideal sensor (see appendix). The basic function only requires 3 pins: VSS, VDD and the IR output. An additional output can be used to monitor the temperature of the sensor itself.
For this output pin(s), various options are available, to allow the customer to optimal connect the IR module to the rest of his system:

- Analog Output with 8 bit precision. The IR output can be relative to VDD, or absolute (0-4.5V). An internal voltage reference better then 15ppm over the full temperature range is included.
- PWM output: 10 bit precision. Extensive error handling built in (signalling overflow conditions on ambient and object temperature range).
- SPI interface: 12 bit precision. In this mode, the microcontroller of the customer can have full access to all asic internals if desired.

4 Module Specifications and Design-Ins

4.1 Calibration and Configuration

A dedicated IR calibration line has been designed, where customer specific requirements for accuracy, temperature ranges, and signal output configuration are implemented into the modules. The line is flexible enough to allow customers to choose an appropriate accuracy vs. cost level for implementation of IR sensing.

4.2 Accuracy

Depending on temperature range specified, and cost level desired, an accuracy of up to 0.5°C can be achieved by the module for normal climate control functional temperature ranges.

4.3 Quality Tracking

Every module, right from wafer level (for the interface chip), gets a unique EEPROM code. This code is traced through the full assembly and calibration cycle. As such, the full history of every module can be traced.

4.4 Reliability of the Module

The assembled devices pass all relevant automotive qualification testing:

- Life test: Temp. cycling -40°C/+115°C, under power and humidity cycling 2000 h - 500x.
- Thermal shock (unbiased): -40°C/+115°C, 500 cycles.
- Temp humidity cycling (unbiased): -10°C/+65°C, 240 hours, RH>90%.
- Low temperature storage (unbiased): -40°C, 1000 hours.
- High temperature storage (unbiased): +125°C, 1000 hours.
- ESD protection (AEC Q100 002): 4KV.
- EMC Radiated Immunity (according IEC61000-4-3 (1995), TEMcell, 150kHz - 1GHz): 100V/m.
- EMC emission (ac. European Requirement for Automotive Subsystems 95/54/EC A7&A8).
- Drop test.
- Load dump.

4.5 Production Status

The IR Module as been described, has been accepted for automotive production. Automotive production began in March 2003. For 2003 it will be used in two vehicles: one vehicle model is for the Japanese market only, the other vehicle model is for the US market only. The Japanese market vehicle is a Single Zone system. The US market vehicle is a Dual Zone system. Additional vehicle model implementations are being planned.

5. Other IR Sensing Applications

5.1 Automatic Windshield Anti-Fog Use

Some advanced climate control system designers are also exploring the use of humidity sensors to further enhance cabin comfort and energy efficiency by controlling humidity [1].

When humidity sensing is adopted, the addition of an IR sensor, aimed at the windshield, allows for implementation of Automatic Anti-Fog sensing and control. The humidity sensor measures the RH (relative humidity) in the cabin. From the RH measurement, the dew point temperature is calculated. The IR sensor measures the interior surface temperature of the windshield. The climate control system can then decide how much heat and dry air to supply the cabin and windshield to keep the interior windshield temperature above the dew point. This can decrease energy waste and increase passenger comfort. The IR Module, as designed, can already accurately track inside windshield temperature for the full range of automotive temperatures.

5.2 IR Arrays

Building on the base sensing and signal conditioning chips developed for the IR Module, Melexis is also currently developing automotive grade infrared arrays of up to 10 x 10 pixels. IR arrays will allow future vehicle climate control systems to not only have as many control zones as desired, but will also allow climate control systems to identify how many passengers occupy the vehicle and their placement. Such smart climate control systems will allow future vehicles to be more energy efficient in delivering heating and cooling to the cabin.

This development is based on some unique on-chip signal treatment for the arrays, for which patent application is pending.

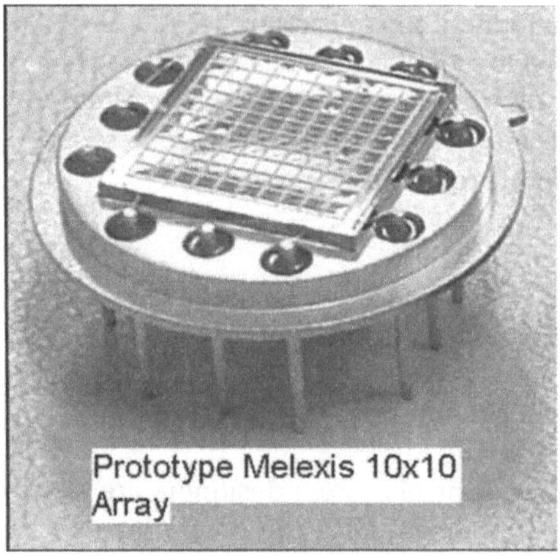

Fig. 6. Prototype Infrared Array Sensor With On Chip Multiplexing.

Melexis has chosen to implement arrays using thermopile technology rather than bolometer technology that is currently being investigated by some researchers. The advantage of thermopile arrays over bolometer arrays, is that thermopiles arrays can work under normal atmospheric pressure, while classical bolometers require a vacuum. Fabrication with a vacuum package is expensive for mass production automotive use. In addition, maintaining the integrity of a vacuum over the expected 10 year life of a motor vehicle would be extremely difficult. Any breach, whatsoever, in the integrity of the vacuum package can cause a sudden "catastrophic" signal error in the device.

5.3 Others

IR Sensors are already well integrated in gas sensing applications. Every gas has its unique absorption spectrum for IR radiation, detectable by an IR sensor. This market started with industrial applications, but automotive applications as CO, CO_2 and H_2 sensing are under investigation.

Racing – Melexis is supplying IR modules to multiple racing teams for use in tire temperature monitoring for suspension set-up. Three sensors per tire are used to sense temperature distribution over the tread surface. Even surface temperature equates to maximum tire grip and proper suspension set-up for a particular race circuit.

6 Further Evolution

To sustain the future needs of IR applications, Melexis is designing the next generation IR interface chip. This new chip will enhance the capabilities of the existing interface chip.

The most important upgrade is towards array sensor handling: the new interface will be able to convert 10 pixels simultaneously, reaching an array refresh rate up to 100 pixels per second. By software, customers can select emphasis towards more accuracy vs. speed, or vice versa.

This chip will also incorporate a LIN interface, based on a dual task microprocessor.

7 Conclusion

It has been shown that IR technology, based on Si sensors and signal processing, has become mature and cost-compatible. The concept of pre-calibration within the production line has simplified the design-in of IR sensors. The maximum use of cmos integration, both for the sensor as the signal treatment, has secured the reliability of the product. As such, first IR sensors are now going into production in Climate Control applications. Other applications will follow.

Acknowledgements

The authors wish to thank the IR Melexis development teams in Tessenderlo and in Sofia, as well as Jens Urban and Uwe Schwarz of XFAB, for their contribution to this development.

Appendix: Technical Details of the Signal Conditioning Chip

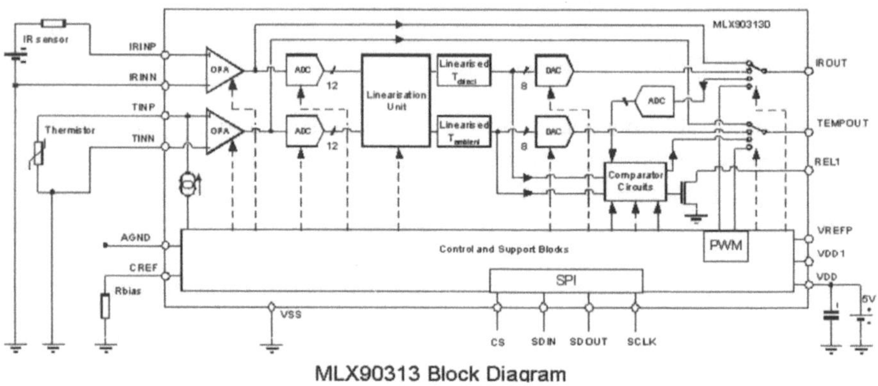

Fig. 7. MLX90313 Block Diagram.

A.1 Supply Regulator

A supply regulator is on-chip and can be powered by an automotive ignition supply line (7V-80V). The chip can withstand SAE standard ignition transients. The resulting voltage of the regulator is available on VDD (5V ±300mV). VDD can source up to 20mA to external circuitry. The chip can also be supplied directly with a 5V regulated supply on pin VDD.

A.2 Band-Gap, DAC and ADC References

The on-chip trimmable, curvature compensated band-gap circuitry provides a stable reference level (less than 10ppm per deg C) for several derived reference potentials used for normal operation in MLX90313.

The analog ground at AGND is directly derived from this band-gap voltage. The output voltages from both amplifier chains are relative to this potential. The AGND reference can be trimmed internally to (2.5V ±20mV). The regulator at AGND pin can be switched off to minimize the current consumption. The pin can be also used as external input for the internal 12-bit ADC.

The reference voltages for ADC and DAC are also derived from the band-gap. The DAC reference is available at pin VREFP. The 90313 DAC reference voltage can be programmed on chip to one of the following values: 2, 2.5, 3, 3.5, 4 and 4.5 V.

Depending on the customer application, one can program the linearised analog outputs for object and ambient temperatures providing absolute voltage/temperature dependence (see figure below) . In its simplest form, the 90601 module is a 3 pin devices (vss, vdd, IRout) representing an ideal (linearised) temperature sensor.

The internal regulator for Vrefp voltage can be switched off to minimize the consumption (if linearised analog output is not in use) or to use externally supplied reference for DAC reference in range 2-5V.

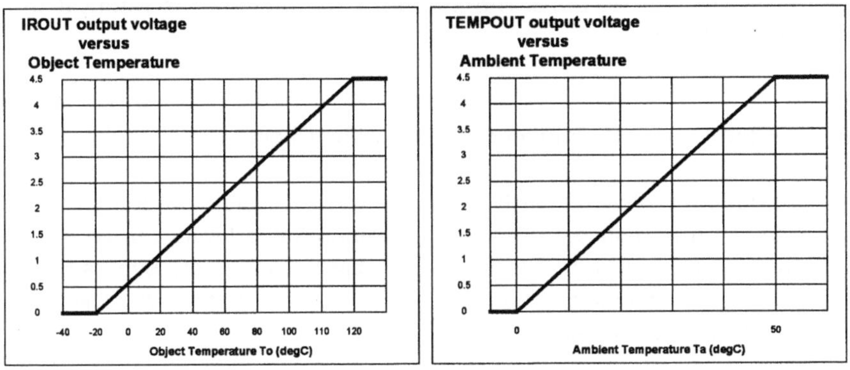

Fig. 8. IR and TEMP analog outputs, using the internal voltage reference.

A.3 IR-Amplifier Chain

MLX 90313 uses two chopper amplifiers for IR chain and Temp. chain.

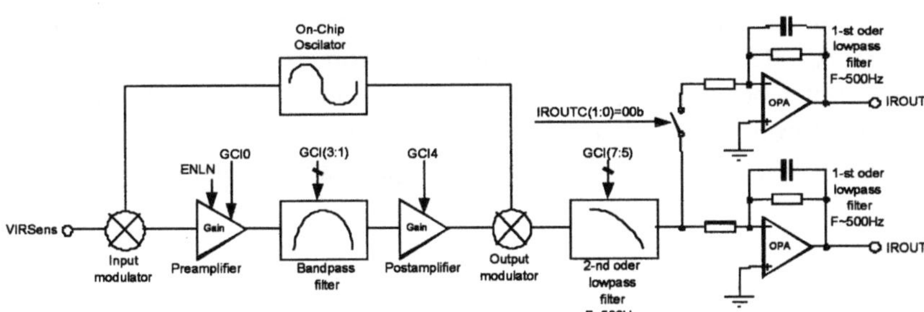

Fig. 9. IR analog amplification chain.

The chosen chopping architecture eliminates the 1/f noise of the input amplification stage. The tracking bandpass filter limits the folding back of the white noise portion.

The MLX90313 is available with gain settings for the IR-amplifier chain ranging from 55 to 5500. Any gain between the above mentioned limits can be obtained within an accuracy of ±6.5%. The amplifier input-referred white noise level is below $23nV_{rms}/\sqrt{Hz}$. In the application with IR-sensors, with output resistances of 50kW typical, the total system noise, however, will depend mainly on the noise of the sensor and will rise up to $45nV_{rms}/\sqrt{Hz}$

The common mode input range of the amplifier is -100mV to VDD-3V. As such, it allows the IR sensor to be referenced to vss, solving esd, emc and high temperature leakage problems (Melexis patent pending). The output range of the amplifier is 0V to VDD-0.2V. The output of the amplifier is referred to the potential on AGND.

A.4 On-Chip Thermistor Temperature - Amplifier Chain

The MLX90313 is available with gain settings for the Temp-amplifier ranging from 5 to 50. Any gain between the abovementioned limits can be obtained within an accuracy of ±6.5%. It is also possible to completely bypass the temperature amplifier and force the input signal directly to the ADC. The amplifier input-referred white noise level is below $400nV_{rms}/\sqrt{Hz}$.

The common mode input range of the amplifier is -100mV to VDD-3V. The output range of the amplifier is 0V to VDD-0.2V. The output of the amplifier is referred to the potential on AGND.

A.5 Current Reference

The thermistor (or any other sensor) connected to TINP can be biased with a current source. The bias current is mirrored from the current through the external resistor between CREF and VSS. The voltage maintained at the CREF pin is derived from internal band-gap voltage. The typical value of the voltage at CREF pin is 2V. The current mirror ratio can be set between 1/7 and 1.

A.6 A/D

The MLX90313 contains a 12-bit internal analog to digital converter. Real 12 bit conversion is achieved by a fully differential analog part of the converter. The input amplifier of the ADC has a fixed gain of 3. Automatic calibration is implemented in the background, which allows precise conversion in a very wide temperature range. The ADC sampling rate is 7k samples/second. The reference voltage for the ADC is normally a scaled version of the internal band-gap reference and is fixed to be 2.5V. Alternatively it can be configured to work with an external reference potential at the AGND pin by setting the appropriate bit in the configuration register. Internal ADC can work with references down to 1V keeping the 12-bit resolution.

The ADC contains an interface circuit to scale and offset the analog signals in order to make the most efficient use of the available resolution. After amplification the IR and Temp sensor signals are referred to AGND (typ value 2.5V). The additional offset is scaled version of the AGND.

A.7 D/A

An 8 bit digital to analog converter can be used to output the data for the linearized Tobject- and Tambient signals. The DAC can work with internal programmable reference voltage, which is available at pin Vrefp as well as external one. The result from DA conversion is stored on hold capacitors and buffered. The signals are available at IROUT and TEMPOUT respectively, if the appropriate bits are set in the configuration register.

A.8 PWM Output Implementation

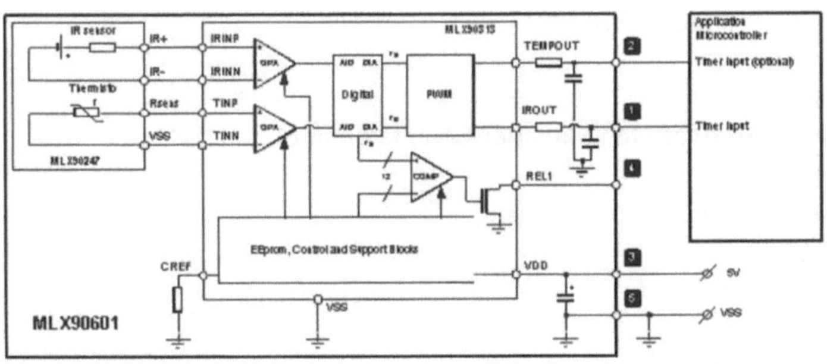

Fig. 10. PWM Block Diagram.

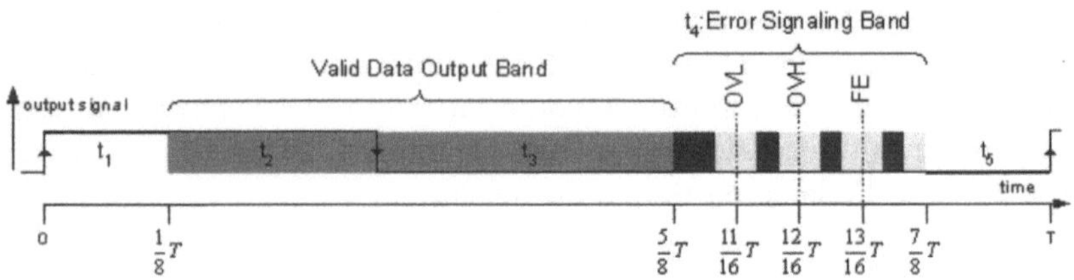

Fig. 11. PWM Output.

As an alternative to analog output, the 90601 offers PWM output. This approach also only requires a 3 pin connection (vss, vdd, pwm_out). The advantage of PWM is that the user does not need an internal AD convertor. Although some error flagging is available with the analog output mode, the PWM mode offers more extensive error signaling (out of range position of ambient temperature, object temperature, and more).

References

[1] Ryuji Hayashi, Yuichi Kajino, Carlos Christopher Arias, Denso Corporation 2002-21-0053 Advanced Thermal Comfort in the Passenger Vehicle Compartment Paper presented at Convergence 2002 Proceedings)

[2] Website Melexis related to IR products: http://www.melexis.com/IR

Roger Diels
Melexis NV -Business Unit Manager Climate Control
Transportstraat 1
3980 Tessenderlo
Belgium
rdi@melexis.com

Domenic Pompei
Melexis Inc - Sales & Marketing Mgr IR
41 Locke Rd, Concord
NH 03301
USA
dpm@melexis.com

Keywords: Infrared, Climate Control, Automotive, IR, Temperature, Comfort

Miniature Servo Drive System for Air Condition Controls and Test Rig for Dynamic Testing of Small Torques

F. Poehlau, Oechsler AG

Abstract

WAVE DRIVE®, is a microsystem gear unit made by injection molding of thermoplastics – as opposed to the more commonly used silicon - and at the same time presents a novel gear principle which makes it possible to achieve extremely high reduction ratios with very few components in a very compact unit. Utilizing WAVE DRIVE®, a fluid duct regulating flap has been developed that completely contains all its drive elements within the shaft of the flap, thus eliminating the need for an external actuator. A sensor determines the angular position of the flap and completes closed-loop control. In order to enable testing of the small gears for performance characteristics and durability, a special test rig needed to be developed. Special emphasis was put on testing of torque while the gear system is in operation. Most other test systems for small torques can only determine them while the system is at standstill.

1 Introduction

Today's automobiles are being equipped with a rapidly growing number of sensor and actuator systems. Besides a broad variety of sensors, a modern luxury car has about 100 to 150 electric motors and other actuators that drive comfort accessories (seats, mirrors etc.) as well as safety (like adaptive front-lighting, chassis adjustment) and engine related systems (e.g. variable camshafts). Weight and compactness are of prime concern, so that miniaturized systems are becoming increasingly important.

Small electric motors achieve their power at very high speeds, but output only low torque. In order to make the motors usable for performing adjustment functions, a gearbox is needed to adapt them to each specific application. For small motors, very high reduction ratios are necessary, which are usually achieved by a series of spur gears, worm or multistage planetary gears. Each of them has their specific drawbacks, as shown in figure 1.

Spur gear	*Planetary gear*	*Worm gear*
High number of gears Large space needed Input and output usually not concentric Low torque capability	Several stages necessary Long High number of parts Planet carriers necessary Complex assembly	Input and output not parallel (usually 90°) Low efficiency

Fig. 1. Comparison of gear principles used in small actuators.

Depending on the requirements, a more compact gear unit is highly desirable especially for small actuators. Additionally, concentric in- and output and a high torque capability may be sought after. All these can be realized by a gear system based on the strain wave gear principle [1].

2 The WAVE DRIVE Gear Principle

WAVE DRIVE®, is an innovative gear unit that makes it possible to achieve very high reduction ratios (up to 5000:1) with very few components (only three to five moving parts depending on the executed version of the principle).

It consists of an elliptical core, called Wave Generator, which is directly coupled to an input shaft. Because of the elliptical shape of the core, plungers of a Plunger Wheel (shown in white in figure 2) surrounding the wave generator are caused to move outward and inward in a wave-like pattern. The Plunger Wheel in turn conveys its elliptical deformation to a flexible ring bearing an outside gear, called the Flexring (blue in figure 2). This gets in and out of mesh with a cylindrical fixed ring (red in figure 2). Because the Flexring has a smaller circumference as the Fixed Ring, it must perform a slight angular displacement with every engagement cycle in order to maintain rolling contact with the Fixed Ring. Thus, a very slow rotation is achieved.

The reduction ratio can be determined by

$$i = z_{flex}/(z_{fix}-z_{flex})$$

with

z_{fix} = number of teeth of Fixed Wheel
z_{flex} = number of teeth of Flexring.

Torque is led out of the system by an Output Wheel (not shown) in mesh with the Flexring. The Flexring and the Output Wheel can also be combined to form a flexpot, further reducing part count and increasing efficiency, albeit at the expense of compromised length.

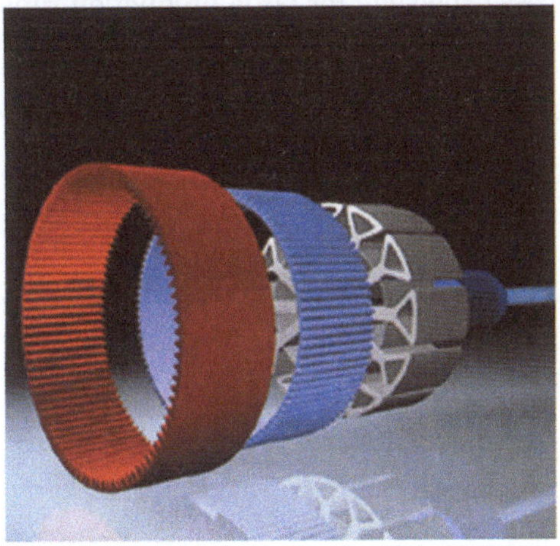

Fig. 2. Principle of Wave drive.

The concentration of high-speed rotation at the comparatively small Wave Generator leads to slow circumferential speeds and makes it possible to use thermoplastic materials for all components of the gear. The heat generated in the frictional contact between the Wave Generator and the Plunger Wheel, despite careful selection of suitable materials (partly modified for tribological purposes) limits the allowable duty cycles, thus predetermines the gears to use in actuating applications. Below certain power levels, the gears may be used continuously. Life expectancy then is around 500h. Typical temperature curves are shown in figure 3. As no values for the microsystem gears have been determined yet, a larger gear is taken as an example.

Because of the wave-like action of the Flexring and deflection of the thermoplastic material under load, a high number of teeth is in engagement simultaneously. This has two beneficial effects: a high power density and high transmissible torque can be achieved because the stress is distributed over a large number of teeth. Up to 30% of all teeth are involved in transmitting torque at any given time, as opposed to usually one or two teeth in conventional gear contact scenarios. Secondly, the large number of teeth in engagement very effectively limits play and thus backlash. A 49:1 gear unit has about 0.2° backlash with no special measures.

The components are manufactured by precision injection moulding. Thus the system is lightweight and can be produced economically. As opposed to similar gears manufactured from metal using complicated processes, this gear is manufactured by injection molding. Even with the necessary precision, this is a very economic process especially for large production runs.

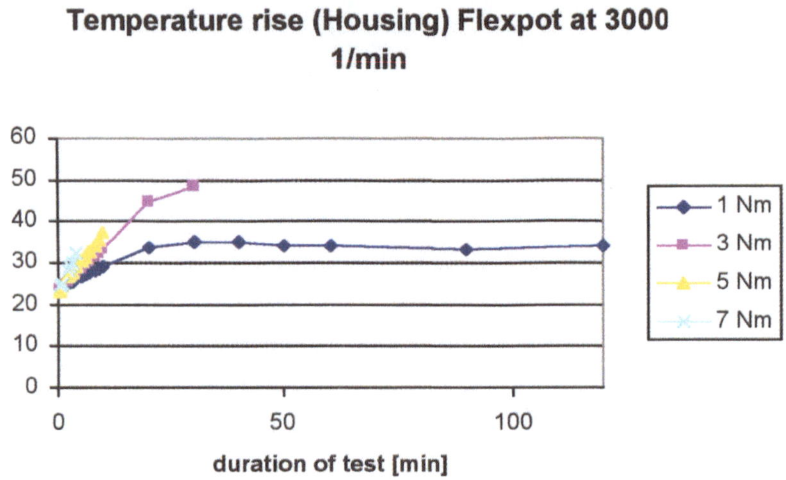

Fig. 3. Temperature curves at the wave generator.

Fig. 4. WAVE DRIVE®, 10mm (without housing).

WAVE DRIVE®, can be combined with electric motors and position sensors to form a compact, disk- or tubular shaped servo drive unit. The system is very easy to assemble because of its low part count and its concentric design. Thus it lends itself to miniaturization and is extremely suitable for use in small actuators and drives where very high reduction ratios need to be realized. The smallest gear units realized to date have a gear diameter (Flexring) of 10mm and a reduction ratio of 20:1 or 400:1. They weigh less than 1 gram with housing or 0,57g without. External diameter is ca. 12 mm. Further miniaturization is planned.

3 Application for Air Conditioning Controls

In most automotive dashboard applications, space is at a premium, as more and more actuators need to be incorporated for air conditioning and other purposes. Conventionally, an actuator for a flap controlling air flow in an air duct is placed on the outside of the air duct, driving the shaft of the flap. The actuators normally comprise a spur or worm gear, thus consuming a lot of space.

The miniaturized WAVE DRIVE®, makes it possible to integrate the drive into the shaft of the flap itself. A 12mm diameter conventional DC motor is combined with a 400:1 gear unit to achieve the rotational speed and torque needed for operation.

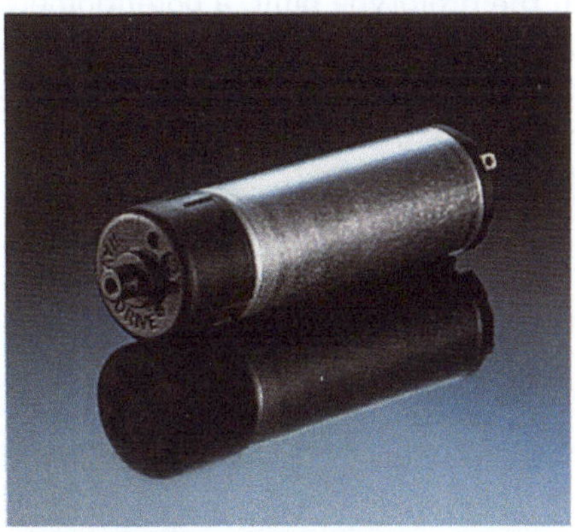

Fig. 5. Miniature Drive with DC motor and WAVE DRIVE®pt.

The diameter of the shaft will increase slightly, but only minimal compromise in flow characteristics of the air duct is necessary. The output wheel of the gear unit is designed to become part of the flap itself. Motor and Fixed Ring are supported against the air duct wall and secured against rotation.

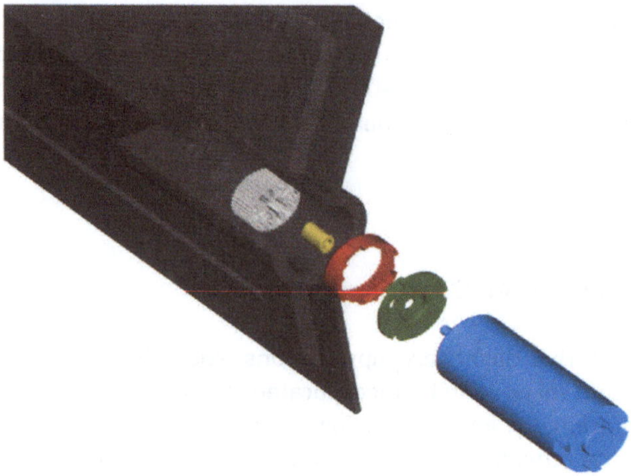

Fig. 6. CAD sketch of drive principle.

Valuable space is saved on the outside of the air duct and a much more compact design of the air conditioning unit becomes possible. The flap also bears a sensor element to determine the angular position, so that exact control becomes possible. In the prototype built, a potentiometer was chosen for ease of design. Other sensors, like inductive or optical units, are of course also possible. Together with a control unit (in this case also a potentiometer), the complete feedback loop of a servo drive has been realized.

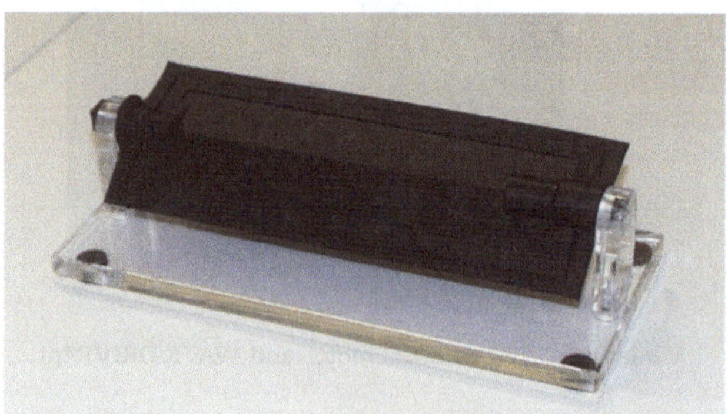

Fig. 7. Air duct flap prototype.

Very similar drive units can be used for actuating passenger vents or other comfort accessories like cupholders. A variation of the gear principle using rollers instead of plungers makes it possible to feature a hollow shaft for passing power wires or glass fibres for optical data transmission through the gear. Also a version as a linear actuator is under development.

Fig. 8. Passenger vent prototype.

4 Development of a Test Rig

A designated test rig was developed and installed. Off-the-shelf equipment could not be used, particularly because the input torques are very small. For long-term efficiency measurements, measuring was specified to be while the motor-gear-system is in motion as opposed to the available static (mainly reaction-torque [2]) methods.

The test rig consists of a brushless motor with attached encoder, a measuring shaft for input torque, a clutch to connect to the gearbox to be tested, another clutch on the output side, a second measuring shaft for output torque and speed, and a brake. Each of the measuring shafts is equipped with a speed output. The brake is of the hysteresis type, excited by current.

Custom-modified torque measurement shafts were used to be able to measure values down to 500µNm. It is possible to perform torque measurements at input speeds of up to 20.000rpm. Long-term measurements are possible up to 15.000rpm. This was only possible by carefully modifying the various components of the optical-bank-like set up in order to reduce vibrations and dirt effects as much as possible.

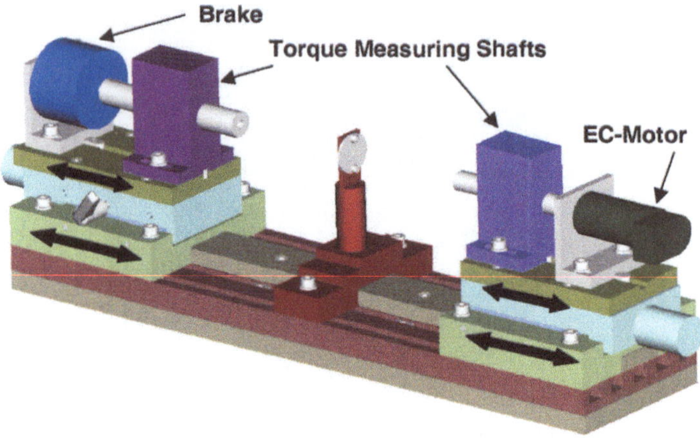

Fig. 9. Concept drawing of the test rig.

High-frequency analyzing software makes it possible to determine average values for speed, reduction ratio, torque (input / output) and efficiency as well as to generate reliable information about running characteristics and noise. This can be used to test the small gears for uniformity or other specific characteristics. The test rig can also be used under controlled climate conditions (-50°C to +150°C, humidity up to 100%) in a specially modified climate chamber. Thus, environmental conditions like in a car's dashbouard or even engine compartment can be simulated.

Fig. 10. Test rig.

	Input side	*Output side*
Min. torque [mNm]	0,5	10
Max. torque [mNm]	30	300
Max. speed [min^{-1}]	20000 (short term)	n.a.
Brake torque [nMn]	n.a.	350
Inertia [$kgcm^2$]	0,044	Ca. 0,199

Fig. 11. Technical data of test rig.

Following the experience collected with this first test rig, a series of further, simplified, rigs for life cycle testing has been developed and installed and is currently undergoing evaluation and calibration.

5 Conclusion and Outlook

WAVE DRIVE®, is a novel approach to solving the problem of achieving high reduction ratios with high precision in small available spaces. Its design with entirely thermoplastic components makes it highly economic. The specially developed test rig makes measurement of very small torques and running characteristics of microsystem gears possible with very high accuracy. It functions with the gears running.

Future applications of the WAVE DRIVE®, can be derived from its specific features. For example, its plastic parts are transparent to magnetic fields and X-rays. This enables the use of the thermoplastic strain wave gears in medical applications like CT machines. The light weight and low cost of the gears makes them very suitable for future use in portable applications, like medical diagnostic or therapeutic devices.

References

[1] W. Musser: Strain Wave Gearing, U.S. Patent Nr. 2.906.143, 29.9.1959

[2] P. Wurmsdobler: Ein Drehmomentenmesssystem für mikro-elektromechanische Rotationselemente in GMM-Fachbericht 33: Innovative Klein- und Mikroantriebe, Vorträge der GMM-IMM-Fachtagung 15./16.5.2001, Mainz, VDE-Verlag, Berlin/Offenbach 2001

[3] P.W: Gold: Micro Gears in VDI-Berichte 1665, International Conference on Gears, 13.-15. 3.2002, Munich, VDI-Verlag, Düsseldorf 2002

[4] H. Meerkamm, K. Paetzold: Gleitreibung und thermische Eigenschaften in einem WAVE DRIVE,-Getriebe, Seminarveröffentlichung

[5] F. Pöhlau, K. Paetzold: WAVE DRIVE,-Getriebe in Konstruktion 6/2001, S.53-55

Frank Pöhlau
Oechsler AG
Matthias-Oechsler-Str. 9
91522 Ansbach
Germany
f.poehlau@oechsler-ag.de

Keywords: actuator, micromolding, thermoplastics, gear, HVAC, air conditioning, testing, wave-drive

Rapid Control Prototyping of Automatic Climate Control Systems

A. M. Strobel, DaimlerChrysler AG

Abstract

Automatic climate control systems in vehicle applications are very complex systems. There is a variety of connections to and from other vehicle systems (engine and engine cooling, cabin, passengers and HMI, environment). The control algorithms are currently based on classic control concepts with a huge set of expert rules. The design and test of these expert rules during simulated and real driving conditions is a key factor for a successful development (time, costs and quality) of an automatic climate control system. This paper presents experiences that have been made during the design of a rapid control prototyping (RCP) system for an automatic climate control system. The advantages and drawbacks of a model-based control system specification are discussed. The concept of bypassing on CAN is shown. Suggestions for future improvements of the applied standard tools are made. Further more, the importance of real-time plant models (RTL) for hardware-in-the-loop applications, model-based control concepts and online diagnosis tools are discussed. Concepts for the derivation of RTL models from more detailed, but non real-time software models are presented.

1 Automatic Climate Control Systems

1.1 System Layout

A climate control system is a collection of different components that are used to heat, cool and ventilate the air for a comfortable climate inside a vehicle cabin, to prevent fogging of the surrounding windows and allow the deicing of the windshield. Figure 1 shows the layout of a state-of-the-art climate control system in a luxury car.

The thermal state of the vehicle cabin and the thermal comfort of the passengers inside it can mainly be influenced by the adequate conditioning of the cabin intake air. This is done inside an HVAC unit (heating, ventilation and air conditioning) that is part of the overall climate control system. Figure 2 shows the air paths and major components and subsystems of the climate control system.

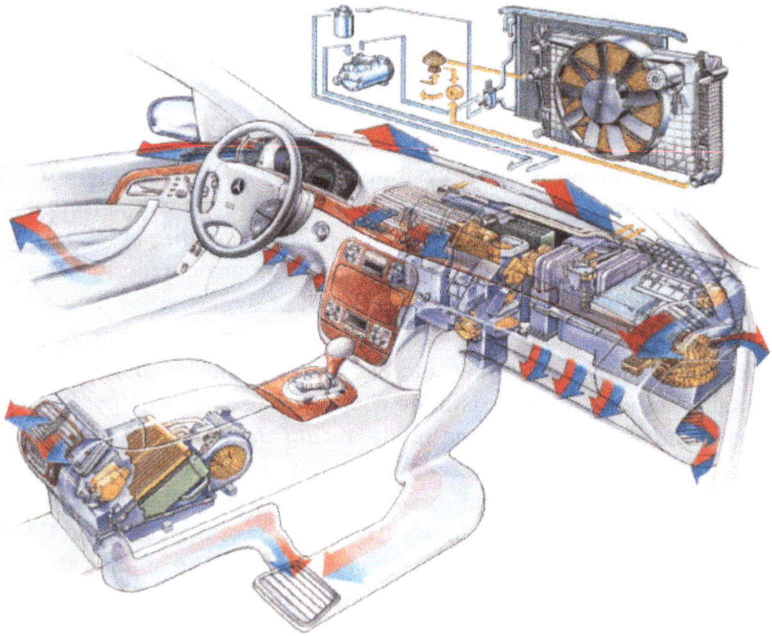

Fig. 1. State-of-the-art climate control system in a luxury car.

In a first step water is separated from the fresh air right after the air suction inlet. Then the fresh air can be mixed with circulated air from the cabin. The amount of circulated air is set by the position of the circulation flap. After that the air is cleaned by different filter systems. Some of them can be activated or deactivated. Then the air is accelerated by the fan. In the next step the air can be cooled at the evaporator. The evaporator is a part of the cooling system (refrigerant cycle). After that the air can be heated at the heat-exchanger which is a part of the heating system. The heating system is connected to the engine cooling system. In a last step the air is distributed to the different air ducts that lead to the cabin air inlets. The distribution can be influenced by the positions of several flaps. The number of air inlets varies. Most vehicles have defrost outlets, left and right floor outlets and left and right side and middle outlets as shown in figure 3.

The thermal state of the vehicle cabin is also influenced by a lot of disturbances. Therefore, it is necessary to continuously adapt the temperature and mass flow of the cabin intake air in order to maintain a constantly comfortable thermal environment for the passengers inside the vehicle cabin. The following list shows some of the most effective disturbances:

- Solar radiation (direct and diffusive).
- Exterior air temperature.
- Exterior air humidity.
- Exterior air quality.
- Engine cooling temperature.
- Vehicle speed.

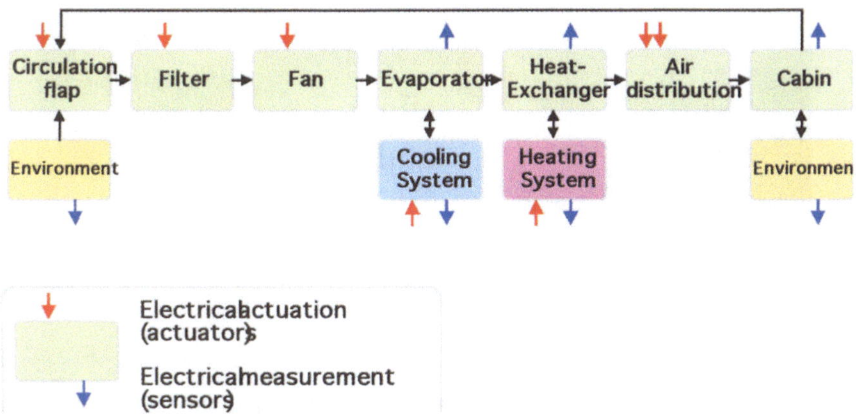

Fig. 2. Air paths through the climate control system.

The comfortable thermal condition may be different for all passengers inside the vehicle cabin. Therefore, more and more multi-zone climate control systems are used. These systems allow each passenger to individually adjust the thermal state of the assigned cabin zone.

The continuously manual control of the temperature and mass flow of the cabin intake air would need high attention of the passengers inside the vehicle cabin especially in case of changing boundary conditions. This is neither user-friendly nor safe. The driver would be distracted from driving the car. Therefore, automatic climate control systems are used for the automatic control of the temperature and mass flow of the cabin intake air. Normally, the control functions are separated into temperature control, fan control and air distribution control. The basic function is the temperature control.

The other functions are often optional. In the following it is assumed that all functions are present, a multi-zone system and the maximum number of state-of-the-art actuators and sensors are used.

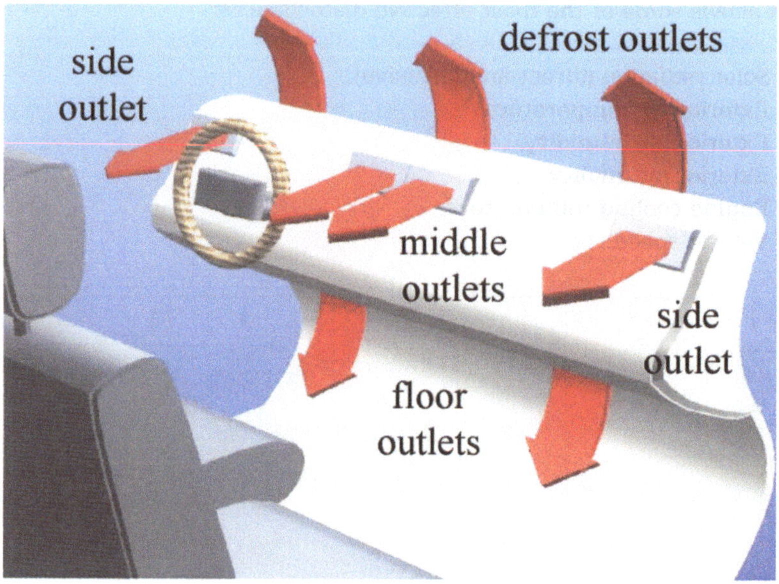

Fig. 3. Cabin air inlets.

The following actuators are used for the conditioning of the cabin intake air:

- Circulation flap.
- Filter actuator.
- Fan speed.
- Actuator for cooling system power.
- Actuators for multi-zone heating system power.
- Air distribution flaps.

The following sensors can be used for measuring the current state of the climate control system and the disturbances:

- "Average" interior air temperature sensors.
- Interior air humidity sensor (fog sensor for windshield).
- Temperature sensors for the cabin intake air of each cabin zone.
- Engine coolant temperature sensor.
- Evaporator air outlet temperature sensor.
- Refrigerant pressure sensor.

- Multi-zone solar radiation sensor.
- Exterior air temperature sensor.
- Exterior air humidity sensor.
- Exterior air quality sensor.
- Engine speed sensor.
- Vehicle speed sensor.

1.2 Standard Control Algorithms

The temperature control is generally done by a classical closed-loop control algorithm with proportional or proportional and integral gain. There will be one desired air temperature value for each cabin zone, if multi-zone temperature control is applied. As in most cases only the average interior air temperature is measured, the average interior air temperature deviation is defined by the difference between the average interior air temperature value and the average desired air temperature value. A top level temperature control calculates the average cabin intake air temperature. This value is adapted subject to the desired air temperature value for every cabin zone. An underlying, cascaded zone temperature control for each cabin zone ensures that the desired cabin intake air temperatures are adjusted. This can either be done in closed-loop or open-loop mode depending on whether the cabin intake air temperature is measured or not.

The measured disturbances are taken into account by feedforward control the desired air temperature values are corrected in a way that the disturbance is balanced. The exterior air temperature is used for correcting the average desired air temperature. For each cabin zone a value for the direct solar radiation can be used for correcting the desired air temperature value for the related cabin zone.

As the air is first cooled and then heated, it can be dehumified. This mode is called reheat operation. The ratio of reheat can be controlled by either a fog sensor signal (closed-loop) or comparing the drew point temperature of the exterior air with the exterior air temperature (open-loop).

The fan control is mainly done in open-loop mode, because the measurement of the cabin intake air mass flow is very difficult and cost-intensive. The average interior air temperature deviation can be used as an indicator for the need for more or less air mass flow. A lot of boundary conditions must be taken into account. In winter, the engine coolant temperature must be high enough, before the fan speed should be increased.

The air distribution control is mostly an open-loop control concept without direct feedback. Depending on an estimated thermal situation, a special air distribution is selected and the according flap positions are set.

2 Rapid Control Prototyping (RCP)

In general the development of an electronic climate control unit (ECU) is not a straight forward process. In general a lot of iterations are necessary between the OEM and the supplier of the ECU, because the specification of the desired system behavior is not only based on physical data, but also on subjective perception. The well-known V-process model [1] with the two branches control software design and ECU testing and calibration has to be executed several times.

In order to reduce the number of iterations that are necessary due to subjective, incomplete or misinterpreted system specifications it is necessary to rapidly generate a variable prototype from the given system specification. Therewith the system specification and the subjective perception can be tested even before the first prototype of the real ECU exists.

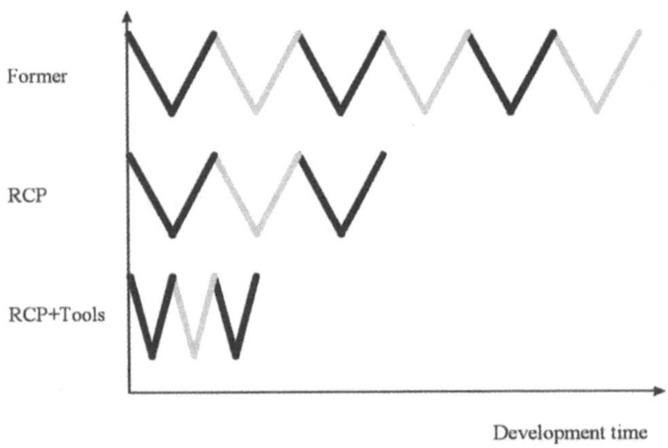

Fig. 4. Reduction of the development time by the use of RCP and other tools.

The remaining iterations should be executed very fast. This can be achieved by the intensive use of tools for all steps in the V-process model. Figure 4 schematically shows the reduction of the development time by the use of RCP and other tools.

It is very important to select a suitable tool chain that is available throughout the whole development process. Data of previous steps should be available and accessible in the following steps. As many steps as possible should be automated.

The release version of the tools should be fixed during a single (short term) project. Only tool functions that have been available and ensured at the start of the project should be used. Otherwise, data integrity might not be ensured throughout the whole development process.

In this paper the control system specification has been delevoped with Matlab/Simulink/Stateflow from The Mathworks. A dSPACE Tandem-Autobox with the dSPACE RTI and ControlDesk software packages has been used as RCP system. The layout of the considered RCP system is shown in figure 5. The RCP system is connected to an existing ECU with bypassing on CAN. The RCP system considered in this paper has been developed for research purposes and not for the development of a new automatic climate control system.

Fig. 5. RCP layout for an automatic climate control system.

2.1 Advantages and Drawbacks of a Model-Based Control System Specification

A model-based control system specification is a functional description of the desired system behavior. Therefore, a model-based control system specification can be executed. It can be integrated in a simulation of the overall system. This allows the test of the specified control algorithms even before any hardware exists. A model-based control system specification is unambiguous. There are no misinterpretations (assuming a perfect interpreter tool).

Normally, a graphical representation of the different functions of a model-based control system specification is used. The graphical representation can be a block diagram, a state chart or something similar. As block diagrams and state charts are used in existing high level programming languages like Matlab/Simulink/Stateflow from The Mathworks, these tools can also be applied for a model-based control system specification. During the last years these tools have been improved a lot for this kind of application.

It is essential for the tool chain that the model-based control system specification can automatically be transferred to real-time structured C code. This step is normally called automatic code generation (ACG). For RCP applications the size (and quality) of the generated C code is not crucial. For the generation of series code that is going to be implemented in the real ECU it is very important that the generated code size is small and the code quality is very high. This method may be different from ACG and can be called automatic series code generation (ASCG). In the ideal case the ASCG method is certified and therefore the generated code can also be considered as certified. Unfortunately, the certification of an ASCG method is very difficult and with any change in this method the certification process must be done again. Due to that there will probably not be a certified ASCG tool in the nearer future. In consequence of that the ASCG methods are currently developed with respect to certified development methods. Therefore, a high quality can be expected. But there still is not a lot of experience in the application of these tools.

The number of applications for this new approach of a system specification is increasing rapidly. But in most of the cases it is only used for special functions and not for the overall control system specification. Any existing, working and validated function will not be redesigned with a model-based approach. Such functions are existing hardware I/O drivers. Other functions like graphical user interfaces are designed with different tools that are not part of the applied tool chain.

The software of the RCP system considered in this paper has been fully specified in Matlab/Simulink/Stateflow from The Mathworks. The control algorithms have been designed in Simulink. The behavior of the user interface has been designed in Stateflow. All the hardware I/O has been modeled with the RTI block set from dSPACE.

The complete control system specification comprises 58273 blocks, 10 state charts. 27 CAN messages are read on three CAN controllers. One CAN message is sent to the bypass ECU on a fourth CAN controller. 15 digital I/O channels and one serial port are used.

On a 1 GHz Pentium machine it takes about 15 minutes to compile the whole code. About 1 GB RAM is needed during the compilation process. Any optimization level of the compiler must be turned off, because the compiler would crash with hash table overflow.

The fastest sample time of the control algorithm is 10 ms. The turnaround time on the given RCP system varies between 1 and 6 ms. The 27 CAN messages are read with a sample time of 100 ms. Each CAN message has a random time offset. This ensures that it is very unlikely that too many CAN messages should be read at the same time on the same CAN controller. A lot of sub-functions like parts of the user interface are time triggered, because they do not need to be executed at the system sample time of 10 ms. The triggers are implemented as software triggers. All software triggers have a random time offset. This ensures that the different calculations are spread over all time steps. The concept of random time offsets is used to simplify the development of the RCP system. For a real series ECU application the priority and sequence of the different functions must be defined accurately.

The given control system specification had to be "simplified" in order to be handled by the tools used with the RCP system considered in this paper. One of the major drawbacks for an overall model-based control system specification therefore is the limitation defined by the applied tool chain.

2.2 Bypassing on CAN

The goal of RCP is to create a prototype for a new control function in a very short period of time with as less effort as possible. The concept of bypassing allows the use of an existing ECU with a few modifications. Only the part that is replaced by the new control function is going to be bypassed. All the connections to the existing hardware can be used. There is no need to develop these connections.

There are two requirements for bypassing. First of all the software of the existing ECU must be changeable. This means that the part of the software that has to be changed should be placed in a chip that can either be replaced or flashed. Otherwise the hardware of the ECU has to be redesigned. Second, the existing ECU needs a bilateral communication channel to the RCP system. In most of the cases an existing serial port like a diagnosis interface or CAN is used. In the future even wireless ports may be useful.

The concept of bypassing on CAN is suitable for an ECU that already has CAN read and write access and that is not time critical. In this case only a few changes are necessary in the existing ECU software.

In the RCP system considered in this paper a single CAN message is used to send data to the bypass ECU. The so-called bypass message has the structure shown in figure 6. With the first byte of the bypass message the RCP system defines whether the bypass ECU should be in bypass or standard mode. In standard mode the ECU must work like it has not been changed. This is necessary to implement a safety function that brings the ECU back to normal operation. The second byte defines the bypass function (e.g. fan voltage, flap position). The third byte defines the status of the bypass function. It can either be in bypass mode (ON) or standard mode (OFF). The fourth and fifth byte define the desired value for the bypass function. The bypass message will only be sent, if it is necessary. This can either be a change of the bypass mode or a change of the status or value of a bypass function. A special bypass logic is used to decide whether the bypass message must be sent or not and if yes which bypass function has to be sent. This bypass logic is executed at 10 ms. The RCP system considered in this paper has 74 possible bypass functions. Therefore, it takes less than 740ms to send a complete set of updated bypass functions. For safety reasons it might be necessary that the bypass ECU will return to the standard mode, if the bypass message is not sent for a special period of time. This has to be ensured by the bypass logic.

Byte 1	Byte 2	Byte 3	Byte 4	Byte 5
Bypass mode ON/OFF	Bypass function	Status of bypass function ON/OFF	High-Byte value	Low-Byte value

Fig. 6. Structure of the bypass message.

A second CAN message is used by the bypass ECU to send data to the RCP system. The bypass ECU must send all the data to the RCP system that is needed as input by the new control function and that cannot be received by the RCP system directly. The so-called bypass data message has the structure shown in figure 7. It is a mode-dependent CAN message. The first byte of this message defines the mode or data block. The second byte is reserved for a sub-block definition. The third to eighth byte contain the data that has to be interpreted with respect to the current bypass data block number. In the RCP system considered in this paper 314 values are sent by the use of 36 data blocks.

The data blocks are continuously sent in sequence by the bypass ECU with a sample time of 100 ms. Therefore, it takes less than 4 seconds to receive a completely updated set of bypass data.

Byte 1	Byte 2	Byte 3 - 8
Bypass data block number	Reserved for sub-block definition	Block value

Fig. 7. Structure of the bypass message.

It has been proved that the concept of bypassing on CAN is very useful for RCP applications that are not time critical. The additional code that is needed to implement the bypass functionality is very small. Therefore, we generally recommend to consider the possibility of bypassing on CAN during the development of any ECU even if there is no application right now.

2.3 Experiences and Suggestions for Further Improvements of the Applied Standard Tools

As mentioned before the RCP system considered in this paper has been designed with Matlab/Simulink/Stateflow from The Mathworks and dSPACE systems.

Matlab/Simulink/Stateflow from The Mathworks has become a standard tool for the development of control systems. It is very easy to use and the support service by The Mathworks has a very good response time (about one day) and offers always helpful solutions. All tools offer a plenty of functions and they are all integrated into the same simulation environment that also has very good post-processing capabilities. There are a lot of interfaces that connect this simulation environment to other third party tools. Therefore, Matlab/Simulink /Stateflow from The Mathworks seems to be the best choice for a top level simulation tool. Hence, it is advisable to create the model-based control system specification in the same simulation environment.

Since Matlab Release 12 it is possible to operate with matrices in Simulink. For control system specifications that also include the description of a graphical user interface with the text that should be displayed it would be favorable to have the possibility to handle strings in Simulink. Figure 8 shows a simple example for such a favored display message.

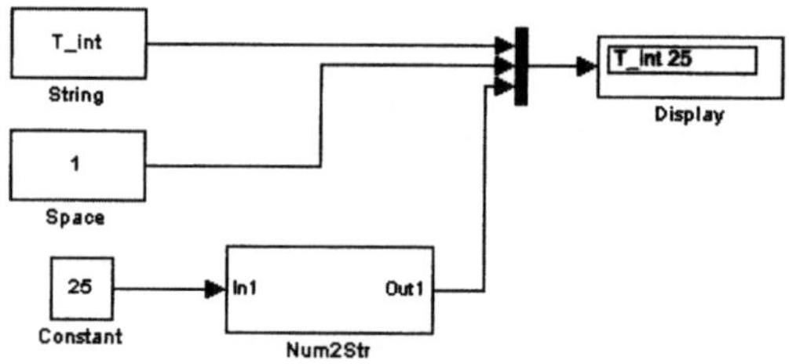

Fig. 8. Strings in Simulink.

With the new version of Simulink that is part of Matlab Release 13 it is possible to model the basic programming constructs like if-then-else and iteration loops. This is a basic requirement for engineers that have to decide whether to switch from C to Simulink or not. However, Simulink operates on blocks functions and unlike C++ not on objects. In order to be competitive to C++ it would be necessary to implement a new block called object in Simulink. This block should be defined in a library. Figure 9 shows a simple example for a subsystem inside a library that would define an object type, because it contains the desired ObjectType block. Relations to parent objects could be defined. The public or private methods could be implemented as subsystems that contain standard Simulink blocks. The object would also contain data that can be either public or private.

Figure 10 shows a simple example for how this object block can be used. It is dragged from the library to the model. By default it is an object definition (like the two blocks at the bottom of figure 9). A new instance of the defined object type is created with this block. The object can be either global, local or scoped. The object definition can be copied inside the model. The target block is an object method block by default. The method can be selected from a list of all possible methods. The block can be changed to an object data block. The data can be selected from a list of all possible data. Using objects would simplify the definition of large control system specifications.

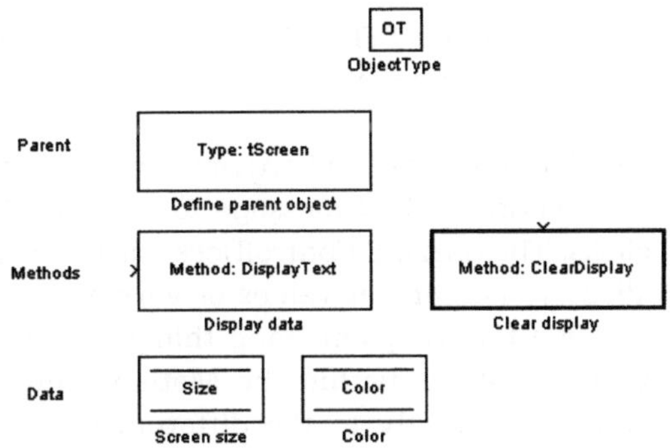

Fig. 9. Object type definition.

Simulink is integrated into the Matlab simulation environment. In this simulation environment parameters can be defined in a so-called workspace. These parameters can be used in a Simulink model. At the end of a Simulink simulation the simulation results can be written to the Matlab workspace. Sometimes it would be favorable to have a standard Simulink block that reads and writes data during a simulation to a special data structure (shared memory) in the Matlab workspace. This data structure could be defined by special initialization blocks. The data structure would also be available in C code that would be generated from the Simulink model with the Real-Time Workshop. There should also be a tool to replace all read, write and init blocks for the shared memory with from/goto or data store blocks and vice versa.

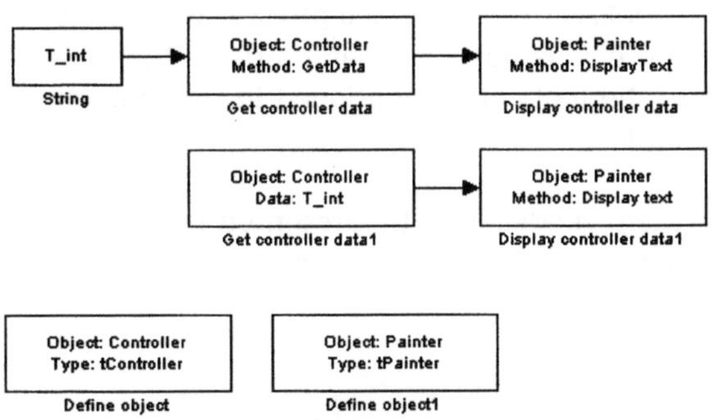

Fig. 10. Object usage

dSPACE systems have been proved to be very suitable for automotive applications, because of their continuous tool chain and a very good support. As far as we could test it, the dSPACE hardware is very good and very reliable. It is not cheap though. Our vehicles normally operate with 12 V. Even the additional hardware (switches, LEDs) for our test vehicles is based on 12 V.

Therefore, we would like to get a TTL to 12 V and vice versa conversion board for the digital I/O directly from dSPACE.

The dSPACE Simulink blocksets for the hardware connections are very easy to use. For example, connecting a CAN message to a control input is done with just a few mouse clicks. The dSPACE ControlDesk software offers a wide range of possibilities to display the current values of variables on the RCP system. Nevertheless, there are a few very annoying things about this tool. It should be possible to start ControlDesk within the Matlab simulation environment and drag Simulink blocks directly to the different display elements inside ControlDesk. Navigating through the variable description file of very complex models is impossible. The definition of parameter sets needs some time to get accustomed to and recording and evaluating test data is not as easy as it should be. Some of these problems will be solved in the upcoming new product releases.

3 Real-time Plant Models (RTL)

Real-time plant models are necessary, whenever a plant model should be used in connection with real hardware. They have to fullfill the following requirements: execution in real-time, dynamic/transient models, very high modelling language for the integration in a model-based control system specification, reuse of existing software models/results. Our general approach for creating real-time plant models is to use as much physics as possible for a real-time simulation. Everything else is put into characteristic diagrams. It will be easy to integrate them in the tool chain mentioned in the chapter before, if they are modeled in Simulink.

Three real-time plant models will be presented in this paper: an air distribution model, a cabin model and an AC-cycle model.

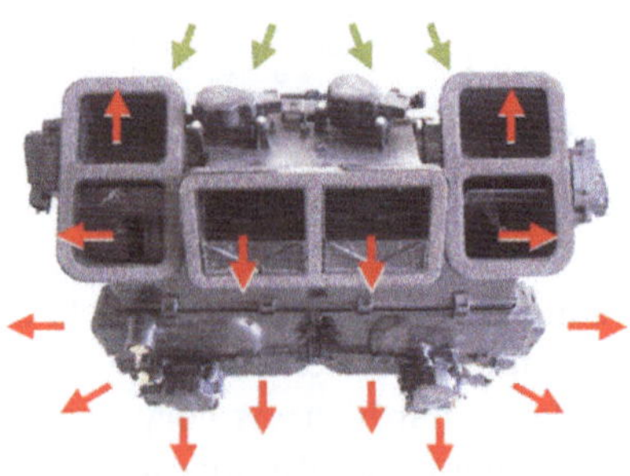

Fig. 11. Main part of the air distribution system.

The air distribution model is able to calculate 30 air mass flows in less than 1ms on a state-of-the-art real-time system. The air distribution is influenced by 12 step motors that are connected to 25 flaps. This leads to 127 possible air paths in the air distribution system shown in figure 11. The parameters for this model are derived from 3D computational fluid dynamics results.

The total air mass flow inside the vehicle cabin is driven by two pressure sources. One is the fan of the climate control system. The other is the pressure difference between the air suction and the air outlet that depends on the vehicle speed. The pressure losses in the air distribution system also influence the total air mass flow. Figure 12 shows the paths for the fresh and circulated air.

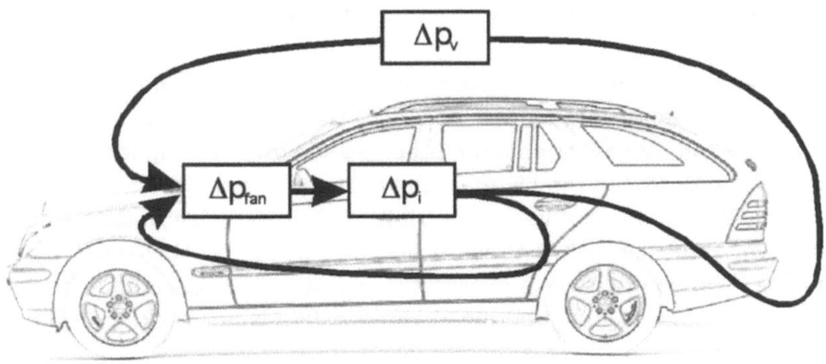

Fig. 12. Air paths.

The air distribution model calculates the circulated, fresh and total air mass flow with respect to the two pressure sources and all pressure losses. The pressure losses are described by conductances G. Figure 13 shows the nonlinear conductance network solved by the air distribution model. The individual air mass flows in the air distribution system are calculated by the total air mass flow and all individual conductances.

Using this approach a model has been created for the air distribution system shown in figure 9. A simple simulation will show the behavior of this model. Figure 14 shows the input variables for the simulation. In the first half the inputs are set according to a cooling condition. In the second half of the simulation the inputs are set according to a heating condition. The fan speed is shown in the upper left corner. In the middle of the simulation it is decreased from 73 to 47%. The vehicle velocity is shown in the upper right corner. After a short idle period it is increased to 60 km/h. The flap positions of the circulation and air-mixing flaps are shown in the lower left corner. The positions of the air distribution flaps (defrost, floor, middle outlet) are shown in the lower right corner.

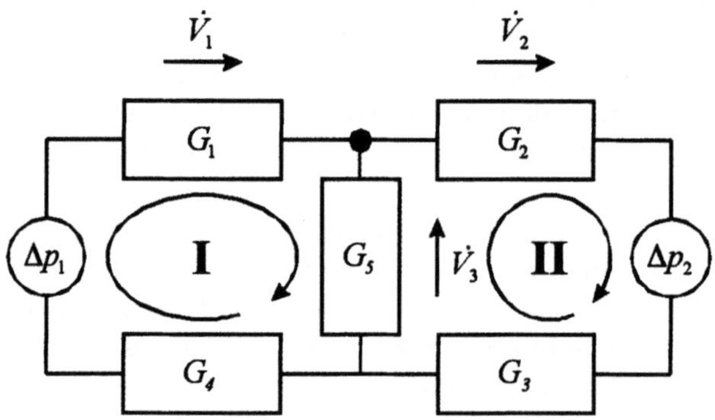

Fig. 13. Conductance network solved by the air distribution model.

Figure 15 exemplary shows the simulation results. In the upper left corner the air mass flows at the left defrost and floor outlets are shown. The air mass flows at the right defrost and floor outlets are shown in the upper right corner. In the lower left corner the air mass flows at the left middle and side outlets are shown. The air mass flows at the right middle and side outlets are shown in the lower right corner. The results are very good and they fit to measurement data. The comparison between the measurement data and the simulation results is not shown as the measurement data is not very accurate.

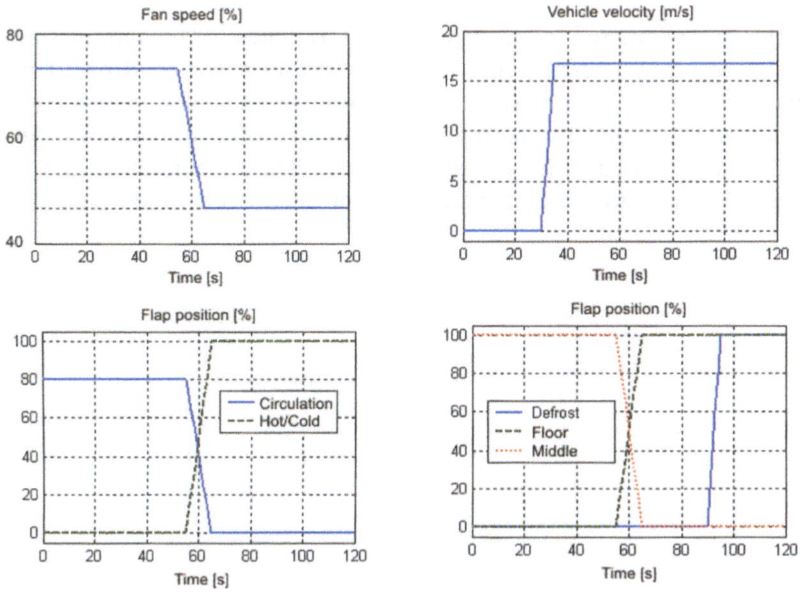

Fig. 14. Inputs for the air distribution model.

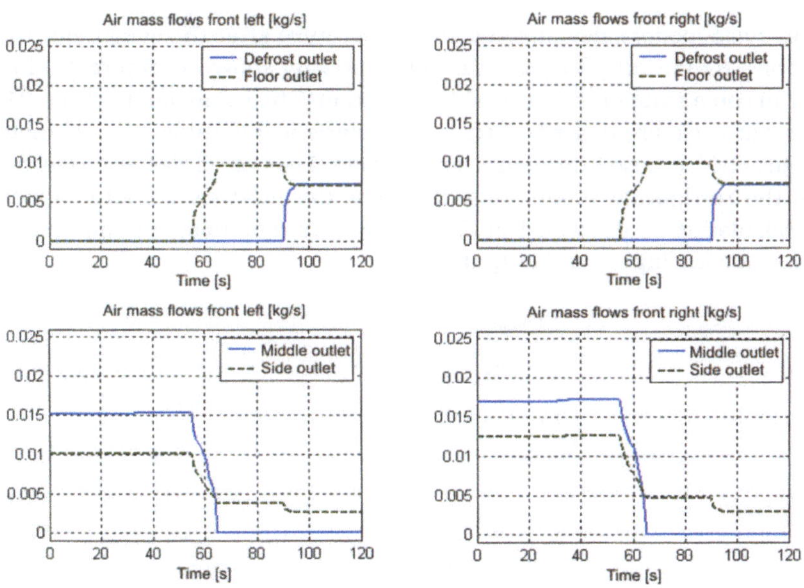

Fig. 15. Results calculated with the air distribution model.

The cabin model (schematically shown in figure 16) is devided into two parts. The first part is mainly based on energy equations and thereby calculates the average interior air temperature. The energy equations are based on enthalpy, convective heat-transfer, solar and thermal radiation. The parameters are mainly based on geometrical and material data. In the second part transfer functions are used to calculate the air temperatures at special sensor positions inside the vehicle cabin. The parameters for the transfer functions are derived from 3D computational fluid dynamics results. The calculation of the whole cabin model takes less than 1 ms on a state-of-the-art real-time system.

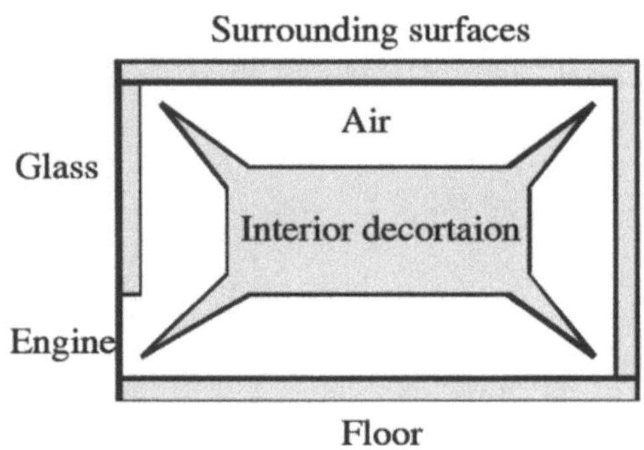

Fig. 16. Cabin model.

The AC-cycle model calculates the air temperature after the evaporator that is a part of a R134a refrigerant cycle (schematically shown in figure 17) in less than 1 ms on a state-of-the-art real-time system. The components of the R134a refrigerant cycle are described with characteristic diagrams. The characteristic diagrams are based on measurement data. Look-up tables are used to calculate the refrigerant media data. Basically, the AC-cycle model is a steady-state calculation as described in [2]. The system dynamics are approximated by time-delayed inputs and outputs.

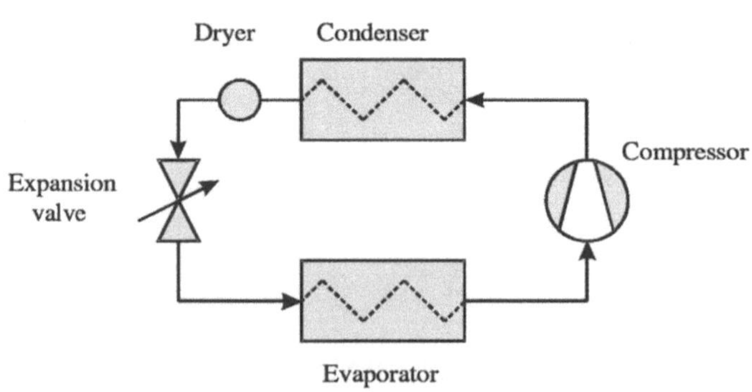

Fig. 17. Typical refrigerant cycle.

Three applications for real-time plant models will be described in the following chapters: hardware-in-the-loop tests, online diagnosis tools and model-based control concepts.

3.1 Hardware-in-the-Loop Tests

A lean development process with less iterations requires testing at an early development stage. An ECU prototype has to pass several tests before it is put into a car for the first time. These tests can be done either in open-loop (e.g. re-run of measured test data) or in closed-loop mode on a hardware-in-the-loop testbed as shown in figure 18. During closed-loop tests the plant has to be simulated by a real-time plant model. The specified functional behavior of the ECU prototype should be tested in a step called function verification. After that several ECUs can be put together and tested in an ECU network. This step is called integration test.

Fig. 18. Hardware-in-the-loop testbed.

The verification of the ECU functions can be quite difficult or even impossible, because full test coverage would take millions of years in real-time for very complex ECUs. Therefore, test coverage is limited to customer relevant and safety critical functions.

3.2 Online diagnosis tools

If the real plant in form of a test or series vehicle already exists and if there also exists a real-time plant model for that vehicle, it will be possible to compare the real system with the simulated real-time plant model online. Assuming accurate real-time plant models any difference between measured and simulated values must be related to changes or defects in the real system.

3.3 Model-based control concepts

Real-time plant models can be used for an online optimization of the control outputs. Such a model-based control concept (see figure 19) is more transparent than a control concept based on many nested rules and look-up tables calibrated during test drives, because the real-time plant models do not need to be limited to a set of special situations. Existing sensor hardware could be simplified or replaced by the use of real-time plant models. New sensors could be used in combination with real-time plant models.

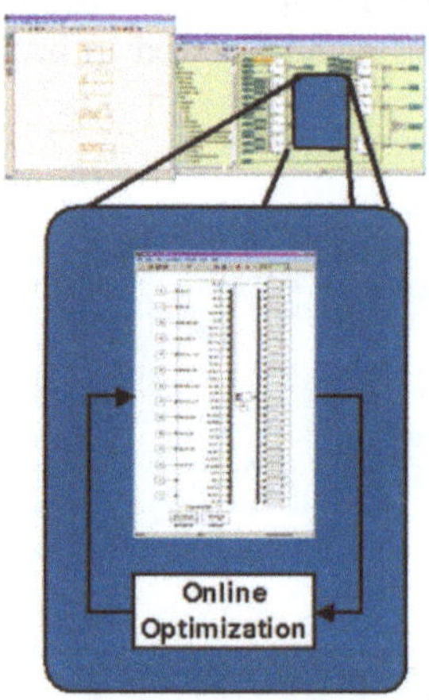

Fig. 19. Model-based control concept.

References

[1] Dröschel, Wolfgang / Wiemers, Manuela: Das V-Modell 97. Oldenbourg 2000.

[2] Davis, Gerald L. / Chianese, Frank Jr. / Scott, Timothy C.: Computer Simulation of Automotive Air Conditioning – Components, System, and Vehicle. SAE Technical Papers Series 720077.

André Manfred Strobel
DaimlerChrysler AG
HPC E223
70546 Stuttgart
Germany
andre.strobel@daimlerchrysler.com

Keywords: rapid control prototyping automatic climate control

Autodoor. Microsystem for Automotiove Door Module

A. Ferré, R. Jiménez, M. López, European Technological Center
S.A. Bota, J.L. Merino, J. Samitier, Universitat de Barcelona
B. Niessen, Austriamicrosystems AG
E. Cabruja, J. Bausells, Centre Nacional de Microelectrònica

Abstract

In this paper a multichip module system for automotive applications is described. The MCM substrate is a IMS (Isolated Metal Substrate), which has been used for integration of analog signal processing stages, complex logic circuits and even power switching transistors. The main functionality of the door module implemented with this MCM, is the control of the power window motor with antipinch feature, although the prototype has been completed with communication (LIN-bus).

1 Introduction

In the automotive industry the use of electronic systems to implement control tasks has increased significantly in a short period of a few decades in order to provide a variety of customer comfort and safety improvements [1].

To be competitive in the marketplace, automotive manufacturers are requiring their systems to have reduced cost, more performance, more flexibility, more configurability, reduced time to market and easy customisation.

The advantages of MCM technologies can be very attractive, taking into account that we deal with complex systems with huge amount of interactions between them [2]. For maximum reliability all blocks in an electronic control unit should be monolithically integrated on one chip or at least accommodated in a multichip module. This strategy has been adopted in order to design and implement an automotive electronic module. The main goal is that microsystem will be able to support power window control with anti-trap functionality although the system has been designed to be scalable to include extra sensors for high performance, as accelerometers. It may include also other functions like central locking and courtesy light control or even mirror control and adjustment

In order to perform this functionality, the multichip system must include sensors, custom ICs, standard ICs and power devices [2]. The implementation of all these components in a single multichip module provides advantages in cost, weight, power consumption, MTTR (Mean Time To Repair) and an increase of functionality, reliability, life time and MTBF (Mean Time Between Failure).

AutoDoor has been funded by EURIMUS (Ref. EM-48). EURIMUS (EUReka industrial Initiative for Microsystem USes) is a European program to improve the research and development of microsystems for industrial and automotive markets. The consortium includes the main contractor (LEAR), who is the system integrator, a supporting university/research institute (UB-CNM) and a silicon design centre and foundry (Austriamicrosystems) able to produce the sensors and circuits required for the microsystem development.

2 System Overview

AutoDoor Beta version (Motor Memory Controller) main characteristics are:

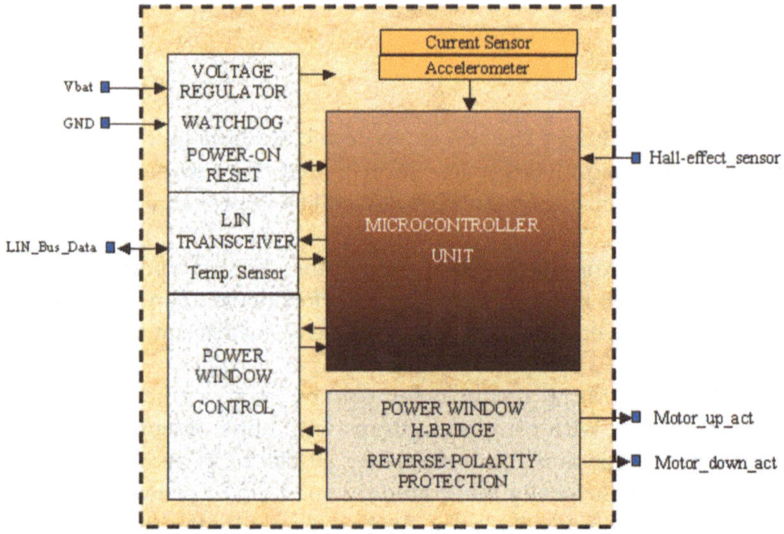

Fig. 1. AutoDoor System description.

It controls motor movement, activated by power transistors, which act as output interfaces. This activation is done according to the input signals read from their respective direction buttons and/or commands received trough LIN-bus, allowing two movement directions.

It controls the position of the motor by means of the input signals providing from some external/internal sensors; in order to know their position at any time, allowing the controller to stop their movement when the window/ reaches any mechanical end or a trap situation occurs.

It is provided with a self-diagnostic program, which recognises system faults, such as stuck switches, damaged outputs, etc. and with the possibility of reporting them to an external tester through bus LIN.

3 Control Devices

3.1 Microcontroller

The microcontroller used is a CMOS FLASH-based 8-bit microcontroller with 200-nanosecond instruction execution. This microcontroller features 256 bytes of EEPROM data memory, self programming, in-circuit debugging, 8 channels of 10-bit Analog-to-Digital (A/D) converter, 2 additional timers, 2 capture/compare/PWM functions, a Serial Peripheral Interface (SPI) and a Universal Asynchronous Receiver Transmitter (USART) to implement the LIN controller. All of these features make it ideal this application.

We have introduced this microcontroller in package form because the QFN package (7 mm x 7 mm) produces no impact on module size while makes easier and cheaper the mounting.

3.2 ASIC

Voltage regulator with added functionality: The ASIC contains a low-power, low-drop voltage regulator to provide internal power and external power for the MCU. The on-Chip regulator consists of the main regulator and additional circuits. The VDD regulator accepts a unregulated input supply and provides a regulated VDD supply to all digital sections of the device. The output of the regulator is also connected to the VDD pin to provide the 5V to the µC. The nominal input voltage is designed for operation from 8V systems to 16V systems. The additional circuits provide the following functions:

- Oower on reset and watchdog.
- Battery voltage monitor.
- VDD voltage monitor.
- Standby circuitry.

LIN I/F: The LIN I/F comply with the LIN standard and is able to wake up the whole circuit out of standby mode if any LIN BUS activity can be detected.

H-Bridge drivers with integrated charge pump: The power window block controls four conventional power FETs to drive a power window motor (H-Bridge) circuit. H-bridge driving technique (non-overlapping drivers, current recirculation through Power MOS) is used. The LS switch is controlled by the PWM signal, while the HS switch at the same side by the inverted signal PWMN (it is used as synchronous rectifier). The ramp (gate current) is controlled by an register, that can be read and written via SPI I/F. The dead time for secure synchronous rectification is controlled by a register too and can vary from 3 to 5 us. The PWM generator is based on a PWM frequency of 20 kHz. The HIGH-to-LOW ratio can be changed by 31 steps from 3.125% to 100%.

Current measurement: A current measurement subcircuit (voltage drop at an external shunt or voltage from the non-contact current sensor) provides the analog voltage signal VIMEAS and sets an over current flag. If this flag is set, all HS and LS drivers are switched OFF. The overcurrent threshold is also programmable. The external resistor has a voltage drop of 300mV to switch OC to 1. The analog output value VIMEAS is amplified by a variable voltage gain, controlled by a register via SPI I/F.

SPI: Data transfer from the microprocessor to the ASIC and vice versa is accomplished by means of a SPI interface.

4 Sensors

The following types of sensors has been considered for this automotive door application:

- Speed sensors, such as Hall effect sensors, to detect the window position and/or velocity. These sensors are external to the module but in a future version may be also introduced.
- Current sensors (non-contact magnetic sensor) to monitor the status of the electrical motor and the status of the power devices. The current sensor has been also tested for an anti-pinch algorithm based on ripple

counting.

- Vibration sensor (accelerometer), to detect the car's vertical movement in order to avoid false anti-pinch detection.

Non-contact current sensors and accelerometer sensors were fabricated in the standard CMOS process of Austriamicrosystems., that has experience in working with such devices.

Non-Contact Current Sensor

The first version of AutoDoor was built introducing a non-contact magnetic sensor. The magnetic field around a current conducting path is measured using Hall Sensors in the CMOS process. Sensor signals processing ensures effective cancellation of offset and temperature drift. Placing the current path close to the sensor ensures maximum sensor sensitivity.

The sensor was measured mounted over the Insulated Metal Substrate. The sensibility achieved in this measure is 23,9mV/A, with a very high linearity transfer function. A critical point is the sensor alignment with the current path during chip mounting. The specifications requirements told us about a maximum position deviation between the sensor and the line of 50μm.

The approach for current sensing changed during the project development because the same sensing performance was achieved using a shunt and the internal amplifier of the ASIC. This results in a much lower cost of the whole system.

Accelerometer

Although it has been tested in a conventional device, the accelerometer allows the vertical integration of a micro-machined structure on a standard ASIC. The main characteristics of the device are:

- Versions from 1g to 300g possible.
- Measurements of positive and negative linear dynamic and static acceleration including tilt angle.
- Bandwidth up to 1 kHz (50g closed loop sensor).
- Low noise: 0.2g peak to peak noise with 50g full scale and 200 Hz bandwidth.

5 Power Devices

The power circuit consists in an H-Bridge converter with four Power MOSFETS mounted on die over the IMS substrate. H-bridge driving technique (non-overlapping drivers, current recirculation through Power MOSFETs) is used.

6 Multi-Chip Module

To cost effectively drive large power window motor, a multichip solution combining logic and power has been developed. The multichip module has been implemented using Insulated Metal Substrate high-density mounting technology. This allows power transistors, small-signal transistors and ICs, and other semiconductor devices to be mounted as bare chips. Even if only considered from the aspect of mounting space, this feature provides significantly increased mounting density over the use of packaged devices.

Different packaging options have been also evaluated including low-pressure overmolding and a full mechatronic solution.

6.1 Substrate Technology

The Alpha prototype has been fabricated in an Insulated Metal Substrate (IMS) by Cire (FR) as shown in figure 2. IMS uses aluminium, i.e. metallic, plate as the base. Manufacturing processes are applied to both sides of the aluminium base plate. In particular, copper foil is attached to the aluminium plate with an epoxy resin which itself forms the insulating layer. This allows the creation of hybrid ICs that take advantage of two superlative features of the metallic material aluminium, namely aluminium's high thermal conductivity allows devices with superlative heat sinking capacity to be created, and aluminium's high mechanical strength makes it easy to create these devices. Optimal characteristics can be created for a wide range of application products by modifying the structural elements of the IMST substrate, namely the material and thickness of the insulating resin, the thickness of the copper foil used, and the thickness of the aluminium plate itself.

The specifications of the substrate technology are the following:

- 2 copper layers, with thickness of 70 microns +/-15 µm.
  Dielectric: FR4, with thickness of 2/10 mm.
- Adhesive (PCB to Aluminium): 3M type F9469PC, thickness of 130 µm.

- Aluminium substrate: 2mm thickness, type 1150A.
- Finishing of metal surface: Ni/Au with anti-solder mask.

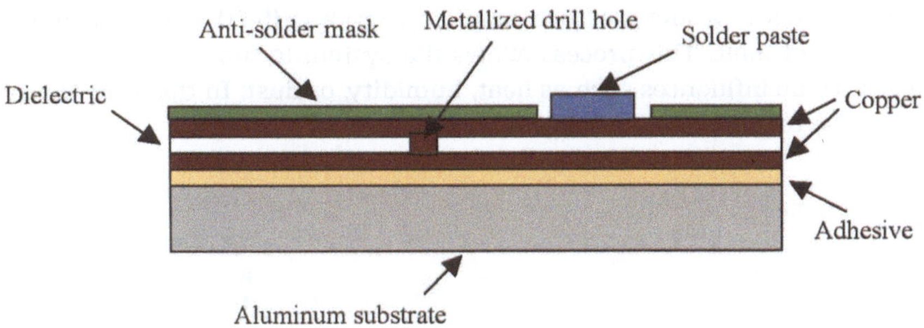

Fig. 2. Insulated Metal Substrate Layers.

6.2 Mounting

The components have been assembled on the boards in the following way:

A screen-printing machine has been used for the deposition of the solder paste, using the stencil metal mask. Bare die components have been manually placed on their locations using a precision component-positioning machine. The solder paste has been melted on a computer-controlled oven using a pre-defined temperature profile program.

Bare chip devices (ASIC and power MOSFETs) are connected to the surrounding circuits using ultrasonically bonded aluminium wires. Since it is possible to pass the wiring pattern under these aluminium wires (crossover lines), wiring patterns that would be implemented with multi-layer printed circuit boards can be acquired. Wire bonding technologies have been established for aluminium wires with diameters of between 25 and 500 mm, depending on the amount of current carried and the purpose of the wire itself, e.g. jumper wires and ground wires.

Bare chip devices (ASIC and power MOSFETs) and their bonding wires have been protected using the glop-top technique (deposition of black polymer plus curing).

6.3 Packaging

A new package created through the use of full moulding is under development (figure 3). This is a low-pressure moulding process (5-20 bar) with molten polyamide hot melt. This process allows the system to function reliably even under extreme influences such as heat, humidity, or dust. In this way, the system may be mounted as a smart connector.

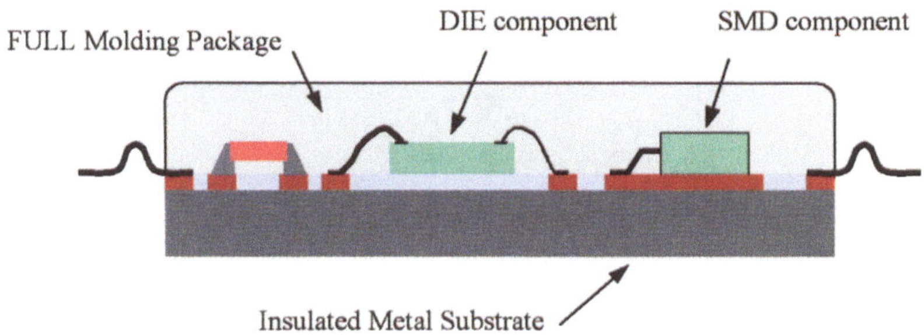

Fig. 3. Full moulding package.

Another option is the integration of the module directly on the end of the motor [3]. We have redesigned the mechanical part so that the IMS acts as cover of the gearbox (figure 4). In this way, problems associated with interconnecting wires are minimised. With proper redesign, even the hall-effect sensor may be included in the MCM and placed directly over the magnetic ring so this results in a very small system package with only three-input wires (LIN-bus, VBAT, GND).

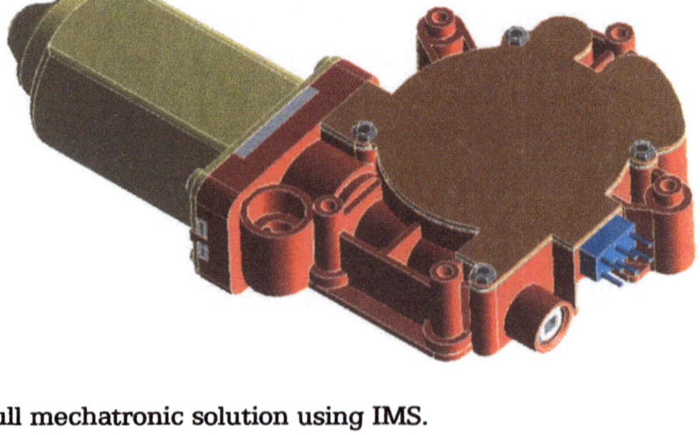

Fig. 4. Full mechatronic solution using IMS.

6.4 Electrical & Temperature Performance

Different prototypes of the MCM have developed. Beta version (figure 5) has been tested in a real car door under different conditions using an Antipinch algorithm developed by the AutoDoor team. It has been experimentally shown that the system is capable to move the window, like the actual standard solution and stop it if a trapping condition is detected, automotive standards are accomplished.

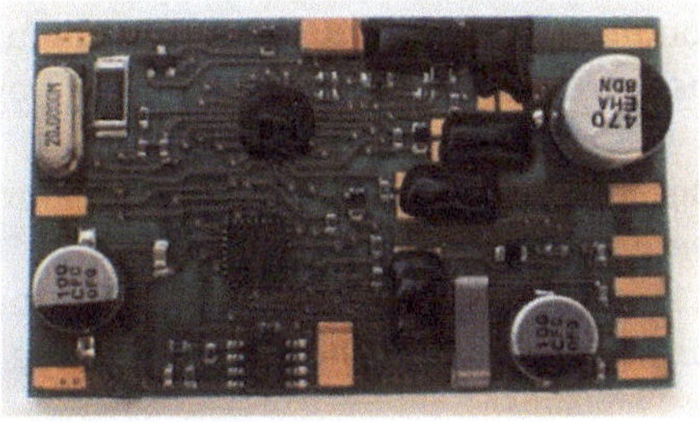

Fig. 5. AutoDoor Beta Version MCM.

The rest of features of the AutoDoor Beta version MCM have also been tested including:

Battery compensation designed to compensate battery voltages higher than 12,5V. When the battery voltage exceeds 12,5V a duty cycle lower than 100% is applied. In this way, we are able equalise the window movement in the range between 12V and 12,5V when the voltage is higher and it performs all the expected system functionality.

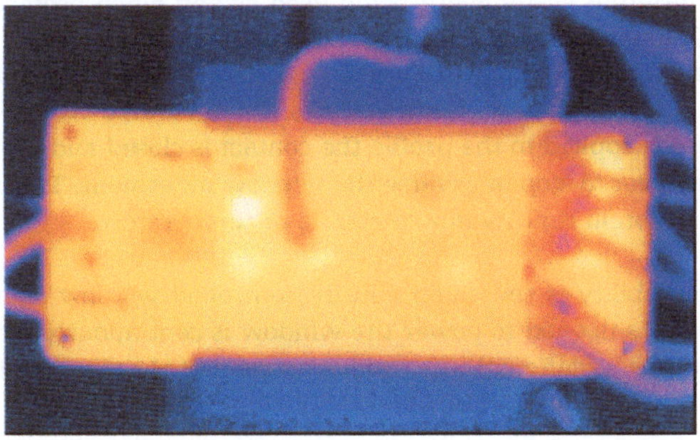

Fig. 6. Thermograph AutoDoor Alpha Version MCM.

Soft-start: The battery voltage applied to the motor is controlled to limit the starting peaks of current during the first 100ms. The system applies PWM with a duty cycle increasing from 0% to X% in steps of 10% each 10ms, where X% is the duty cycle calculated for battery compensation.

Another very important issue is the thermal performance of the system. An especial version of the MultiChip Module (called Alpha version) was developed at early stages of the project to perform the thermal test. The power devices were tested in both current during switching at 25 kHz switching frequency and DC operating conditions up to 20A in order to assure that the IMS substrate satisfied the heat transfer requirements for the power devices.

The infrared thermography pictures of the module (Figure 6) showed that the temperature increase of the power devices reaches 51 °C in the worst case, while the substrate attains an almost homogeneous temperature of 43 °C, if free convection cooling is allowed on the aluminium side of the module.

7 Anti-Pinch Algorithm

Power windows are now standard in new cars and the automated function is becoming increasingly popular. The more comfort the car driver requires, however, the greater the need for safety as all automated functions are latently dangerous. Although not a common occurrence, the news of a child having been throttled by an electric window, highlights just how important an efficient and reliable anti-pinch system is.

Usually, two different systems are being used as anti-pinch protection in the automobile. One possibility is the tactile sensor that consists of two electrically conductive surfaces. Usually, the sensor is integrated directly into the window seal. In its normal state, both surfaces are separated from each other. When pressure is applied to the profile the contact is closed and the controller is informed in order to stop or reverse the window movement. The main problem is the variation in rigidity due to ageing or temperature.

The second possibility is the electronically controlled window opener [4]. In this case, the force needed to power the window is continuously measured. If this force deviates significantly, the motor stops and the window is put into reverse.

The antipinch is based on sensing two parameters:

- Power Window Motor Speed. Sensing the speed motor an obstacle can be detected if the speed decrease (the system is able to use hall-effect sensors and/or ripple-counting).
- Power Window Motor Current. Sensing the current supplied to the motor an obstacle can be detected if the current increase.

A typical "express up" window movement is shown in the Figure 7. The instantaneous current (in blue) is sampled and measured in order to extract the speed and the "mean current". Using these inputs and a software algorithm based on the prediction of the speed and mean current sequence combined with a self-learning function allows the system to compute the force (in red) fulfilling the FMVSS118 norm [5].

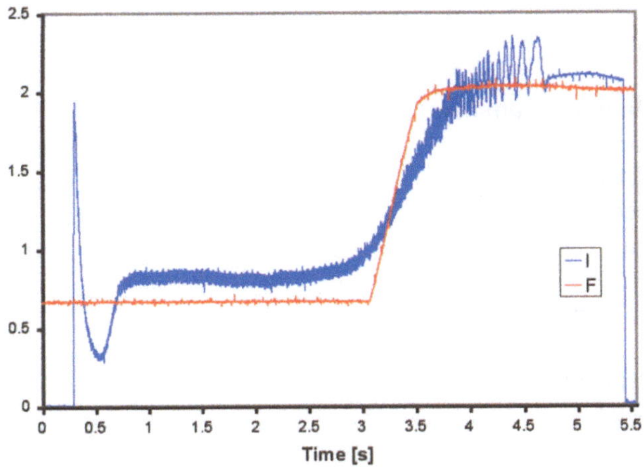

Fig. 7. Current and computed force in an "express up" window movement.

8 Conclusions

In this paper a multichip module system for automotive applications is described. The MCM substrate is a IMS (Isolated Metal Substrate), which has been used for integration of analog signal processing stages, complex control circuits and even power switching transistors. The main functionality of the door module implemented with this MCM, is the control of the power window motor with antipinch feature.

References

[1] R. Ohba, Intelligent Sensor Technology, John Wiley & Sons, 1992
[2] M.H. Westbrook and J.D. Turner, Automotive sensors, Institute of Physics Publishing, 1994
[3] PCT No W0239799
[4] S. J. Chapman, Electric Machinery Fundamentals, McGraw Hill, 1999 (Third Edition)
[5] Federal Motor Vehicle Safety Standard 118, Power-Operated Window, Partition, and Roof Panel Systems

Antoni Ferré, Rafael Jiménez, M. López
European Technological Center, Lear Automotive (EEDS) Spain,
c/ Fusters 51
43800 Valls
Spain
aferre@lear.com

S.A. Bota, J.L. Merino, J. Samitier
Universitat de Barcelona
Departament d'Electrònica
c/ Martí i Franquès 1
08028 Barcelona
Spain

B. Niessen
Austriamicrosystems AG
Via Leone Tolstoi 64
20146 Milano
Italy

E. Cabruja, J. Bausells
Centre Nacional de Microelectrònica (CNM-CSIC)
Campus UAB
08193 Bellaterra
Spain

Keywords: automotive, microsystem, door module, antipinch

Cabin Air Quality Management in Automotive Practice

T. Kardinahl, M. Richter, R. Mönkemöller, K. D. Frers, paragon AG

Abstract

Air quality management in the passenger´s compartment is important for comfort but also for more safety. The complex functionality is based on an intelligent network of microsystems like sensors and actuators. The development of an air quality management in vehicles from the early state of dust control to today performance is reviewed. As an example an air quality sensor is described and characterized. This smart sensor device monitors the outside concentration of gases like NO_x and closes the recirculation flap in case of increased pollution. To improve a given air quality it is necessary to build up an active element: an air quality conditioner for micro-oxydation. Based on the principle of dielectric barrier discharge it produces active oxygen, oxygen-ions and ozone. These gases are known for their sterilization effects, and the number of germs and bacteria in the air is reduced.

1 Introduction

More and more influences must be taken into account to improve the air quality in the driver´s cabin. At first the intake of harmful gases and odor from the outside must be controlled. To improve air quality management the cabin must be protected from the surrounding. On the other hand the sources of contamination are located in the cabin or the climate system. Identification is necessary to eliminate them. New requirements are in discussion [1,2].

All these influences can be handled by an intelligent sensor actuator system.

2 Historical and Future Development – From the Outside to the Inside

Cars are manufactured to drive through our nowadays environment daily. The driver should be save in his cabin. The development of air quality management is shown in figure 1.

All efforts started with filters to exclude dust. Improved filters absorb low concentrations of some odors. Of course the best way to prevent pollution is to shut the recirculation flap befo re the gases enter the ventilation system. This is possible with an intelligent sensor system which is described in the next chapter. Doing this the life time of filters increases.

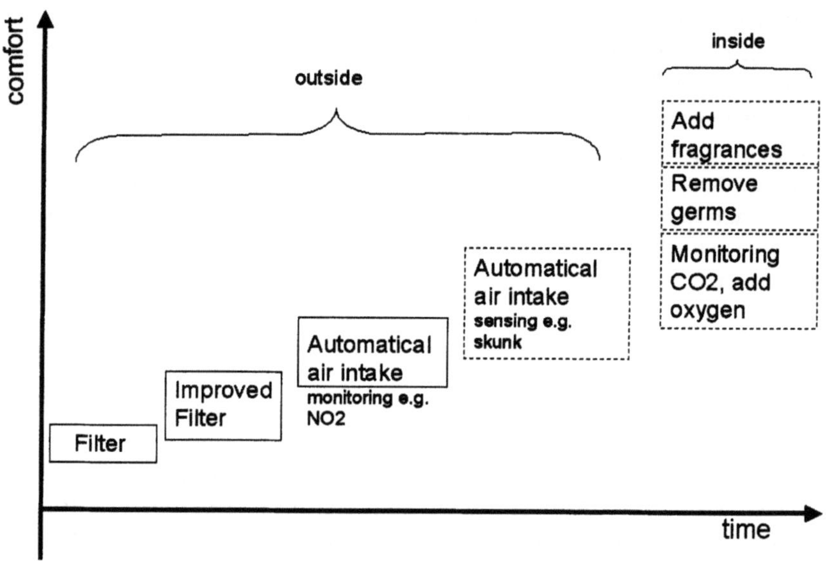

Fig. 1. Development of air quality management.

Parallel, climate control systems have been developed to control temperature and humidity. Part of this system is the condenser, which gets wet and warm under some conditions. This is an optimal biotop for germs which spread harmful and toxical gases. They are polluting the climate system and the cabin air. A concept to avoid this is presented in the last chapter.

The development in cabin air quality is turning from the outside to the inside at this point. Here an external parameter is controlled (the temperature) and the ventilation system produces a new contamination (germs).

Inside the car are sources, too. The applied materials like plastics are evaporating volatile organic compounds (VOC). Human beings themselves produce CO_2 while breathing and its concentration must be monitored. In some cases oxygen will be add in future. Smokers pollute the cabin and are concerned about their health.

In a last step sometimes fresh odors or fragrances are added for the drive´s convenience. This masks the odors but does not neutralize them.

Future developments will be:

- Still incoming strong odors like trash, liquid manure or skunk are eliminated. They will be detected by new sensors and the air intake is switched off.
- The growth of germs at the condenser as a source of harmful gases must be prevented. One possibility, the micro oxidation, is discussed later.
- Odors from the inside (smoke, VOCs) must be reduced. One technique is the photo catalysis.
- The CO_2-concentration must be monitored to avoid depletion for the driver´s safety. Oxygen will be added.

3 Air Quality Sensors in Automotive Practice

Air quality sensor are a good example to visualize the connection between a microsystem and a macroscopic device. This smart sensor device monitors the outside concentration of gases and closes the recirculation flap in case of increasing pollution. The proportion of fresh an recircled air depends on the monitored value of the sensor device.

For various reasons thick film based MOS (metal oxide semiconductor) are chosen for this application in rough surroundings. The research on understanding these materials is very deep. The principle of operation is a change of resistance in dependence on the air quality, see for example [3,4]. Their long time stability fulfils the automotive requirements.

A state of the art electronics evaluates the sensor signals. An intelligent software controls the monitoring of the air quality. Dependent on the signal and its dynamic the values refer to a level of air contamination. This level is transmitted to the climate control computer which decides to close the air intake.

All components are capsulated and housed as shown in figure 2. The maximum dimensions are approximately 39mm x 35mm x 25mm. The sensor system has a weight of 15g.

Fig. 2. Smart sensor device Air Quality System by paragon AG.

Figure 3 shows how the air quality is monitored during a typical test drive on highways, country roads and in the city. The change of sensor signal is shown with time, or because the car is moving with place. At a high content of pollution the sensor signal drops. If the air quality improves the sensor signal rises again. Of cause the values depend on the environment. On a windy day the pollution is spread over larger areas and diluted. Never the less the significant changes are detected by the intelligent software.

In figure 3 significant changes are indicated by arrows. There are several clusters which lead to a dramatic decrease and these data can be connected to the source of pollution. A first group are trucks, see 1,2,5. The signal goes down while driving beside or overtake them. A real disaster are tunnels. Inside the tunnel the ventilation system of the tunnel is not strong enough to keep the pollution due to the cars and trucks in certain limits. The graphic clearly points out that the sensor signal changes when entering (3,7) and – in the opposite direction – when leaving (4,8) the tunnel. At traffic lights and at crossings pollution is high, number 6.

These data are evaluated and translated into levels of air quality in accordance with the car manufacturer. In principle a high decrease of the sensor signal by entering the tunnel shows a bad air quality and therefor the air intake is closed.

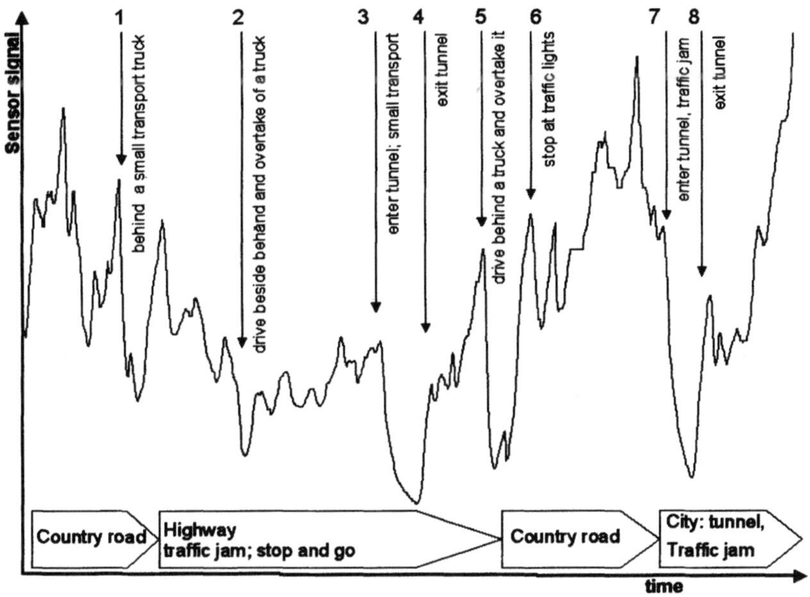

Fig. 3. Typical sensor signal during a test drive.

Sensors usually have location requirements and there are some sensors that can be conveniently clustered to take advantage of the synergies. Packaging is one of the more expensive parts of sensor manufacturing. The single housing reduces costs and weight. The eliminated connectors improves reliability and the assembly process and reduce vehicle weight.

This investigation leads to a multifunction device with microsystems measuring the air quality, temperature and humidity to control the air intake.

4 Improving Air Quality Due Micro-Oxidation

To improve a given air quality it is necessary to built an active element for micro-oxydation: the Air Quality Conditioner. This technique is also used to clean noxious fumes.

Based on the principle of dielectric barrier discharge it produces active oxygen, oxygen-ions and ozone. The principle is shown in figure 4. A dielectric material is sandwiched by two electrodes. In this case the lower electrode is planar and the dielectric material is covered by a grating used as the lower electrode. A high voltage alternating field is applied to accelerate the electrons. They make elastic and inelastic impacts with the present gas atoms. If the energy is high enough ionization occurs. The process is complex, especially when second and third order reactions are discussed.

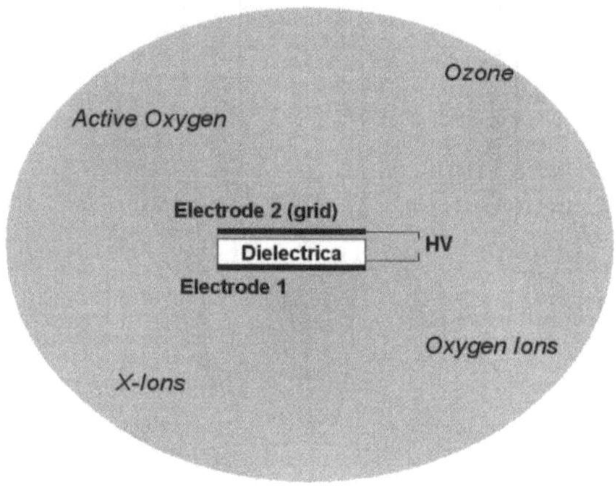

Fig. 4. Principle of operation.

Possible devices are the Siemens-tube and their deviations. Two prototypes are shown in figure 5. The step towards a suitable microsystem is small.

The arising gases are known for their sterilization effects, and the number of germs and bacteria is reduced. For the automotive application it is of importance to control the process carefully. The ozone concentration in the drivers compartment must not compromize the health of the passengers, and must be kept under certain limits.

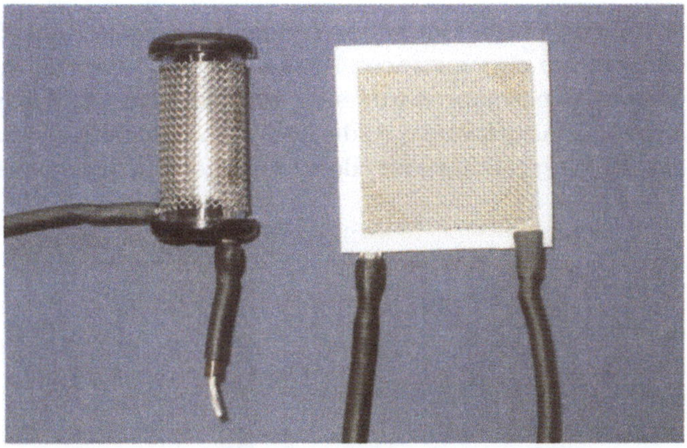

Fig. 5. First and second generation prototype of the active element of the Air Quality Conditioner.

An optical ozone analyzer based on uv-photometry monitors the dependence of the ozone generation rate on the applied voltage in a given (small) volume. Figure 6 clearly points out that the ozone generation starts at a fixed value and increases then rapidly. From theory we know that the kinetic energy of the electrons must be higher than the binding energies.

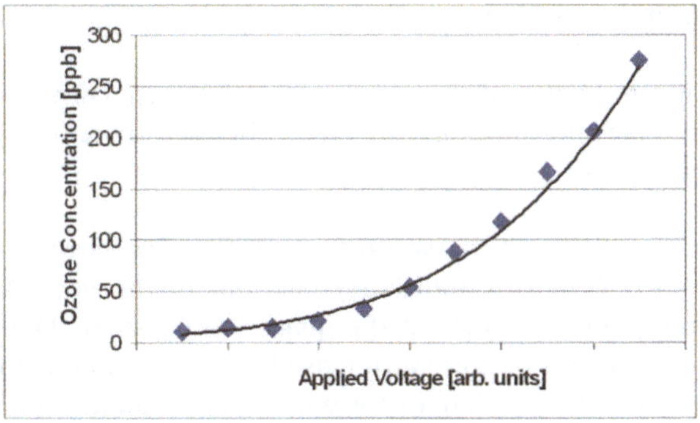

Fig. 6. Dependence of the Ozone generation on the applied voltage in a give small volume $<0.01 m^3$ (example).

We focus on the special problem of germs at the condenser. Figure 7 illustrates the situation. From outside we have dust, particles and odor. The permeation of the dust is prevented by the filters. Odors from outside are partly refused by the filter and – as discussed – by a sensor system which controls the automatic air intake.

Inside the climate control system the condenser becomes a perfect biotop for germs under certain circumstances. At the end of a drive on a sunny summer day the condenser becomes warm and wet. The motor is switched off and in the next hours germs grow up quickly. By switching on the ventilation systems on the next morning the harmful odors will enter the drivers cabin.

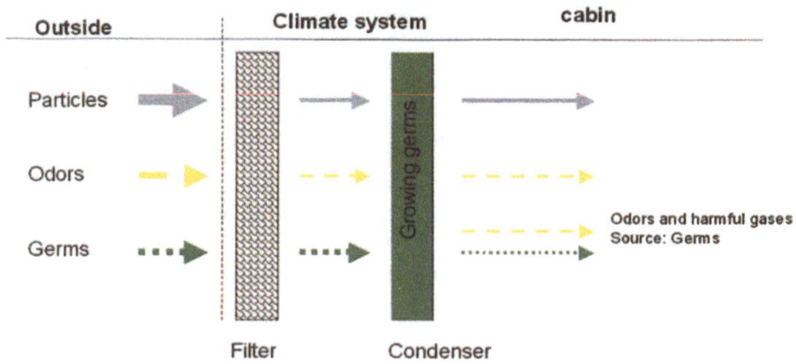

Fig. 7. Scheme of the sources of odors in the cabin.

A possible solution may be a permanent disinfection with low concentrations of ozone and the related ions. There is a direct impact of ozone on growing of the germs. Therefore an active element is positioned in front of the condenser. At the condenser ozone reacts with the water to H_2O_2 which is used in many hospitals as a disinfectant.

Most important is the health of the driver. The ozone concentration shall not reach values above certain limits in any case. This will be done by intelligent software and electronics.

For estimation of the parameters we performed first measurements in a car, see figure 8. The Air Quality Conditioner was put in the ventilation system with a fixed ozone generation rate at the given temperature. One louver was open at the drives side. All others were closed. The ozone concentration was monitored directly at the open louver for various air velocities (air ventilation system step 1 to step 12).

Figure 8 shows the ozone concentration for three different applied voltages. As discussed above each voltage varies the ozone generation rate. The air flow was increased step by step.

The achieved ozone concentration is below 45 ppb in any case. In the cabin the ozone is diluted and the value even lower.

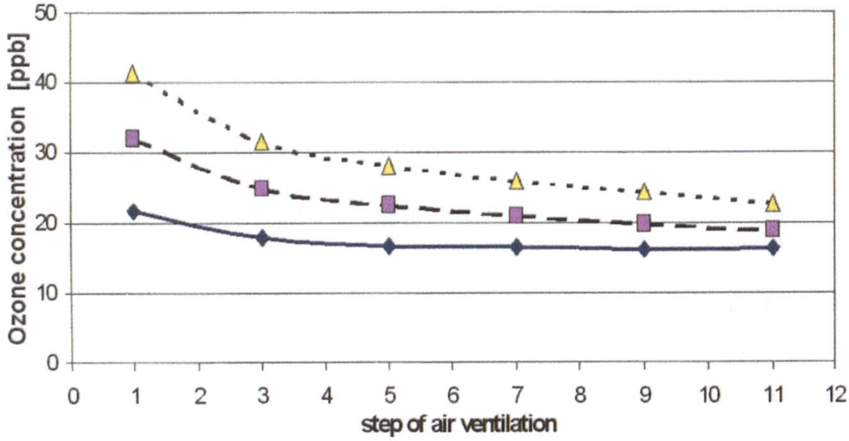

Fig. 8. First results of the ozone concentration in the cabin at the louver for three ozone generation rates.

5 Conclusions

The perspective of modern cabin air quality management turns from outside to inside the cabin. The development to avoid incoming pollution improves with the development of the sensor with new sensitivities. A step to future may be a multifunctional device bases on the Air Quality System.

Inside the car ventilation systems new challenges occur due to germs and the produced harmful gases. Micro oxidation is a chance to improve air quality substantially.

To summarize the development, drivers cabin is still one of the cleanest places for his safety and health – in most cases cleaner than the surrounding air outside.

References

[1] VDI 6033

[2] VDI 6022

[3] Kohl, D.: Function and applications of gas sensors, J. Phys. D: Appl. Phys. 34 (2001) R125-R149

[4] Hauptmann, P.: Sensoren - Prinzipien und Anwendungen, Carl Hanser Verlag, München, Wien, 1991

Thiemo Kardinahl, Marcus Richter, Ralf Mönkemöller, Klaus Dieter Frers
paragon AG
Schwalbenweg 29
33129 Delbrück
Germany
thiemo.kardinahl@paragon-online.de

Keywords: air quality, sensors, micro-oxidation

Advanced Driver Assistance Systems for Increased Comfort and Safety – Current Developments and an Outlook to the Future on the Road

R. Adomat, G.-O. Geduld, M. Schamberger, ContiTemic
Dr. Peter Rieth, ContiTeves

Abstract

Adaptive Cruise Control and Parking Aid Systems are already introduced in the automotive market for years. Based on environmental detection sensors all systems available today are comfort systems. The driver can overrule the system at any time. Other systems supporting the driver in the lateral control of the vehicle are under development. For the future a common direction on all driver assistance systems is the opening out to safety. This seems to be one of the key factors to penetrate the market sufficiently. Almost 95 percent of all car drivers regard safety as being more important than comfort. The second important target for future driver assistance systems is the multi-use of sensors to cut the over-all system costs.

1 Introduction

The „seeing" car is not a future scenario any more; it is already on the road. Advanced driver assistance systems are introduced in various vehicles during the last four years. Today Adaptive Cruise Control (ACC) systems based on radar or infrared sensor technology are available in different cars, including Mercedes-Benz, BMW, Jaguar, Volkswagen, Lexus, Mitsubishi and Nissan.

ACC provides supports to the longitudinal control task of the driver with the intention to minimise the workload and optimise the condition of the driver even on long distance travel. Manipulating the acceleration, brakes and gear control, the system is able to maintain an adequate distance depending on the present traffic situation. ACC is called a comfort system, because the driver can overrule ACC at any time.

Another area of interest on driver assistance is the lateral control of the vehicle. Systems that will warn the driver and/or will steer the vehicle to prevent an unintended lane departure are already available. Using the well designed infrastructure of the present roads video image processing will determine the road geometry and the present driving path by detecting the lines on the road surface. By active steering and/or braking of individual wheels the Lane Keeping Support (LKS) system assists the driver in his lateral control of the car.

Fig. 1. The seeing car.

Beside these applications additional driver assistance systems will show up within the next years. Advanced functionalities using existing sensor and actuation technologies (e.g. for ACC) will help to reduce fatalities or will mitigate crash results within the next future. Systems to provide assistance to the driver on parking manoeuvres or on its ride in bad whether conditions or during the night are in development, and will be on the market within the next three years. All types of driver assistance and safety applications based on environmental sensing can be grouped into four categories: longitudinal support, lateral support, low speed manoeuvres (incl. reversing) and vision enhancement.

According to the EU commission [1] the road traffic accidents and the fatalities should be reduced to 50% by 2010. The vehicle technologies to reach this goal are advanced driver assistance systems that are mainly designed to increase the driver's comfort, but will be enlarged to safety applications. These systems will take into account not just the driver and the vehicle, but also the environment around the vehicle. At the end of this process the same technologies will be exploit for driver assistance and intelligent active safety systems.

2 Sensor Technology for Advanced Driver Assistance

One of the several factors defining the introduction scenario for driver assistance systems is the availability and realization of sufficient sensor technologies. The large variety of driver assistance applications call for a wide span of sensor specifications, which require different sensor technologies. In long term it can be expected, that ultrasonic, active infrared, radar and 2-dimensional as well as 3-dimensional computer vision sensors are used in the vehicles.

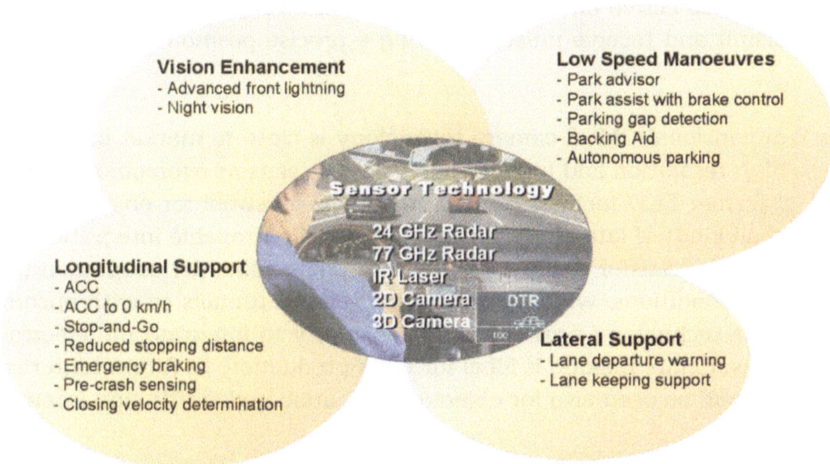

Fig. 2. Driver Assistance Systems on the road or under development.

The ultrasonic sensor is very attractive in cost and size; it covers a wide field of view and is well known due to its long ago market introduction in 1991 [2]. Its disadvantages are the detection performance, which allows shielding a range up to 2m around the vehicle. This limits the operation to low speed applications where only short detection ranges are required, typically parking aid assist systems.

The 24GHz short range radar technology will be introduced into the market in 2004/2005. Their detection performance exploit these sensors for most kind of applications, where a robust and precise object detection up to 30m with a high data up-date rate is required. Precise lateral object measurement and the full coverage around the vehicle can be reached by the use of multiple short range sensors, connected in a sensor network. The 24GHz short range process is ongoing to harmonize the use of the 24GHz frequency band for vehicle sensors [3].

The 77GHz long range radar technology is used primarily for ACC applications. It covers a range up to 150m. Starting in 1999 with designs and production technologies not adequate for the high volume automotive market, the radar sensors produced today allow a sufficient pricing. Radar sensors enable obstacle detection in most driving situations, and they are widely independent from weather conditions [4].

The active infrared sensors are operating with a time-of-flight measurement in the range between 0.2m up to 180m. A short transmitted pulse is reflected by all objects in front of the vehicle. The distance to the objects is determined by the time between the transmit and the receive pulse. Compared to the radar technology the active infrared sensors provide a very sharp lateral control of the transmit and receive pulses, allowing a precise positioning of the objects [5].

The 2-dimensional CMOS camera technology is close to market introduction. These high resolution and high dynamic passive sensors overcome the restrictions of former CCD technology and create the essential for object classification and all kinds of lateral control applications. Controllable integration times or logarithmic transfer curves of the sensor chips allow the adaptation to difficult light conditions, which occur on the road in tunnels or with oncoming traffic. The technology provides the opportunity to integrate the electronics on the sensor chip, making it ideal for mass production. In future high resolution chips will be used also for object classification and traffic sign interpretation.

Sensor technology	Range	Field of view	Applications	SOP	Example
Ultrasonic	<2 m	about 60°	Parking aid	in production	
24GHz short range radar	<30 m	about 60°	Stop-and-Go-Assist, pre-crash, advanced parking aid, blind spot detection, lane change support	2004/05	
77GHz long range radar	150 m	about 16°	ACC, longitudinal control, reduced stopping distance, crash mitigation, FSRACC (with 24GHz short range radar)	in production since 1999	
Active infrared	150 m	about 45°	ACC, visibility estimation, advanced front lightning, crash mitigation , FSRACC	in production since 2000	
2-dim CMOS camera	120 m	about 45°	Lateral vehicle control, LDW, LKS, object classification, traffic sign interpretation	CCD: in production CMOS: 2005	
3-dim PMD camera	<30 m	about 45°	Advanced LKS, object and pedestrian detection, safety applications	2006/07	

Fig. 3. Sensor technology for driver assistance

The 3-dimensional PMD camera is an active technology for the fast detection and analysis of traffic scenarios up to 30m. The scene is illuminated by a modulated non-visible infrared light, and the reflected signals are received by a PMD chip. The PMD camera provides distance and grey picture information for each pixel of the PMD matrix [6]. The fast analysis of traffic scenarios is a key factor for further safety applications.

Today based on a study of Frost&Sullivan [7] 99% of all sensors are ultrasonic, and are used in parking aid systems. This will change within the next years, when other sensor technologies become more important and other assistance and safety applications show up in the market.

3 Longitudinal Control of the Vehicle

The longitudinal vehicle control contains systems like ACC, Stop-and-Go support and automatic braking systems , all responding to environmental traffic situations.

3.1 ACC Today (ISO 15622)

Like in a regular cruise control the driver sets the speed, he wants to drive. ACC regulates the own vehicle to this set speed, when there is no vehicle ahead. When detecting a slower vehicle in the own lane, ACC measures the distance continually and adapts the speed to follow the vehicle ahead by controlling the brake, the engine and the gearbox. The deceleration is due to legal and comfort reasons limited to a maximum of 0.3g. If the own lane is clear, ACC accelerates to the set speed again.

If the maximum deceleration is not sufficient to keep the right distance, may be because the vehicle ahead is changing the lane suddenly, ACC generates a warning and requests the driver to activate the brake in addition to the ACC braking.

The driver is responsible for the vehicle at any time, even in ACC operation. But ACC relieves the driver from the routine work to control the vehicle speed when cruising, so that he can concentrate on other important tasks.

Beside this comfort feature ACC offers a safety aspect as well: The ACC sensor identifies the traffic conditions immediately and reacts on changing situations faster than the driver can do. The own vehicle is decelerated to adapt the speed, when ACC is in operation and a critical situation occurs. When the driver activates the brake after his reaction time, the vehicle is already decelerated to a slower speed than without ACC. Also, the brake is pre-conditioned to the ideal operating point by the ACC, so that no additional swell time occurs, when the driver activates the pedal. This supports the reduction of the stopping distance.

3.2 ACC to 0km/h (FSRACC) and Low Speed Follower (LSF)

The enhancement of the current ACC will be the full speed range ACC, coming to the market in 2005. Short range detection capability help to cover the entire area in front of the vehicle, so that all moving objects – even in urban road conditions – are detected. The vehicle is decelerated down to standstill when following a vehicle ahead. A drive-off assist keeps the pressure in the brakes, so that the vehicle stands still reliable, independent from the gradient of the road. This condition is kept without the driver's interaction.

Fig. 4. Stop-and-Go traffic scenario.

The full speed range ACC (FSRACC) notices, if the vehicle ahead starts to drive again, and informs the driver. After confirming by the driver, the vehicle automatically accelerates to the set speed or to the speed adequate for the current traffic conditions.

This feature supports the driver in stop-and-go situations, but still requires the driver to confirm the re-acceleration of the vehicle. A fully autonomous stop-and-go-system requires additional development work to cover all the complex scenarios on the road.

3.3 Reduced Stopping Distance

The reduced stopping distance (RSD system) is an extended brake assist function, where the brake is adapted to environmental traffic situations. The system will operate in all ACC equipped vehicle, even if ACC is not in use. After the detection of a dangerous situation by the ACC sensor the brake is pre-filled immediately. With this conditioning the vehicle reaches the full brake performance earlier in a braking manoeuvre. As a result the accident can be avoided, or the crash energy can be reduced by up to 50%. Especially for smaller vehicles having a short crash zone this can mean an advantage for the drivers safety.

Fig. 5. Dangerous traffic situation requiring RSD support.

The RSD system determines the position, the distance and the range rate of all objects ahead. During the approach to a preceding object a brake pressure will be applied like in ordinary ACC functionality automatically. If the time-to-collision falls short, the algorithms of the brake assist will be adapted to a faster apply. Releasing the accelerator pedal rapidly by the driver a higher brake pressure will be implemented (like ACC max brake). If the driver uses the brake pedal in panic, the brake assist will be activated in addition.

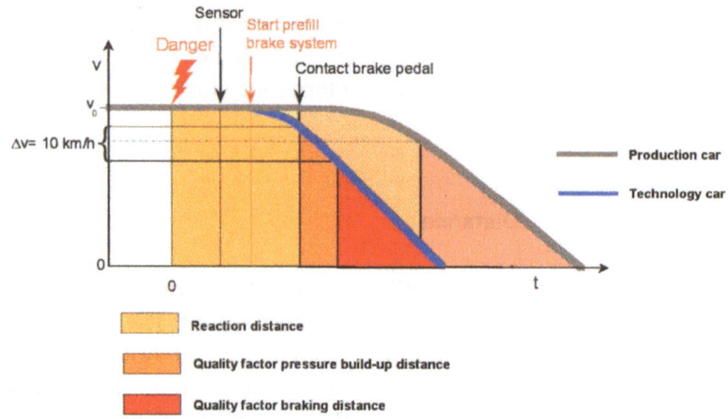

Fig. 6. Braking behaviour with and without RSD.

4 Lateral Control

The Fatality Analysis Reporting System (FARS) published by the NHTSA analysed the root causes of run-off-road accidents in the US. It reports 15.000 annual deaths.

Similar figures showing about 20% of all accidents caused by driver fatigue are available for Germany [8].

Fig. 7. Three types of typical run-off-road accidents

A technical solution to avoid this kind of accidents is the electronic stability system ESP. By applying single wheel braking the driver keeps the control on his vehicle during the steering maneuver.

To increase the road safety forehand, lane departure warning systems and lane keeping support systems can help to avoid the occurrence of the skid situation itself.

4.1 Lane Departure Warning and Lane Keeping Support

A lane departure warning system detects the vehicle position in its lane and informs the driver, when the vehicle approaches the lane borders unintended. To reach the best benefit the lane departure warning system is available both during night and daylight conditions. Most of the systems available today or under development use camera vision sensors to detect the relative position between the desired vehicle path and the present road geometry (lines). These kinds of sensors scan the road 40m to 60m ahead of the vehicle to grab the necessary information. A robust system approach requires high dynamic cameras and sophisticated algorithms to handle. The critical situation is analysed by calculating the lane position change by generating an acoustical beep or by a haptical feedback on the steering wheel or the driver's seat.

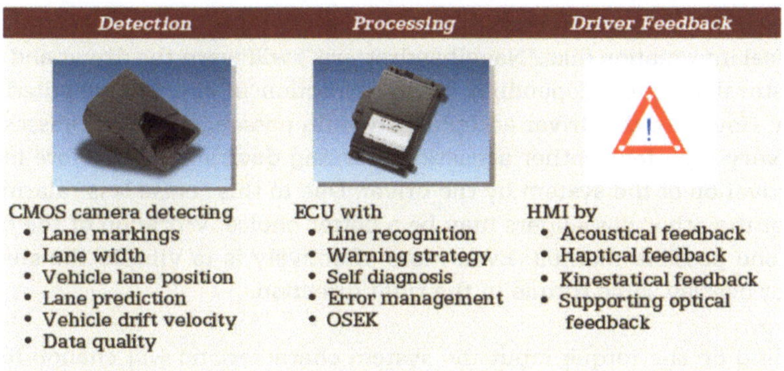

Fig. 8. Components of a lane departure warning system.

Main issue to inform the driver regarding an unintended lane departure is the kind of HMI. Due to the fact, that lane departure will mostly happen during boring long straight road sections, optical information will not be the adequate manner.

Fig. 9. LKS applies brake and steering to keep the vehicle in its lane.

Acoustical information (like "Nagelbandrattern") will warn the driver and causes a natural reaction depending on the direction of the implemented loud speaker. However, the driver acceptance within passenger cars or busses may not be very high for another acoustical warning device and therefore lead to an inactivation of the system by the driver. Due to this "noise less" alarms not to up-set the other passengers may be a better choice. Vibration of the driver seat is one possible method. Even more effectively is to vibrate the steering wheel or overlay some torque in the right direction.

Depending on the torque input the system characteristic will change form a simple warning device up to an autonomous control device.

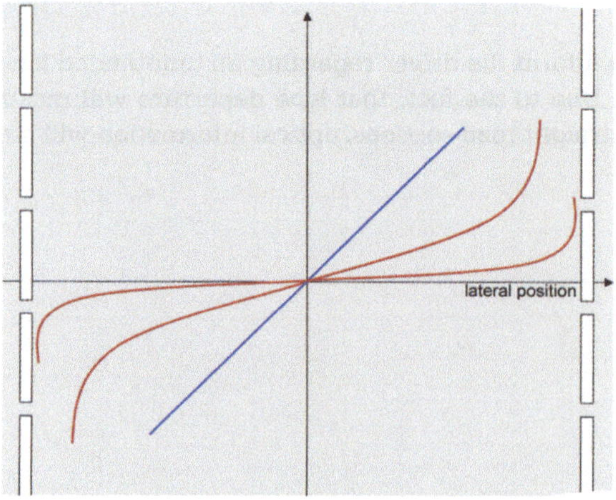

Fig. 10. Transfer functions for warning, lane keeping support and lateral control .

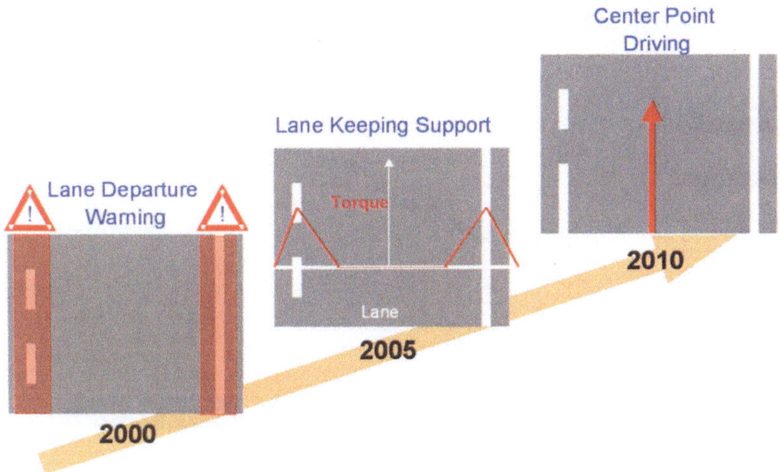

Fig. 11. Roadmap for lateral vehicle control.

5 Integration of Active and Passive Safety

The following definition for active and passive safety is used:

- Active safety means all measures to prevent an accident or to minimize the effects of an accident.
- Passive safety means all measures to protect the occupant and the pedestrian against injuries caused by an accident.

The active safety phase covers the regular driving up to the time, when a crash occurs. This is the moment, when no active system can help to avoid the accident. The passive safety phase is not limited to the time after the crash, but includes the time before the crash, when reversible actuators are triggered. With this definition there is a time overlap of active and passive safety.

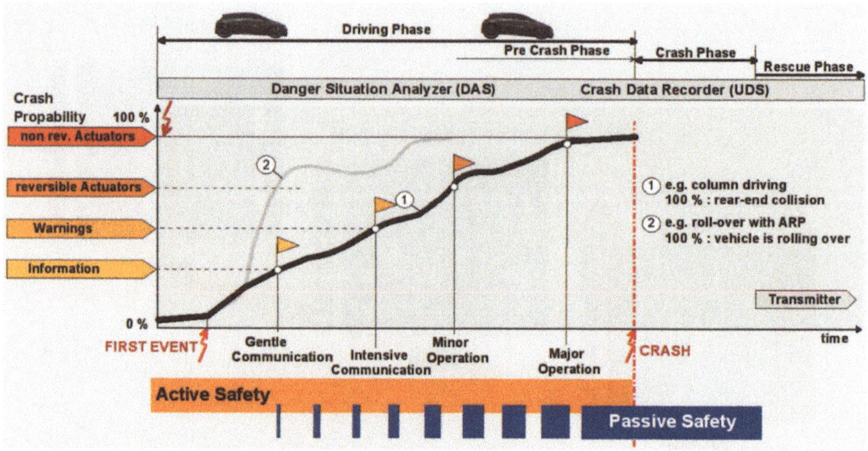

Fig. 12. Definition of active and passive safety.

5.1 Pre-crash

Restraint systems today are activated based on the signals of deceleration sensors located central or de-central in the vehicle. All measures and algorithms can be triggered, when the crash already appeared. Therefore the time to activate protection devices is limited.

With a pre-crash sensor the restraint systems gather information in the time before the crash. Objects around the vehicle are detected and their relative position, their relative velocity and the time-to-impact can be calculated. With this information the behaviour of the passive safety devices in the vehicle can be optimised.

5.2 APIA

The active and passive integration approach APIA combines the advantages of all available systems in the vehicle to optimise the driving comfort and accident behaviour.

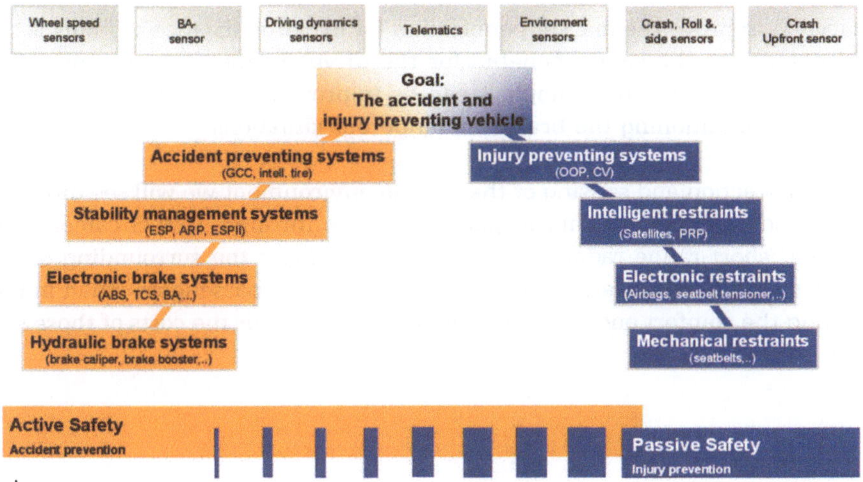

Fig. 13. Applications on active and passive safety.

APIA accomplishes new unprecedented functionality with comprehensive vehicle occupant and traffic participant injury protection through the decisive networking of active and passive system components. To achieve this all measures to avoid the accident and minimize the effect of an accident are initiated with best possible effect. The analysis of the potential danger by a crash probability calculator, the monitoring of active and passive system information and the launching of stepped protective measures are the technical methods for the realisation.

6 The Road into the Future

A common direction on all driver assistance systems is the expansion to safety. This seems to be one of the key factors to penetrate the market sufficiently. Almost 95% of all car drivers regard safety as being more important than comfort. More than 70% are in favor of ABS and ESP being made a legal requirement. These and other findings about the acceptance level of electronic safety and assistance systems are contained in the safety study presented by Continental AG in autumn 2002 [9].

Crucial to improve the market penetration for driver assistance systems is the engagement of the safety benefits for the driver to the end customer. ACC equipped vehicles – for example – can be improved in their braking process when pre-conditioning the brakes in dangerous situations.

For the detection and sensing of the vehicle environment we will see different sensor technologies working in parallel. Active infrared, CMOS camera and long and short range radars will be used to evaluate the surrounding of the vehicles. The huge advantage for the driver and the passengers when implementing the comfort and the safety features will pay for the costs of those systems.

References

[1] EU commission initiative "e-Safety - Research on Integrated Safety Systems", 2002, http://europa.eu.int/information_society/programmes/esafety/index_en.htm

[2] Valeo Product Data „Ultrasonic Park Assist", http://www.valeo.com

[3] SARA "Ultra wide band 24GHz Radar Sensors for Automotive Applications", 2002, http://www.ero.dk/EROWEB/SRD/UWB-2/Presentations/UWB%20Workshop%2002041 1%2005_020404_Ka_2_HH2.pdf

[4] DaimlerChrysler, "Your car will be watching the road – even if you're not", http://www.daimlerchrysler.com/index_e.htm?/company/campaign/mot6/detail6_e.htm

[5] Nissan Motor Co., Ltd „Technologies for reducing the driver's workload", 2000, http://www.nissan-global.com/GCC/Japan/NEWS/20000405_0e.html

[6] ContiTemic Press Release, "Optische Sensoren erhöhen Insassenschutz", 2001, http://www.kfzelektronik.de/b_020023.htm

[7] Frost&Sullivan, 2002, „Moving Into Top Gear: Driver Assistance Systems Propel Driving Experience To A New Level", http://www.frost.com

[8] Fatality Analysis Resorting System, NHTSA, 1999, http://www.nhtsa.dot.gov

[9] Automobilclub von Deutschland AvD, Alfred F. Fuhr, "Sekundenschlaf", 2001, http://www.avd.de/index.html?presse/aivs/aivs_vortrag_sekundenschlaf.htm

[10] EASi und ContiTemic „Sensorstudie für den Frontalaufprall", 2002, http://www.easi.de/company/publications/bag_belt_2002-de/bag_belt_2002-de.pdf

[11] Study of Continental AG, November 2002, "Continental Sicherheitsstudie – Deutsche wollen mehr aktive Sicherheit im Auto", http://www.conti-online.com/generator/www/index_uv.html

Rolf Adomat, Georg-Otto Geduld, Michael Schamberger
ContiTemic - PL Advanced Driver Assistance Systems, A.D.C. GmbH
Kemptener Str. 99
88131 Lindau
Germany
Rolf.Adomat@adc-gmbh.de

Dr. Peter Rieth
ContiTeves - Zukunftsentwicklung
Guerikestr. 7
60488 Frankfurt
Germany
Peter.Rieth@contiteves.com

Keywords: ACC, Active Safety, Adaptive Cruise Control, APIA, Blind Spot Detection, Comfort Systems, Computer Vision, Distronic, Drive-by-wire, Drive-Off Assist, Driver Assistance Systems, Emergency Braking, Full speed range ACC, Infrared Sensor, Lane Departure Warning, Lane Keeping Support, LDW, LKS, Long Range Radar, Multi-Use of Sensors, Night Vision, Photonic Mixing Device, Park Assist, Passive Safety, Pre-crash Detection, Safety Systems, Sensor Technology, Short Range Radar Sensor, Stop-and-Go Support, Stopping Distance, Vision Based Systems, Vision Enhancement

Functions and Applications

Networked Vehicle

Bus System Initiators

T. Görnig, Conti Temic microelectronic GmbH

Abstract

This paper is focused on the cooperation project IBA (Integrated Bus compatible Initiator) and shall give an overview about the results achieved during the project. The cooperation project itself was initiated on January 1st 2000 and last until December 31st 2002. The involved project partners of the cooperation project IBA are the following companies and institutes: Dynamit Nobel Automotive Initiator Systems GmbH, Fuerth; NICO Pyrotechnik Hanns-Jürgen Diederichs GmbH & Co. KG, Trittau; Chemnitz University of Technology, Department ET/IT, Center of Microtechnologies, Chemnitz; Conti Temic microelectronic GmbH, Ingolstadt and TRW - Automotive GmbH, Aschau/Inn. The research project is being funded by the BMFT (Ministry for Education and Research) and supervised by the VDI/VDE - Technologiezentrum Informationstechnik GmbH, Teltow.

1 Introduction

With the development of standards for automotive passenger restraint firing and sensor bus systems, the need for integrated bus compatible initiators has became evident. The firing bus systems like the BoTe System from Bosch and Temic (three pin initiators, Daisy Chain architecture) [4] or the SURFS System from Siemens (two pin initiators; parallel architecture) and the B/S/T firing and sensor bus system based on a development by Bosch, Siemens and Temic (combined Daisy Chain and parallel architecture) [10,11] are to be mentioned in this respect. An advantage of the bus systems is the increased flexibility as they can be introduced on a new carline by just adding the additional components and a reconfiguration of the bus master. An important component for a firing bus system is besides the bus master, the integrated bus compatible initiator.

2 Integrated Bus Compatible Initiator

For the functionality of a future restraint system that is based on a bus technology, it is important that the initiators of the various restraint devices have the required bus interface. All necessary electrical components have to be included in the initiator within the given mechanical outline. The integrated bus compatible initiator developed within the IBA project [5] fullfills all these mechanical and electrical requirements:

- Three or two connector pins for the bus interface (depending if parallel or serial bus architecture is selected).
- Integration of all electrical components into the initiator.
- Bus interface including daisy chain switch.
- ESD protection circuits (25kV@150pF;150W).
- Energy reserve for the bus interface and deployment electronics.
- Energy reserve capacitor for the initiator element.
- High side and low side firing stages.
- Bus protocol decoder and diagnostic circuit.
- Compatible with the mechanical dimensions of a conventional bridge wire initiator.
- All environmental and life time requirements (e.g. -40°C to +125°C; 15 years).

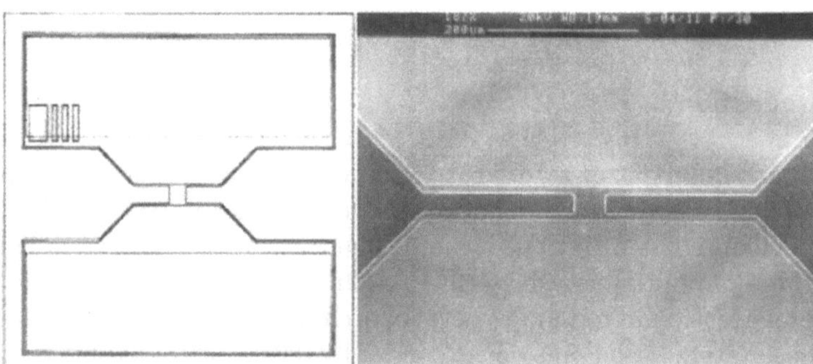

Fig. 1. Layout and REM image of a HfHx initiator bridge.

Due to assembly and technology restrictions the first generation IBA (known as IZE) has been built as a two-chamber design [6][7]. The electrical components and pyrotechnics was encapsulated into two separate hermetical sealed chambers. Further focus of the IBA project was on the consequent integration of the electrical components and pyrotechnics into one hermetical sealed chamber.

3 Initiator Bridge

The IBA project utilizes an earlier development of an initiator bridge that has been conducted within the SIFA/MST project. The HfHx initiator bridge has been further optimised and the structure of the element has been defined for further test series. The size of the HfHx silicon is app. 2x2mm² while the actual initiator bridge element is much smaller (figure 1). Tests which have been conducted within the IBA project showed an energy consumption <20µJ. The initiator bridge is manufactured by using standard semiconductor processes [3].

Firing Energy and Test

To achieve the smallest possible mechanical package for the IBA it is important that the firing capacitor also has the smallest possible mechanical dimension. The required energy that is stored in the capacitor to reliably initiate the initiator bridge has been defined in various test series. The different parameters that have an influence on the result have been taken into account. Typical test curves can be seen in figure 2. An example of the activation of the HfHx initiator element is shown in the images taken from a high speed video in figure 3. The parameters for the test series are defined as: C = 500nF; ESR = 0,3Ω; Ron = 3Ω @ 2µs; V(Bias) = 22,1V.

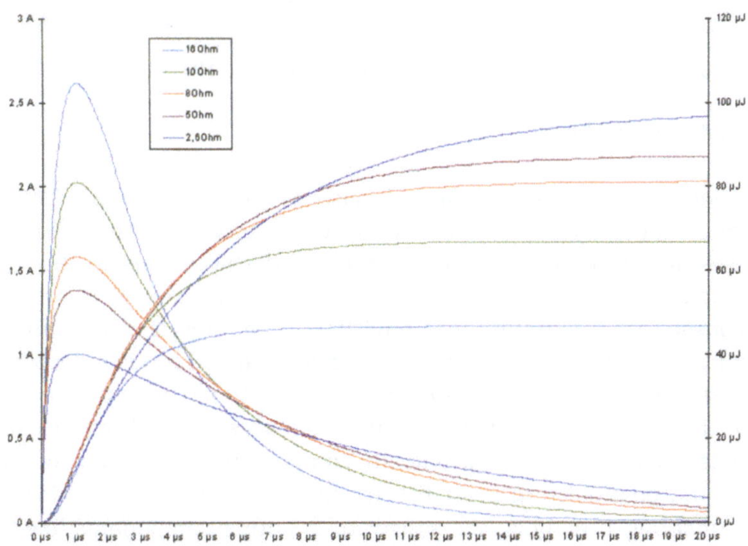

Fig. 2. Typical current and energy at HfHx initiator bridge.

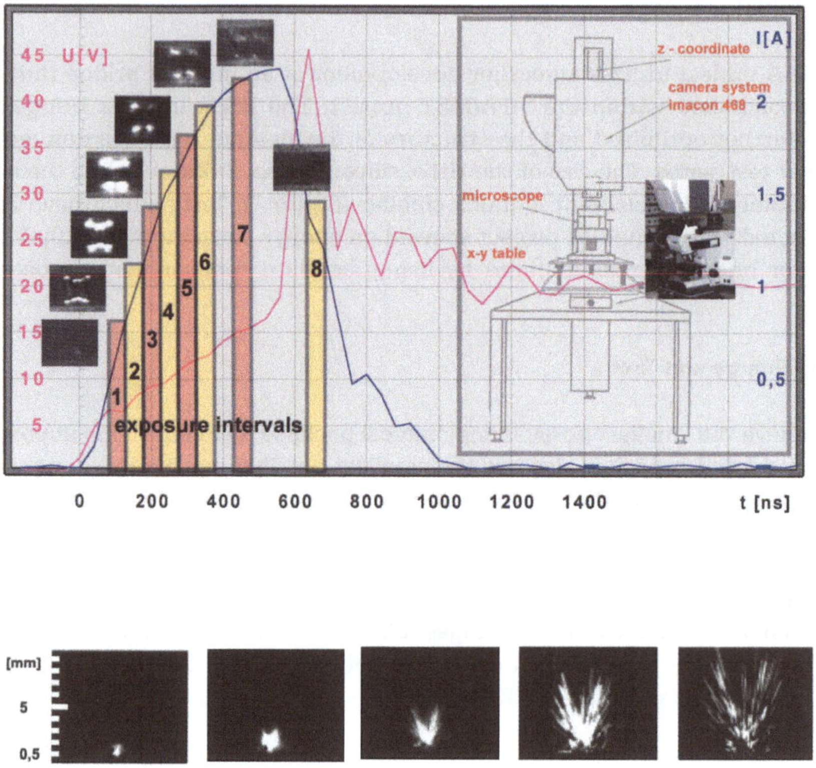

Fig. 3. High speed camera setup and images of a HfHx firing test.

IBA Initiator Assembly

The project also address manufacturing issues related to the IBA. This includes the highest integration, best usage of available and/or further to be developed processes and optimum design for manufacturability. Due to the very tight mechanical dimensions that are available for the electronics, the requirement for a consequent integration was obvisious. A volume of only 7mm by 8mm was available for the electronics. Due to this reason all semiconductors like the ASIC and the ESD protection have been integrated as Chip on Board technology. An Al_2O_3 substrate of 6x8mm^2 with a thickness of 0,64mm was chosen as a suitable carrier for the electronically components. The Al_2O_3 substrate also enabled the possibility to print the required reference resistor for the diagnostic functions directly on the substrate. All other components like the back-up energy capacitor and the ignition capacitor have been assembled by use of a high temperature solder. This is required due to the extensive environmental conditions the IBA is subjected to when it is assembled in the airbag inflators. The life time is specified 15 years, the temperature range is specified from -40℃ to +125℃. The ASIC has a chip size of app $10mm^2$ and is manufactured by using a CMOS BCD3ss process [1,2]. The semiconductors are attached to the Al_2O_3 substrate by conductive glue, the electrically connection is done by conventional bond technology. The IBA assembly is finally protected with a molding compound before the pyrotechnical materials are added. An example of the IBA with a HfHx initiator chip without molding is shown in figure 4. The COB Technology that is used for the mayor semiconductors can be seen in figure 5.

Fig. 4. IBA electronics, 3 pin design with HfHx Initiator.

Fig. 5. IBA electronics, COB technology.

IBA Initiator Tests

Complete IBA initiators have been installed in standard side airbag gas generators. Deployment tests have been conducted in cold and hot climate to determine the performance of the initiator. A typical measurement, pressure curves of a test lot with two different types of IBA initiators can be seen in figure 6.

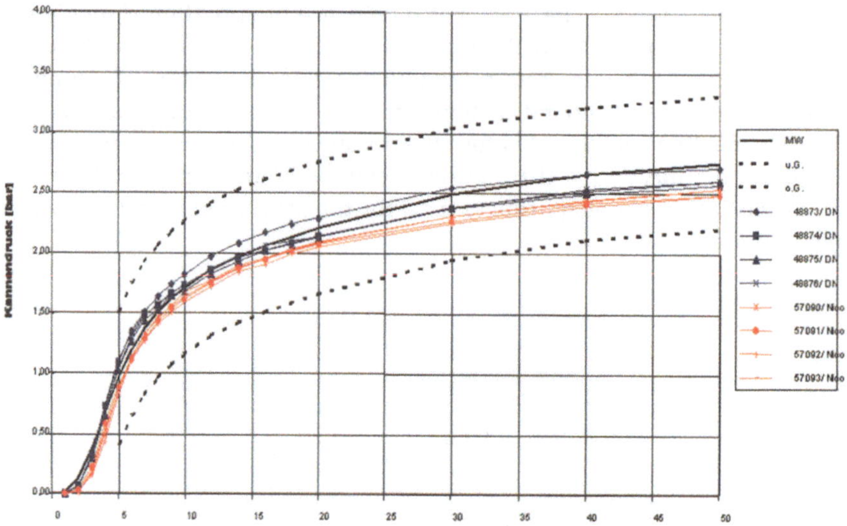

Fig. 6. Pressure curve of gas generator with IBA initiator.

4 Conclusion

With the results that have been achieved within the IBA project [8][9], it will be possible to design and manufacture an integrated bus compatible initiator at economical costs. The consequent usage of all possibilities for miniaturization and the identification of additional synergy effects will lead to further cost optimised solutions and to a wide introduction of integrated bus compatible initiators. Small mechanical dimensions are an important issue for the integration of the IBA into the next generation inflators as it can be seen in figure 7.

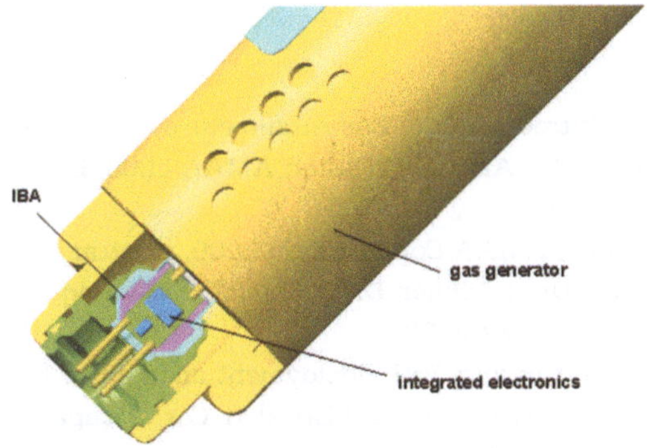

Fig. 7. Side airbag gas generator with IBA initiator.

Acknowledgements

The author likes to thank the involved partners of the IBA cooperation project: G. Kordel (Dynamit Nobel Automotive Initiator Systems GmbH, Fuerth), W. Schmid (NICO Pyrotechnik Hanns-Jürgen Diederichs GmbH & Co. KG, Trittau), U. Weiss (NICO Pyrotechnik Hanns-Jürgen Diederichs GmbH & Co. KG, Trittau), Dr. B. Löbner (Chemnitz University of Technology, Department ET/IT, Center of Microtechnologies), Dr. H. Laucht (TRW Automotive GmbH, Aschau/Inn).

References

[1] H. Flocke, M. Herz, Introduction of Bus Architecture for Vehicle Safety Restraint Systems - a Challange for Airbag IC Design AMAA 2000, Berlin, Germany April 6-7, 2000, www.amaa.de

[2] S. Krueger, W. Gessner, AMAA Compendium 2000, www.amaa.de

[3] G. Kordel, W. Schmid, Dr. B. Löbner, Dr. H. Laucht, T. Goernig IBA Status Seminar, Ingolstadt, Germany, Nov. 27 2000

[4] K. Balzer, C. Zelger, T. Goernig, BST Deployment and Sensor Bus Airbag 2000+, 5th International Symposium on Sophisticated Car Occupant Safety Systems, Karlsruhe, Germany, Dec. 4-6, 2000

[5] T. Goernig, Integrated Electronics for Bus System Initiators mstnews 1/01, February 2001, www.vdivde-it.de/mstnews

[6] T. Goernig, Integrated Electronics for Bus System Igniters AMAA 2001, Berlin, Germany, May 21-22, 2001, www.amaa.de

[7] S. Krueger, W. Gessner, AMAA Compendium 2001, www.amaa.de

[8] T. Goernig, IBA - Integrated Bus Compatible Initiator AMAA 2002, Berlin, Germany, March 21-22, 2002, www.amaa.de

[9] S. Krueger, W. Gessner, AMAA Yearbook 2002, www.amaa.de

[10] T. Goernig, Advanced Sensors for Future Restraint Systems Airbag 2002, 6th International Symposium on Sophisticated Car Occupant Safety Systems, Karlsruhe, Germany, Dec. 2-4, 2002

[11] Common Bosch Siemens Temic Bus Description Rev.: 3.0, October 2002, www.temic.com

[12] G. Kordel, W. Schmid, Dr. B. Löbner, Dr. H. Laucht, T. Goernig, IBA Abschluß Seminar, Ingolstadt, Germany, Feb. 6th 2003

Thomas Görnig
Conti Temic microelectronic GmbH
Ringlerstraße 17
85057 Ingolstadt
Germany
thomas.goernig@temic.com

Keywords: IBA, HfHx, initiator, integration, reliability, firing and sensor bus system, BST

Future Trends in Software Architectures for Automotive Systems

Dr. S. Voget, Robert Bosch GmbH

Abstract

This paper focus on the future integration of software modules into vehicle electronics architecture. Based on an analysis of the challenges coming soon we present common approaches and solutions for the software architectures which are already perceptible. To validate the architecture concepts a demonstration of an example completes the contribution.

1 Introduction

The grown structures of today's electronic control units (ECUs) in the vehicles are characterized by two trends:

- An increasing number of relatively loosely coupled, mono-functional ECUs.
- An increase of the total share of software in each of the ECUs [4, 5, 7, 8, 11].

As a result of this trend, the communication between ECUs increases, too. But the networks of ECUs are oriented on a traditional grown functional division of automotive functions. Following this division the ECUs are grouped into several sub-domains, for example, power train, body, telematic, chassis, etc. The networks of the ECUs are neither standardized according to their interfaces across these sub-domain borders nor developed with respect to the interrelationships between nodes of the network.

Comparable statements are valid for the development processes. The software development processes for different ECUs have evolved based on the individual history of the sub-domains, and are currently quite divergent. In the automotive industry most of the widespread system development processes map the functional requirements one to one into software and hardware components. Software architecture models that provide an appropriate level of abstraction of the systems are often missing.

This paper is organized as follows: first, we motivate the needs for new system structures in the electronic networks of vehicles. Next, we present several surveys for activities which consider and develop future system structures. How the future software architecture division will look like is summarized in chapter 4. A demonstration of an example from the vehicle-environment-supervision domain completes the contribution. This example will be used for the validation of the architecture concepts.

2 Needs for New System Structures

2.1 A Short History

A brief consideration of the history is necessary to understand the grown structures and the challenges of software development in the automotive industry. In this section we concentrate on the history of technologies in this area.

The technologies changed several times during the last 150 years. Up to World War 1 the production was determined by small series that had been mainly craft oriented. About up to the sixties, big series productions get through. In this period mechanical and first electro-mechanical products determined the technology. Up to today there had been a continuous increase of electronic in combination with more and more precision mechanics.

There will be a fourth phase of technology change in near future: the software will get of dominant importance for the costs of the vehicles. For the next ten years an increase of 10%-15% of software in the share of costs of a vehicle is forecasted every year [8]. There are two main drivers for this which are not only reasoned by the technology:

- Software makes electronic systems cheaper.
- New or improved functionality will only be possible with the help of software.

2.2 Problems With These Structures

The outlined historic survey influences the grown structure of electronic systems in the vehicles. It is one of the main reasons for the problems in the software development. These lead to delayed start of productions and an increasing number of call backs because of errors in electronic systems [14].

On Software Level

The increasing total share of software leads to high complexity and costs. This becomes more critical with non standardized development processes and without adequate networks. In addition, the incorporation of third party software increases the complexity of collaboration between companies.

An appropriate level of abstraction in the software architecture modeling and appropriate integration concepts are still missing. Current architectures do not reflect the effects of quality requirements. As a consequence, these often remain vague and unexplored. The architectures grown up by the single solution development strategy do not represent long-term solutions.

On Functional Level

Today a lot of functionality is distributed on several ECUs. E.g., indicator functionality is distributed over up to eight ECUs in high end vehicles. Furthermore, some of the future functionality is not realizable with the current loose side by side of ECUs. E.g., drive-by-wire will need a very close and safe interlocking of ECUs across different domains [6]. The traditional functional division of automotive functions is more and more crossed by the new functionalities.

In the next section we list some of these upcoming functionalities with main influence on the problems mentioned here.

2.3 Challenges Because of New Functionalities

As described above the present situation leads to a lot of challenges for the further development of electronic systems in vehicles. In addition several new functionalities will increase these challenges. This section contains a list of functionalities which will have main influence for the network of ECUs and the associated software architectures. The list is grouped into

- Improvement of Propulsion Technology.
- Substitution of Mechanical and Hydraulic Linkage.
- Intelligent Vehicles.

These groups form the drivers for the evolution of the technologies presented here [1, 3, 12].

Improvement of Propulsion Technology

Technology	Description (D) / Improvements (I)
Improvement of Diesel Engines	D: Diesel engines offer a consumption advantage compared to gasoline engines due to their better thermal efficiency of about 20%. This advantage can be further improved by 15% to 20% by optimizations at the injection system, the combustion control, and the engine control system. I: The particle emission is reduced by high injection pressures and prevented by diesel particulate filter systems.
Hybrid Power-train	D: The rising electrical power demand leads to the application of the integrated starter-generator and the 42-Volt electrical system. The next step in this evolution is to use the integrated starter-generator as an electric motor. Thus a simplified hybrid drive is available. During acceleration the electric motor can support the combustion engine. During breaking it can be operated as generator to regenerate energy. In case of standstill the combustion engine can completely be switched off. Low speeds will be mastered alone with the electric motor. I: This leads to substantial fuel saving, in particular before red traffic lights and in stop and go traffic.

Substitution of Mechanical and Hydraulic Linkage

Technology	Description (D) / Improvements (I)
Electro-mechanical Valve Lift Control	D: The electromagnetic valve lift control replaces the camshaft. The valves are actuated by electromagnets which are controlled electronically. I: Gas change can be optimized such that fuel consumption is reduced or engine power is increased.
Automated Manual Transmission	D: An automated manual transmission operates the switching mechanism and the clutch in electric motors. A mechanical linkage between the switching level and the clutch pedal can be dropped. I: This influences the weight and crash behavior positively.
By-Wire Car	D: Components for braking, steering, shifting and other functions are controlled by electrical/electronic systems. I: Electronic systems can allow a reduction in overall costs during the complete life cycle of a car. The reduced space and weight necessary for electronic systems can lead to reduced fuel consumption. Electronic systems provide new concepts for remote diagnosis and maintenance, and they enable new technologies such as fuel cells and intelligent power-train control systems.
Integrated Chassis Control	D: Drive-by-wire chassis consist of individual corner modules that are segmented further into smart sub-modules for steering, braking, suspension and damping. These will be cross-linked via a control module with an integrated chassis controller. I: The objectives are to improve vehicle stability, to cut braking distance, to optimize the comfort/driving stability trade off and to harmonize steering, braking, suspension and damping systems.

The substitution of mechanical and hydraulic linkages by purely electrical/electronic systems is due to the limited nature of mechanical systems in the areas of safety and comfort as well as the rapidly improving price/performance ratio of microelectronic components.

Intelligent Vehicles

Technology	*Description (D) / Improvements (I)*
Driver Assistance Systems	**D:** Driver assistance systems support the driver in specific situations. Upcoming ultrasound and radar sensors enable new driver assistance systems like advanced Adaptive Cruise Control, Blind Spot Detection, Stop and Go Control and Parking Assistance. The introduction of video-based sensors permits the recognition of traffic signs, lanes and obstacles. **I:** Safety and comfort is improved.
Pedestrian Protection	**D:** Sensors embedded in the vehicle's front bumper detect a collision with a pedestrian's legs and raise the rear of the bonnet around 10 cm, using either gas-filled bellows or electro-mechanical activators. **I:** Pedestrian-protection systems are designed to minimize head impact in the event of a collision.
Usability	**D:** Since the entering of microelectronics the number of functions in vehicles has increased. A simple evolution of the well-known control surfaces would lead to a tide of switches and small lamps. **I:** The user interface has to present all information clearly and enable an easy and intuitively operation of the vehicle functions by different driver types, without diverting the driver.

Each and every of the listed functionalities will increase the complexity in the network of distributed hardware. The need for the control of the complexity – also in future the vehicles have to be realized – leads directly to some needs for new structures. These will be considered in the next section.

2.4 Need For New Structures

Future developments have to control and manage the increasing complexity on a network of distributed hardware. A cost minimum is not achievable on the basis of single ECU considerations. But, a cost minimum for the whole system of ECUs is necessary with the important constraint to be flexible enough for the realization of future requirements, especially with respect to the further displacement of mechanic/electric systems by software intensive systems.

To minimize costs, to win efficiency and quality it is necessary to develop new, integrated system architectures, e.g. platform based [10]. Such architectures have to be standardized to improve the collaboration between vehicle manufacturers, system suppliers and 3rd party software suppliers.

Furthermore, appropriate system architectures can only be realized by associated software architectures which reflect the requirements of the whole network of all ECUs. To enable the communication of single parts in this network a comprehensive, global software architecture framework is needed which contains a set of decisions about the used interfaces.

Several projects already work on the topic of a software architecture framework. Before we go into detail of the structures we introduce some projects dealing with these needs in the next chapter.

3 Common Efforts to Develop the Future Architectures

Several activities, initiatives, and projects work on the challenges and problems described above.

The projects where OEMs and suppliers work together will have a lot of impact on the development of system/software architecture frameworks for the future.

AIL

With the support of the French Under-Ministry for Industry several French OEMs and suppliers worked together in the project AEE (Architecture Electronique Embarquée). The project closed in 2001 but the results influence several subsequent activities. The mission of the project was:

"In order to meet end customer needs in terms of quality and innovation, at the right price, and also to control the resultant growth in complexity, this project designs and validates a fast and reliable process for defining the Systems Architecture and developing the associated embedded software. This process is based on independence between hardware and software, on the use of standard methods, tools and components, and on the commitment by contracts for exchanges between the main players. The project focuses on transport applications, particularly for automobiles." [aee.inria.fr]

One of the main outputs of this project was an AIL (Architecture Implementation Language).

TITUS

As a counterpart to the French initiative some German companies developed TITUS. Originally developed for the body domain the methodology is in the meantime extended to other domains.

"Vehicle body electronic software has reached a level of complexity and cost that methods focusing on re-use and seamless vehicle integration of networked ECUs are being heavily embraced. The TITUS methodology shows that vehicle functions can be re-used over several physical ECU networks, i.e. several vehicle models." [2]

HIS

Since 2000 some German OEMs work together in the HIS (Herstellerinitiative Software) initiative. The goal of this initiative is to develop software modules in such parts of the system which are not relevant for the competition between the OEMs:

"The vehicle manufacturers Audi, DaimlerChrysler, BMW, Porsche, and Volkswagen introduce their joint efforts to standardize so called standard software modules for in vehicle networks. This will also include methods for determining process maturity levels, software test and software tools. The objective is to achieve common standards." [9]

ITEA-EAST/EEA

All the beforehand mentioned activities influence this project. We present it here in more detail. ITEA-EAST/EEA is an European funding project. More than 20 OEMs, suppliers, tool-suppliers, and research institutes from 4 European countries work together to...

"... enable a proper electronic integration through definition of an open architecture. This will allow to reach hardware and software interoperability and re-use for mostly distributed hardware [figure 1]. By developing a European wide accepted embedded electronic vehicle architecture, ITEA-EAST/EEA is a prerequisite for the implementation of these advanced features and functions." [http://www.east-eea.net]

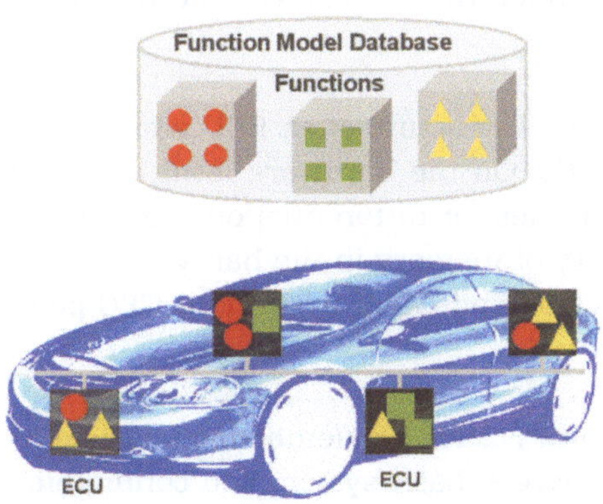

Fig. 1. Various functions have to be distributed over several ECUs.

The project started in 2001 and will end in 2004. The activities are based on two main columns:

- the definition and specification of a middleware and communication concept for an automotive embedded electronic architecture (figure 2a),
- the development and validation of architecture and functional modules where the complete development process is taken into account (figure 2b).

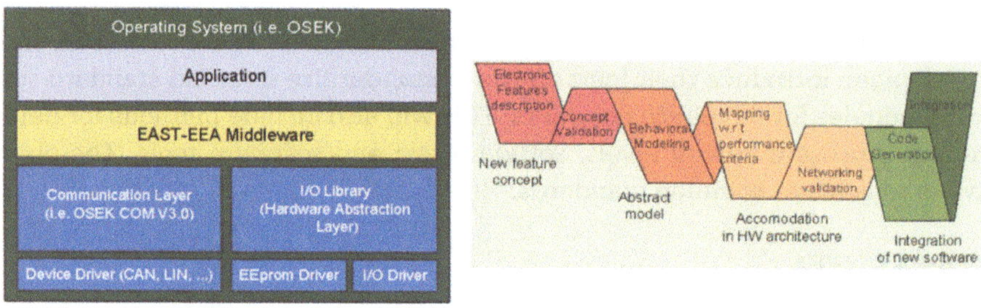

Fig. 2. a) layered architecture.
b) general development process.

4 Trends for Software Architectures in the Automotive Domain

All those projects and initiatives have different goals in mind and follow different approaches to solve the mentioned challenges. Although several projects and companies work on the development of future software architectures with different goals a common trend is perceptible. This trend is driven by the parts of the system in which the OEMs have to differ to contrast their brand from others.

Figure 3 shows the general structure of an ECU which is forecasted as a standard in near future ECUs. On the hardware level the OEMs will differ from each other. The main values for differentiation are the number of ECUs, the CPU-power, and quantity of memory in the hardware. A vehicle in the luxury class will have more ECUs and ECUs with more CPU-power than a car in a compact class.

The next layer – sometimes called EBS (embedded basic software) – consists of the operating system, drivers, basic system and communication functionality, diagnosis services, and hardware abstraction layer. These parts of the software are not relevant for the individual differentiation of an OEM in market.

Therefore, OEMs as well as suppliers are interested to define and to use standards in this area. The details given in figure 3 for the EBS part are more or less common for all the mentioned initiatives and projects.

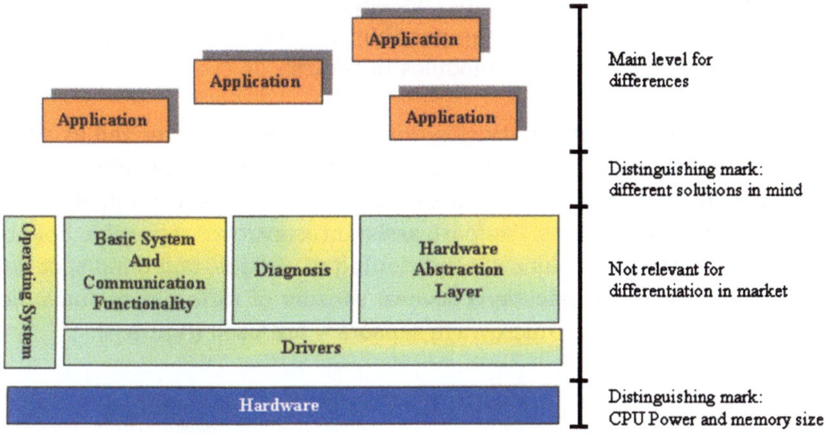

Fig. 3. General layer structure.

A slightly different situation can be found on the next higher layer, the API. This layer is also not relevant for the differentiation in market. But for this part of the system all initiatives and projects have different solutions in mind. Some projects put only a thin layer with interfaces in this part. Others add several set of services (e.g. notification services, mode management, ...) which can be used by all applications positioned above this layer. The thickest versions are constructed by extensive runtime environments which will be automatically generated and embed the applications and their relationships.

The application level is the part of the software which is the most relevant part for the differentiation in market. Today there are no efforts for common solutions or standardizations in this layer noticeable. From a supplier point of view this situation prevents the suppliers from cost efficient development by reusable software modules [13]. In future the establishment of - at least - common component models will be decisive for the success of cost efficient application software modules from the suppliers.

5 Example: Portable Sensor Software

Currently there are several pilot projects which experiment on the concepts presented in the last chapter. In the ITEA-EAST/EEA project Bosch builds up a vehicle to validate the architecture concepts. The vehicle-environment-supervision domain is used to show the possibilities which will be enabled by distributable application software modules in future.

The vehicle-environment-supervision domain can be characterized as a family of applications that are based on sensors installed around the vehicle to monitor its local environment. The sensors enable either the realization or enhancement of several applications, like park assistant, control of reversible hold back systems (e.g. seatbelt tensioner), conditioning of airbags, and blind spot detection. To realize these applications several sensors of different technologies – e.g. ultrasound sensors or microwave sensors – are used (figure 4).

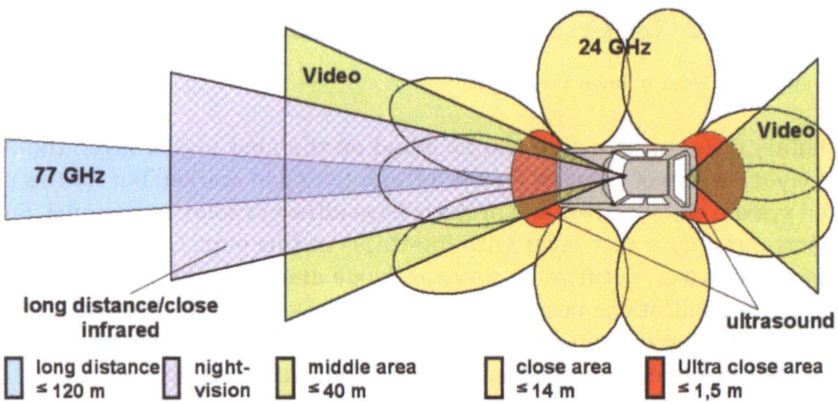

Fig. 4. Overview about distance areas and associated sensor technologies.

A processing in several steps leads to a description of the environment and the obstacles in these areas. The necessary software has to be distributed on several sensors and ECUs. The hardware equipments differ from one OEM to another. But some constraints on the used hardware and topology of ECUs are reasoned in the requirements of the applications on the hardware, e.g. a pre-crash is not realizable with ultrasound sensors.

In the ITEA-EAST/EEA project parts of a park-assistant will be realized independent from the underlying sensor technology – subject to all constraints – and the ECU hardware. As a second scenario the upgrade of sensor processing software via CAN Bus from a telematic unit will be realized (figure 5). Both scenarios together will show that a flexible distribution of an application is possible in (at least) the body domain. In addition, the limits of such a procedure are tested out. The realization enables the inclusion of the park-assistant software modules in function-libraries. Several OEMs plan to install such data bases. This will show how the integration into company specific libraries could look like.

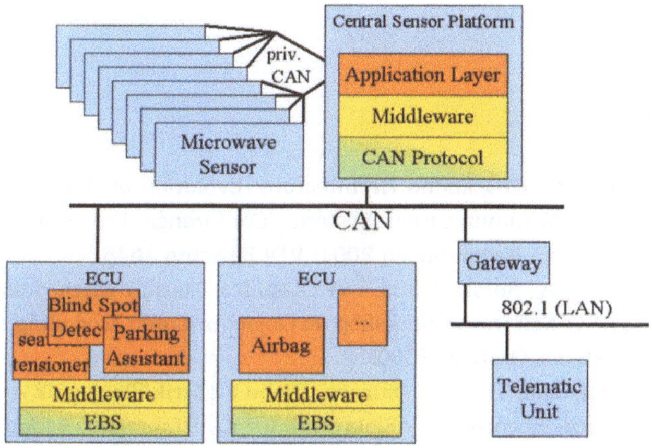

Fig. 5. Network structure for the validation.

6 Conclusions

We have shown that the electronic networks in the automotive domain will change in near future. This will take place on system- as well as on software level. For specific parts of the software architectures common approaches and solutions are already perceptible. They have a good chance to evolve into standards.

But the enormous grows of new functionalities to be included shows the complexity of the necessary solutions. Flexibility with portable software modules is not reachable with cost optimal solutions. Today the two quality attributes "portability" and "cost optimal" are often considered separately. In addition the quality attribute "portability" is often contrary to safety requirements. Future considerations have to take into account a sufficient compromise between all relevant quality attributes.

Acknowledgements

This work was partially funded by the ITEA project No. 00009 EAST/EEA which is part of the EUREKA program. We would like to thank A. Lehn for his assistance in the section "Challenges because of new functionalities".

References

[1] R. Beck et al.; "Backbone Architecture: Evolution of the Central Gateway to an Integrated Communication System" Conference Proceedings - Elektronik im Kraftfahrzeug, Baden-Baden 2001, VDI Berichte 1646.

[2] U. Freund, A. Burst; "Titus – A Graphical Design Methodology for Embedded Automotive Software" Workshop on Object-oriented Modelling of Embedded Real-time Systems, OMER2, 2000.

[3] P.E.H. Hoffman, Thomas Thurner; „Neue Elektrik/Elektronik Architekturansätze" Conference Proceedings - Elektronik im Kraftfahrzeug, Baden-Baden 2001, VDI Berichte 1646.

[4] ITEA; "Technology Roadmap on Software Intensive Systems", ITEA Office Association, Eindhoven, March 2001.

[5] K. Jost; "Electronics demand to grow nearly 7% annually" in Automotive Engineering International, September 2001.

[6] K. Jost; "From fly-by-wire to drive-by-wire" in Automotive Engineering International, September 2001.

[7] ATKerney; "Software-betriebene Fahrzeugsysteme bestimmen die Zukunft des Automobils" Study, 2001.

[8] McKinsey&Company; "Automotive software: A battle for value" Study, 2002.

[9] K. Lange et. al.; „Herstellerinitiative Software" VDI Conference for electronic in vehicles, Baden-Baden, 2001.

[10] H. Postema; "Platform Based Product Development" In F. van der Linden (Ed.): Software Product-Family Engineering; Lecture Notes on Computer Science (LNCS) Vol. 2290; Springer; Berlin 2002.

[11] D. Sallee, R. Bannatyne; "Trends in Advanced Chassis Control" in Automotive Engineering International, September 2001.

[12] P. Steiner, F. Schmidt; „Requirements and System Architecture of future Body Electronics" Conference Proceedings "Elektronik im Kraftfahrzeug", Baden-Baden 2001, VDI Berichte 1646.

[13] . Voget; "Product Line Variability in the Automotive Domain" In Conference Proceedings: International Colloquium of the Sonderforschungsbereich 501; University of Kaiserslautern; 2003.

[14] S. Voget, M. Becker; "Establishing a Software Product Line in an Immature Domain" In G. Chastek (Ed.) - Software Product Lines, Springer 2002.

Dr. Stefan Voget
Robert Bosch GmbH
Corporate Research and Development, Software Technology
P.O. Box 94 03 50
60461 Germany
stefan.voget@de.bosch.com

Keywords: electronic control units, software architecture, vehicle-environment-supervision domain

Melexis LIN Solutions Enable Cost Optimized LIN In-Vehicle Networks

M. Bender, Melexis GmbH

Abstract

The paper will present an integration concept which is based on a LIN optimized 4-bit Dual-Task RISC microcontroller together with the possibility of high voltage integration (12V and 42V boardnet). This enables the complete integration of LIN-transceiver, voltage regulator, ADC, microcontroller and application specific parts on one chip at a cost level of optimized state machines. This concept combines the benefits of powerful, robust and cheap LIN state machines with the flexibility of small conventional microcontroller solutions.

The MLX4 Dual-Task CPU consists of 2 independent register sets which can simultaneously run both the whole LIN protocol on one task and the application on the other task. Both tasks run completely independent from each other and there are no influences of the LIN to the application and vice versa. With this CPU concept we have functions of a RtOs integrated in silicon, which is easy to use, robust and extremely fast for event processing. Therefore it perfectly fits to LIN applications.

1 Introduction

The growth of electronics modules in vehicles providing increased comfort, safety and performance features, makes the use of different in-vehicle networks in dependence of the application a necessity. Vehicle manufacturers (VMs) have agreed upon industry wide network standards in order to reduce costs and to increase performance. The most recent standard is called LIN (Local Interconnect Network), which is the communication protocol for a new low-speed network to which all comfort and convenience applications in the car are connected.

Instead of every air conditioning vent, electrical seat motor and sunroof being individually wired point-to-point to its own switches, only one wire can connect up to 16 separate points using „LIN".

LIN can be used for every application which has no demand for high data rates. Currently three different main application areas exist:

- Intelligent switching modules for window lifter, sunroof, mirror- and seat position control, steering wheel modules,...
- Actuators for climate flaps, wiper motor, positioning motors, blower, ...
- Sensors for light, emission, rain, intrusion, tow away protection, ...

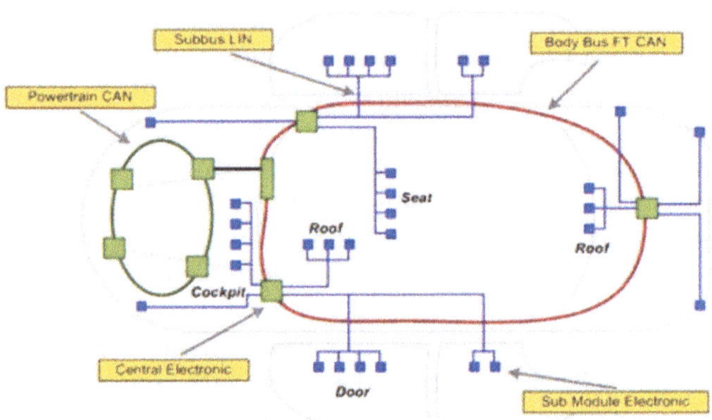

Fig. 1. Future In Vehicle Network.

With the help of LIN we are able to develop distributed networks within the car. LIN has the potential to cover all applications which can't be directly connected to CAN because of to high costs. Therefore it is necessary to bring the costs of LIN modules to a level, which is far below CAN. Up to now this was only possible with state machines. That kind of solution is efficient, robust and cheap, but it has the disadvantage that it's only proven for one application and therefore not flexible. Changes are difficult to implement only with a high effort and development time. Conventional MCU architectures are flexible usable, but LIN and application influence each other. These solutions don't reach for some applications the cost level we need for LIN.

Melexis has developed for LIN a complete new MCU concept, which combines the advantages of both solutions: On one side the efficiency, robustness and cost level of a state machine and on the other side the flexibility of a conventional MCU solution.

2 Current Situation on LIN Integration

LIN is a concept for low cost automotive networks, which complements the existing portfolio of automotive multiplex networks. The standardization will reduce the manifold of existing low-end multiplex solutions and will cut the cost of development, production, service and logistics of in vehicle networks. To reach this target, the costs of a LIN node must be far below CAN.

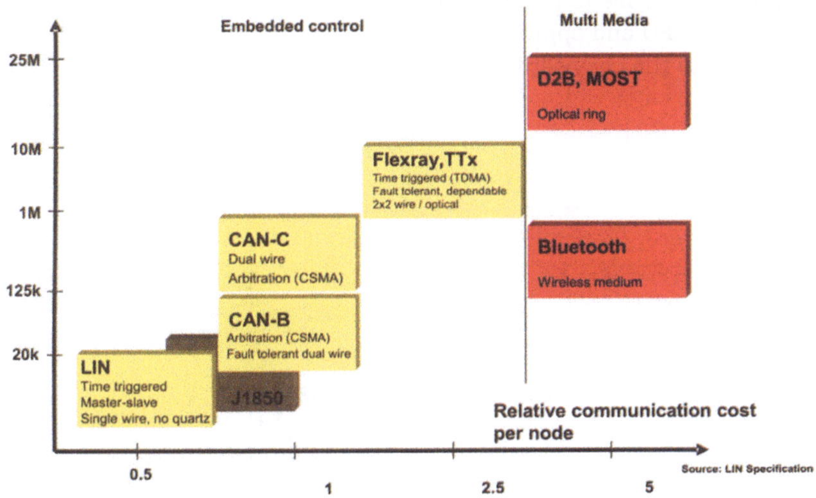

Fig. 2. Cost overview of different bus systems.

Figure 2 show that the cost per LIN node is a very important point. Current implementations within master or slaves are realized with standard MCU's together with voltage regulator and separate LIN transceiver. This realization principle as 3 chip solution doesn't meet the target from the cost point of view, but in the beginning it is a very flexible and secure solution. For further developments we have to develop more integrated products to cut down the cost. This cost optimisation can be done in different steps. The first step is to put the LIN transceiver, voltage regulator and support functions for the MCU together on one chip. This kind of ICs – the LIN System Basic Chip (SBC) – makes it possible to cut down the cost and improve the development time due to decreasing the LIN slave implementation to a 2 chip solution. An example is shown in figure 3.

For further improvements we have to integrate all necessary functions of a LIN node within one chip. Melexis currently develops for different applications complete integrated LIN slave ICs with this new generation of CPU. Possible applications areas are climate control, door applications, dashboard, seat functions and others.

This concept based on a LIN optimized 4-bit dual task microcontroller (MLX4) together with the possibility of high voltage integration (12V and 42V boardnet), makes it possible to integrate a complete LIN slave (LIN transceiver, voltage regulator, CPU and application) on one chip. This kind of ICs can be directly connected to the car boardnet.

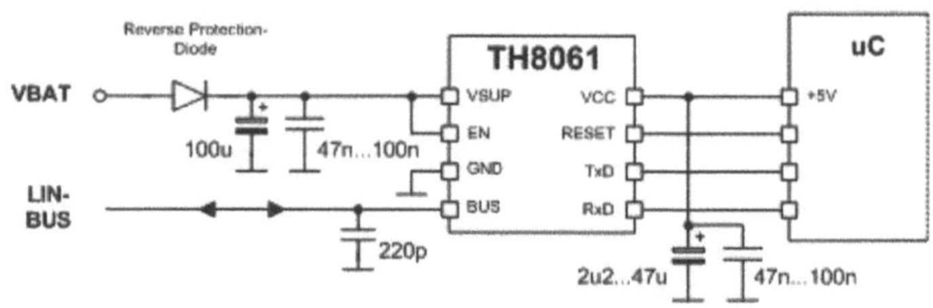

Fig. 3. LIN slave 2 chip solution with SBC TH8061.

3 LIN Optimized 4-bit Dual Task Core for Complete Integration of LIN Slaves

3.1 LIN Implementations

The development of LIN applications can be done currently with different principles:

- Standard MCU: To implement the LIN functions only standard hardware like Timer, Capture, I/Os is used. The software must be implemented down to bit level. The LIN protocol will be transmitted via standard I/O pins to the LIN transceiver. This solution generates a very high MCU load, because of processing the LIN protocol at bit level with standard MCU hardware, but it is very flexible in case of any changes. An additional big disadvantage of this LIN solution is the high interactions between the application itself and LIN because of using a pure software implementation for LIN.
- Optimized "LIN UART": This LIN UART processes some LIN functions in hardware like break detection, synch field functions and bit processing.

The usage of this hardware part decreases the MCU load compared to the previous implementation, but there are still high interactions between LIN and application. This is caused by using the same MCU for processing LIN and the application, but it is still a very flexible solution.

- State Machine: For dedicated applications it is also possible to develop LIN nodes based on state machines. With this principle all functions are realized in hardware (ASIC) without using a MCU. This implementation is very powerful, robust and reaches a cost level, which is not possible with standard MCU's. But it is developed only for one application and it generates high development costs. If there are any changes the complete IC must be redesigned and this generates a high effort in development costs and time as well as design risk. The big disadvantage is the inflexibility. If there are future extensions or changes at protocol level it requires also a complete redesign of this type of LIN ICs.

Melexis has developed for LIN a new CPU concept, which combines the advantages of all solutions: On one side the efficiency, robustness and cost level of a state machine and on the other side the flexibility of a traditional MCU solution.

3.2 MLX4 – The Optimized LIN CPU Core

The Melexis low cost 4-bit CPU Core is optimized for processing the LIN protocol. Different versions of this core are available:

- Dual task, which is well suited for simple LIN slaves.
- Slave CPU, which can be combined together with 8-bit or 16-bit CPU for more powerful LIN applications.

The MLX4 LIN CPU core is based on RISC architecture and has a calculation power of 4Mips@12MHz. It has a guaranteed timing of one CPU cycle for each instruction and has special instructions for direct command of periphery and direct moves between RAM and I/O are possible. The CPU can address 4k instructions per task, 256 nibbles RAM and 128 nibbles I/O.

LIN operations are processed very efficient, e.g. a LIN parity calculation needs only 5 cycles (1.25µs@12MHz) and 11 ROM words and a 16-bit CRC of a byte needs 36 cycles (9µs@12MHz) and 32 ROM words.

3.3 MLX4D – The Optimized LIN Solution for Simple Applications

The Dual Task version of the 4-bit CPU consists of 2 independent register sets, which enables the possibility to handle 2 different software tasks at the same time. With this system it is possible to run on one task the whole LIN protocol while the other task runs the application. Both task running complete independent from each other and therefore there is no influence of one software part to the other (LIN and application).

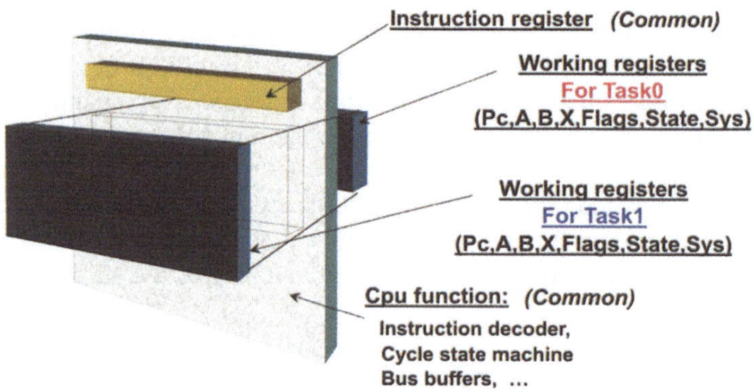

Fig. 4. MLX4D dual task CPU principle.

Each task of this CPU owns a private set of peripherals like timer and UART, as well as a private memory part. The data transfer from one task to the other will be done via a protected memory area. This protection prevents that both tasks write to the same RAM address at the same time.

The switching of the register sets occurs after every instruction. Therefore every task has a guaranteed processing power of 50% of the whole MCU power (2 Mips@12MHz).

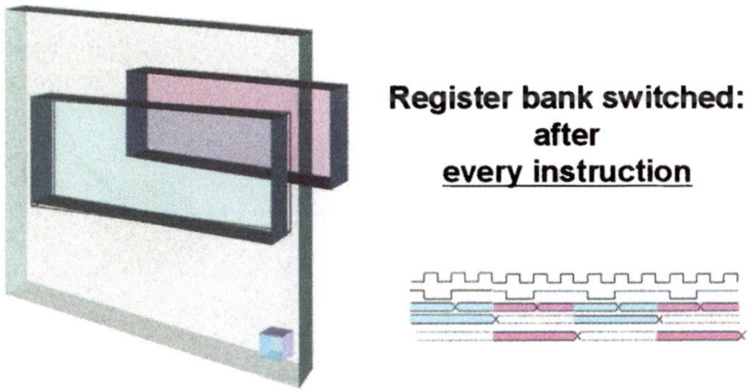

Fig. 5. MLX4D switching mechanism.

The synchronization between both tasks is performed via different flags. For a typical LIN application this is achieved via mutex flag for protection of shared resources. It is also possible, to synchronize both tasks on different signals for timing synchronization or on events for functional synchronization.

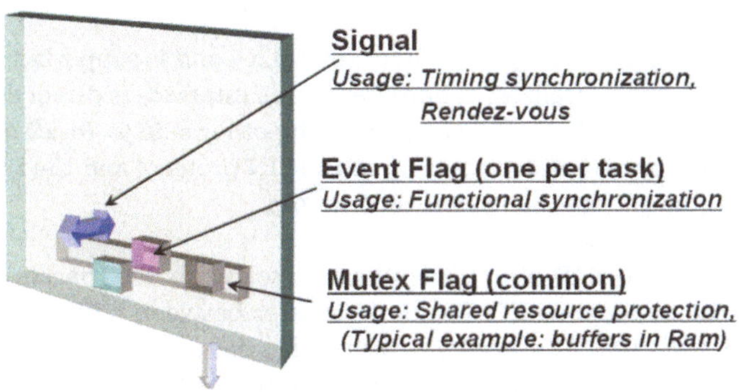

Fig. 6. Task synchronization via CPU flags.

This CPU is a full dual task implementation. It has integrated functions of a RtOS in hardware, of course limited to two tasks. The MLX4 CPU core was developed for integration into complete integrated LIN slave ICs. The working principle of complete integrated LIN slaves is the following:

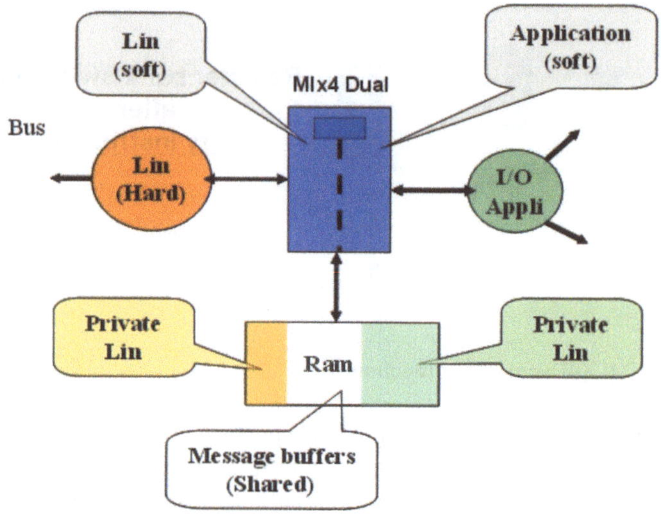

Fig. 7. MLX4D working principle.

The Lin protocol is running on task0 of the MLX4 and is supported from a Class A network interface realized in hardware. This interface is direct connected on one side to the LIN physical layer and on the other side to Task0 of MLX4D. It is responsible for bit-level processing of the LIN protocol and the LIN task have access to all bus states, bus events and errors.

On Task1 of the MLX4D runs the application software. This part is directly connected to the application I/O. The data transfer between both tasks is done via a shared RAM area. The used RAM area for data exchange is protected via mutex flags to avoid that both tasks write to the same RAM address. The general communication between both tasks is done via an API.

Both tasks run complete independent from each other, that means there are no influences from the LIN to the application and vice versa. The application task transfers data's to the shared RAM area and the LIN task transfers this datas to the LIN bus. From the application point of view the transferring of the data's via LIN is done automatically. In the same way the data's are processed which coming from the LIN.

3.4 MLX4S – Powerful Multi CPU Solution

There are also LIN applications which need more calculation power than a 4-bit CPU can support. Melexis has developed a solution which uses the same principle as the MLX4D configuration with the same advantages. It is used the MLX4 for processing the LIN and a powerful 8-bit or 16-bit for the application.

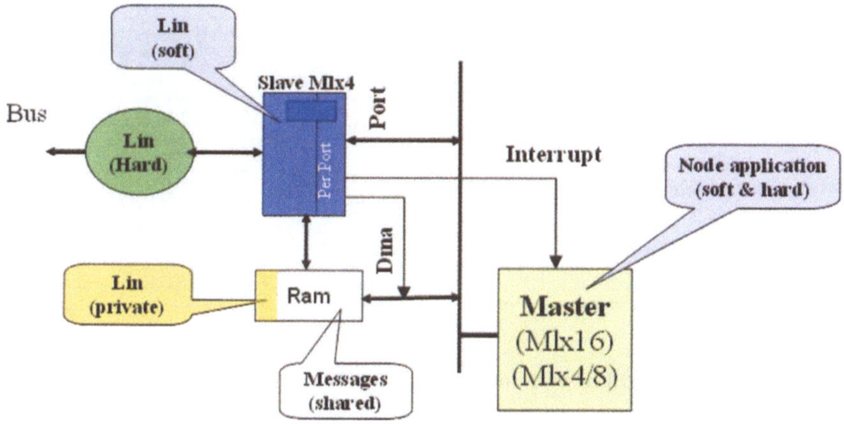

Fig. 8. Principle of powerful multi CPU configurations.

From the MLX4D it is only used one register bank. The second task and the task switching mechanism is replaced by a dual port memory controller with protected RAM areas. This RAM area is shared with the master for data exchange. From the communication port is one side unchanged (connected to the MLX4) and the other side is moved in a peripheral port of the master CPU. With this configuration it is possible to develop powerful multi CPU configurations. The software of Task0 of the MLX4D runs identically on the slave MLX4S.

3.5 MLX4 and LIN Software

The LIN software developed by Melexis running on task 0 of MLX4D or on MLX4S is identical. This software processes the serialization/ de-serialization of the LIN bit stream, the LIN frame itself and frame filtering, makes the data- and error handling and is responsible for the network monitoring and status reporting. In other words the LIN handler does everything regarding LIN. Therefore the application software must only transfer the data's to or from the LIN to the application via shared memory buffer. The message filtering secures that the application will only get the data's which are valid for these slave nodes. The software also supports protocol extensions like auto baud rate detection and auto configuration via dynamic ID assignment.

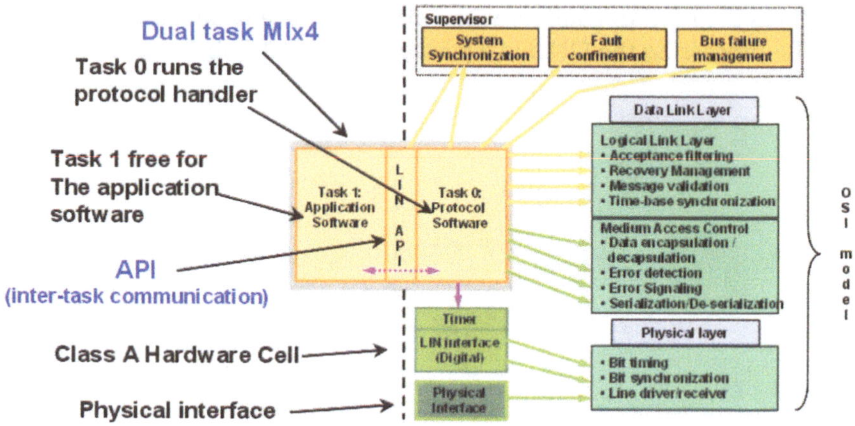

Fig. 9. MLX4 and LIN software.

Because of the realization of the protocol processing via software it is very easy to adapt this software to future protocol extensions or specification changes (LIN 2.0).

The complete LIN software for MLX4 based integrated LIN slaves will be developed by Melexis. This software will also be tested via the LIN conformance test and can be used for every MLX4 product. That means that customers only have to focus on development of the application software itself.

3.6 Development Tools

For development of the application task software a complete environment is available. It consists of:

- Assembler, Linker.
- MLX4 CPU Simulator.
- Emulator E-MLX available as ROM-Emulator and In-Circuit Emulator.
- Flash Programmer with in application programming capability.

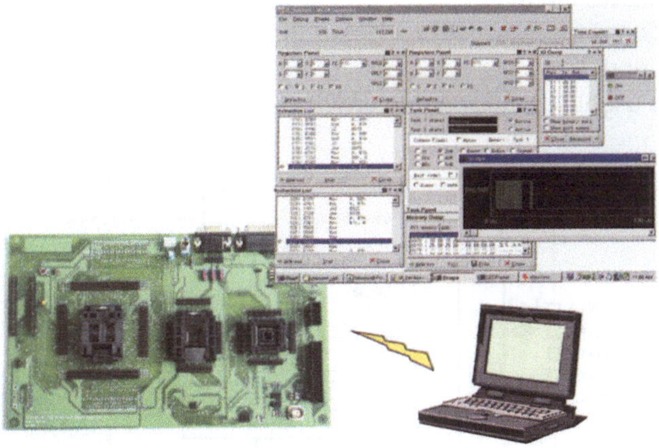

Fig. 10. MLX4 development system.

4 MLX4 Integration Example – IC for Intelligent Switch Modules TH8100

4.1 TH8100 General Description

The first member of our MLX4 complete integrated LIN slaves is the TH8100. This IC can be used for development of distributed intelligent switch modules, which can be linked via LIN together.

The combination of physical layer LIN transceiver and LIN protocol controller in combination with switch matrix or single switch inputs and PWM outputs on one chip make it possible to develop simple, but powerful and cheap switch slave nodes in LIN Bus systems.

The following functions are implemented:

- LIN transceiver and protocol controller based on MLX4D, wake-up capable.
- 17 Switch inputs to connect switch matrix up to 5 x 5 or single switches and encoder reading, wake-up capable.
- Three high current PWM outputs up to 50mA with configurable PWM frequency and slew rate limitation, constant current source mode 2 to 30mA.
- 3 ADC channels, one with high voltage option.
- Very low standby current, 10uA in sleep mode.

- ▶ Internal RC-Oscillator.
- ▶ EEPROM with internal chargepump.
- ▶ Watchdog.
- ▶ Direct usable in automotive environment.

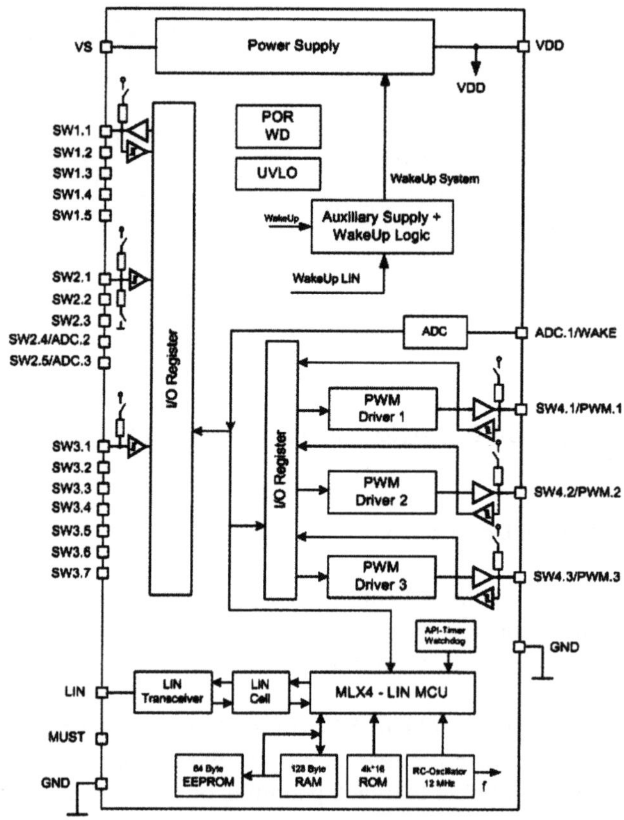

Fig. 11. Block diagram of TH8100.

4.2 Possible Applications

The TH8100 is a universal MCU for distributed intelligent LIN modules. It is well suited for applications like switch modules for window lifter, mirror adjustment, seat adjustment as well as switch modules within the steering wheel or the dashboard. It is also possible to use this IC as LIN gateway for exiting non LIN solutions via the SPI interface.

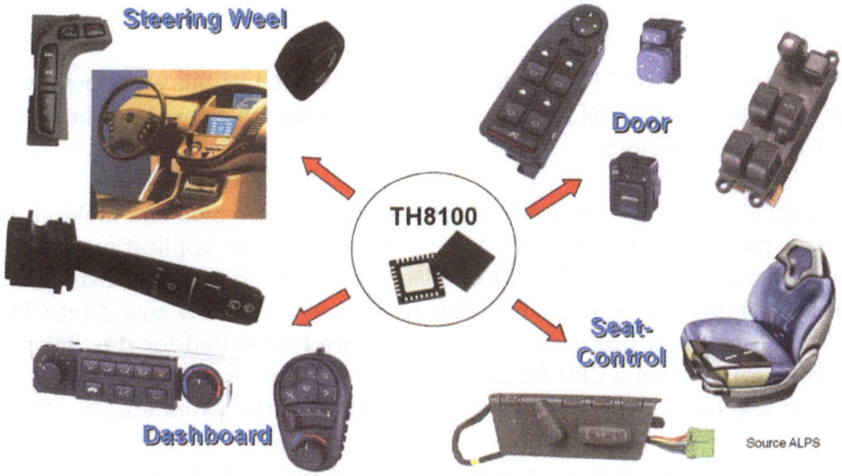

Fig. 12. Possible applications.

The external circuitry is very simple. Just a few capacitors and a reverse protection diode are needed as external components. The switch inputs of this IC can be directly connected to the I/Os. These pins are short circuit proof against battery voltage as well as load dump protected for using in automotive environment. The PWM outputs can be used for realisation of a search and function lighting. These outputs can also be used in different modes that make it possible to connect directly LED's to battery voltage as well as conventional realisations via pre resistor. If higher current is needed than the TH8100 can supply this can be extended with an external load transistor. It is also possible to connect rotation encoder. For reading in analogue values the integrated 10-bit ADC can be used.

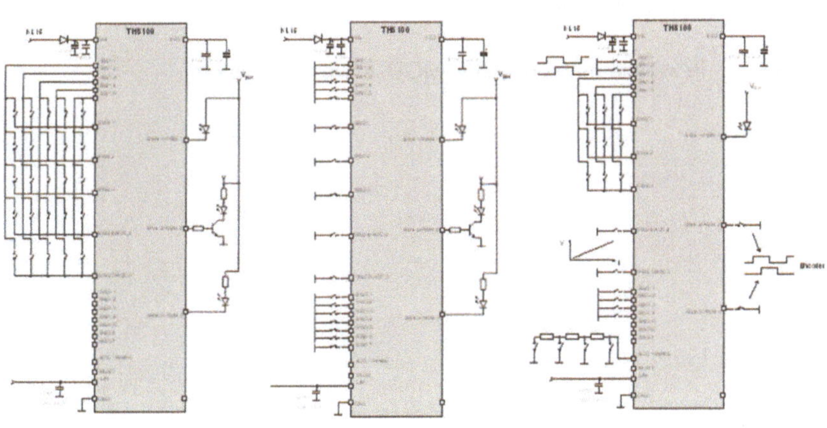

Fig. 13. Example of application circuitries of TH8100.

5 Summary

With the shown 4-bit LIN solution Melexis provides LIN bus architects with a very flexible low cost LIN MCU for slave application which enables the LIN bus to be a successful story.

It is a very safe and robust system with inter task protection and makes it possible to proof the LIN software independent from the application software. This LIN software is supported from Melexis and therefore the customer has only to develop the application software by itself. This low cost 4-bit CPU is optimized for event processing and is therefore best suited for the LIN protocol. Because of using a CPU for processing the LIN protocol it is a very flexible solution and can be adapted to every protocol evolution or extension.

The future of LIN has now started with a new progress of integration of LIN applications. It offers a very low cost solution with a high level of flexibility.

References

[1] LIN Specification 1.3
[2] Datasheet TH8100 Rev. 002b

Michael Bender
Haarbergstrasse 67
99097 Erfurt
Germany
michael.bender@melexis.com

Keywords: in-vehicle networks, LIN, MCU, slave, bus system

An Advanced Network Vehicle Controller (NVC) To Support Future Technology Applications

S. Channon, P. Miller, Ricardo Consulting Engineers Ltd.

Abstract

Future vehicles will require significant changes to their available processing power, network capacity and electrical fault tolerance to allow ongoing improvement in safety, powertrain and comfort systems. This Paper introduces the concept of "Network Vehicle Control" (NVC), shows how this meets the needs of future vehicle systems and details real implementations of NVC. The prototype implementation consists of a number of network vehicle controllers. These controllers include a very high performance microcontroller (>500MIPs), high speed network interface (100Mbps) and a wide range of automotive I/O in a rugged automotive enclosure.

1 Introduction

Ricardo Group are addressing the growth in automotive electronic systems through research projects to provide enabling technologies and evaluate potential solutions for tomorrow's vehicles. This paper describes one of the key projects undertaken during 2002/3 within the Controls and Electronics group.

2 Technology Trends in Future Vehicles

Electronically controlled systems and functions within the vehicle have underpinned a significant growth and application of technology within the automotive environment. This technology trend is continuing at an ever increasing rate as more and more functions are realised. The increased feature content both at a driver comfort and at a system safety level has extended existing electrical networks, which has greatly increased system complexity.

The demand for more feature content and controllability has increased in line with the capability provided by the many electronics control units (ECU's). Chassis control system ECU's developed as stand-alone functional units where the system boundary was identifiable to the mechanical control being replaced e.g. braking, steering. These system boundaries have become blurred as more features are added onto existing controllers. New technologies are now demanding an alternative approach to the vehicle electrical architecture and the traditional system boundaries:

- Drive-by-wire.
- Integrated Safety.

The result is a need for a higher level of systems and functional integration within vehicle electrical systems. This need puts higher demands on electronic technology and increases the complexity. Enabling technologies in automotive electronics computing performance will need to be introduced to support the trend toward smart actuators and sensors systems. An overview of the complex requirements for integrated functionality is illustrated in Figure 1.

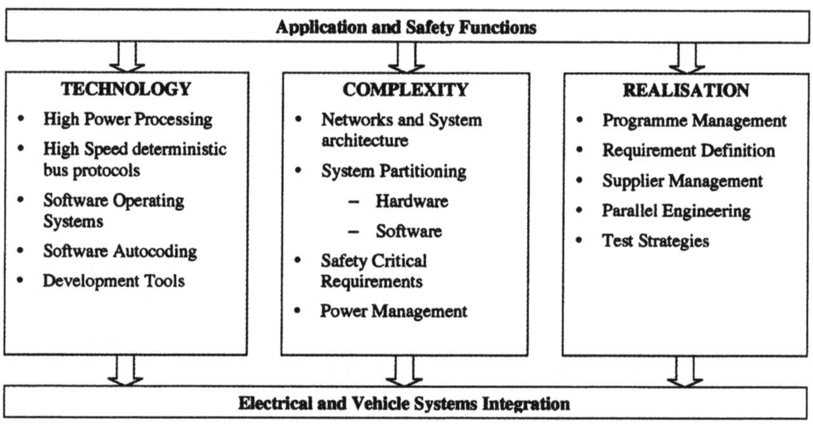

Fig. 1. Demand for integrated functionality increases the complexity.

3 Distributed vs. Centralised Systems

Communications between ECU's have evolved from point-to-point wiring to multiplex serial busses with several different standards now widely used, from LIN and ISO9141 up to CAN. Use of any particular bus protocol is dependent on function requirements, the most widely used for vehicle control being CAN at up to 1Mbps. This capacity has so far satisfied the needs of powertrain and driveline communication as the sub-system controllers have developed higher level features and integral safety functions.

The trend toward functional integration with sub-system interactions is leading to a need to consider alternative network architectures. This reflects a desire to shift from single function control units to distributed functionality as there is no more space for extra ECU's in the car. It can also be said that a smaller number of high capability ECU's is more cost effective so that the number of control units per function can be reduced.

The 1Mbps CAN-bus speed and protocol does not have adequate capability to meet the demands of the next generation applications that require deterministic low latency communication links with higher bandwidth for safety critical real-time vehicle control. New protocols are under development e.g. Flexray, however the extent to which this can meet all safety and data communication needs has yet to be fully determined. There will be many other questions arising from network implementation requirements. While work to develop a standardised protocol is in process it will be important to find an approach to investigate system interactions and data transfer requirements using a high performance ECU platform.

Figure 2 suggests one possible roadmap for the evolution of vehicle electrical architectures. Vehicles launched during the 1990's that are in production today could be said to have a distributed architecture of single function controllers. The challenge is how to move this to a distributed multi-function control architecture that will provide not only functional integration but also support the electrical safety architecture requirements.

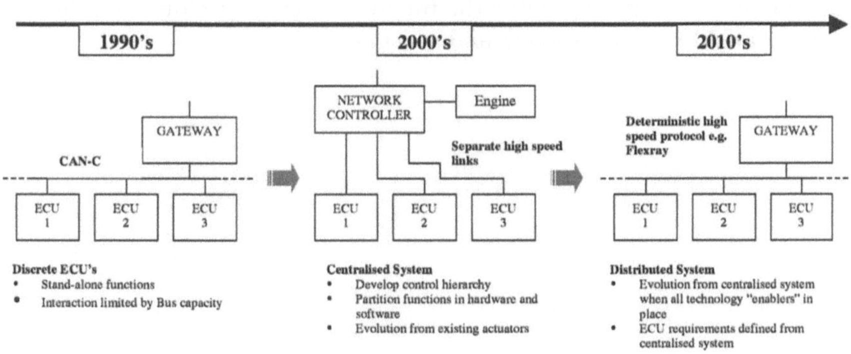

Fig. 2. One potential Network Architecture Roadmap.

Using a high performance network controller as a central communications node provides one solution. Sub-system ECU's can be connected via high speed dedicated links (e.g. CAN) avoiding any bus contention and allowing time-synchronisation protocols to be overlaid for any safety-critical data. Bus communications and network requirements can be explored as a step towards a more ideal fully distributed system. The use of Byteflight [4] in the BMW 7-series is also an example of this type of architecture.

4 Fault Tolerance

As functions become more distributed and as mechanical control is replaced by all electronic control in Drive-by-Wire applications, there will be a need for a robust electrical safety architecture [5]. Electrical safety requirements will move from ECU level solutions to network level implementations, however each ECU will still need to implement hardware and software fault detection and protection strategies on their Input/Output signals and power supplies. The ECU internal protection strategies will need to be augmented with fail-safe and fault tolerant solutions for integrated system-level control functions and network connectivity to provide system-level redundancy.

Figure 3 shows some example techniques that can provide fail-safe and fault tolerant operation. The diagram does not show the sub-system connections that will be necessary to provide the functional system redundancy, for example duplicated links to actuators and sensors.

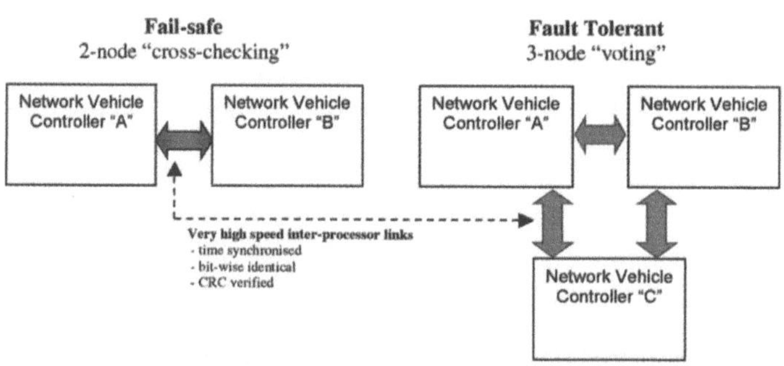

Fig. 3. NVC providing a scalable development platform for Electrical Safety Architecture.

In the case of 2-node and 3-node configurations, cross-checking and voting algorithms in each node perform calculations on exactly the same data. If the data that is cross-checked between two nodes does not agree, then the vehicle electrical systems will reduce to a predetermined base level safe functionality. In the 3-node case, at least 2 nodes must agree on the outcome of the voting before any function is implemented. If one node does not work correctly, it is possible for the other two nodes to identify the faulty node. In this way, the overall system can continue to operate, although at a reduced level of safety, provided there is adequate redundancy in the sub-system connections. This section has only provided a brief overview of a very complex subject. For more details please see references [5-7].

5 Requirements for an Advanced Network Vehicle Controller

Ricardo identified the need to address the increasing system complexity and undertook a project to produce a network vehicle control unit to support and investigate the enabling computing technology requirements. This project was a one year development with the specific intention to deliver a prototype hardware platform as a step to support future research and development. The next sub-sections describe the key technology requirements and how these are addressed by the NVC project.

5.1 Processor Performance

Demand for higher power computing within automotive ECU's has been increasing with new technology developments in engine management and fuel combustion techniques, together with the need for data processing and information integration from safety system sensors and telematic functions. Existing high performance vehicle control ECU's (e.g. engine management systems) provide in the region of 40-60MIPs of processing power.

Next generation devices improve on this level of performance typically by an order of magnitude to provide >300MIPs. Such embedded "supercomputing" devices today have applications in multimedia, networks and communications. This higher performance means more limited operating temperature range and lower integration. Devices with this level of performance designed for use in an automotive environment are under development and will become available over the next 2-3 years.

It is important to start early development with high performance processors in an automotive environment. For the NVC project, potential microprocessor devices were selected based on

- Core processor.
- Communications peripherals.
- Memory support.

After evaluation of all criteria, the Motorola MPC8260 was chosen. It is difficult to assess with any accuracy how well a processor will perform in a given application without some idea of the application itself. However, in the case of NVC which is designed for a range of applications, many of which are currently "undefined", "artificial" benchmarks were used to assess the processors performance. Two benchmarks were used, Dhrystone 2.1MIPs and a matrix inversion benchmark. It is anticipated that for many NVC applications, matrix inversion will dominate the execution time (for example a control system based on a Kalman filter).

The simulation results using a cycle-accurate model of the processor showed that an MPC8260 will meet the NVC objective of exceeding 300MIPs and that it is feasible to run a 60 state Kalman filter with a 10ms update rate (with the processor running at its maximum clock rate). In fact manufacturers data for the MPC8260 state that when running at the fastest clock speed of 300MHz, the processor performance exceeds 500MIPs.

The NVC hardware architecture is shown in Fig. 4. Each of the two processor devices in the design has external RAM and FLASH memory. The fast (MPC8260) processor is designed to run at up to 300MHz, and will generally run application code. Apart from some simple demonstrations, application code is outside the scope of this project. The I/O processor will typically run at 40MHz, and will contain I/O software services which control the hardware within the NVC such as output drivers and serial protocol buffers.

The fast processor can communicate directly to the outside world via 100MHz Ethernet, this feature being used to link NVC units together to create fault-tolerant systems. These links are used in a point-to-point configuration for full-duplex communication to ensure no bus contention with data from other network nodes. Custom software provides fault tolerant time synchronisation and "end-to-end" message integrity checking ("pseudo CRC's") that is both very efficient and very reliable. These are implemented over the Ethernet links to provide a deterministic protocol framework for the cross-checking and voting algorithms. Each NVC includes two externally accessible 42V tolerant Ethernet interfaces to allow for the scalable safety network architecture shown in Fig. 3 and initial measurements show that they can meet all automotive EMC requirements.

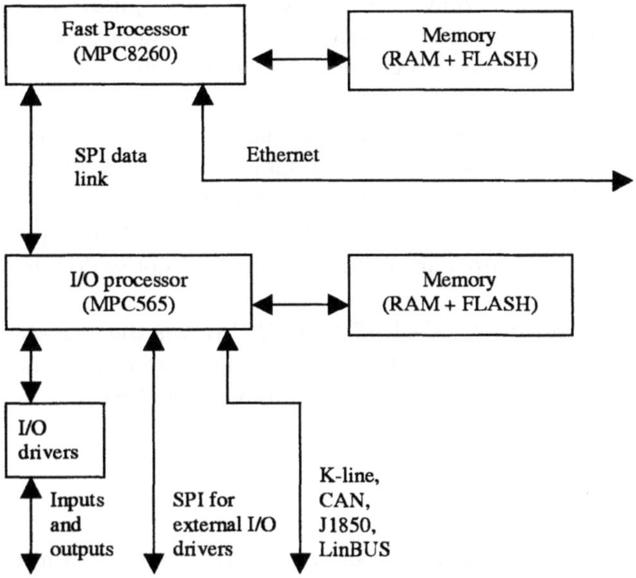

Fig. 4. NVC System block diagram.

5.2 Software Development

The need for future electronic controllers to link directly to software development tools is driven from an increasing emphasis on virtual engineering to address the system complexity issues and to speed up the development process. Simulation and auto-coded software are becoming indispensable.

Current auto-code generators convert a diagrammatic model into (typically) C-based code. The resulting code is much more likely to be functionally correct and generated in a faster time than hand coding. Automatic code generation has been used in the aerospace industry since the late 1980's and code generators are available that are qualified to ED-12/DO-178 level A (the most strict requirements) [8].

The main requirements for the NVC project were therefore to include:

- An RTOS which includes memory management and real-time capability and the means to schedule auto-generated code at the required rates.
- A calibration protocol e.g. CCP, allowing interface with calibration tools (e.g. INCA).

- A method to download code to the fast application processor (ideally via Ethernet), to monitor and control task execution, to debug code and to "blow" an auto-executing image into flash.

A real time version of Linux was selected as the development Operating System. While an OSEK standard operating system was considered, an OSEK OS is not already ported to the MPC8260 (a significant consideration on a time limited development) and the OSEKtime and FTCOM specifications for fault tolerant architectures are relatively new and not widely implemented.

The ETAS tool suite, (ASCET-SD and INCA) was the tool chain selected for initial development, as it met our requirements and in particular, the INCA calibration tool is widely used in the automotive industry.

5.3 Flexible Automotive I/O

One target for the NVC project was to provide a platform that can be used to support the development of advanced engine control algorithms. Therefore a comprehensive IO capability is provided that includes the ability to handle interrupt and crank synchronous tasks. The IO listing also provides a number of software configurable drivers as well a several communications bus interfaces. This range of control interfaces delivers a second NVC project objective to support a number of sub-system load and sensing connections to provide network redundancy for electrical safety architectures. A summary of the principle IO capability is;

- Three CAN 2.0 bus.
- One ISO9141 K-line bus.
- One J1850 bus.
- Two LIN-bus.
- One SPI interface available for external peripherals.
- 24 Digital IO in arranged in blocks of 4, software configurable as inputs or outputs.
- 24 Analogue inputs that may be optionally configured as digital inputs
- 16 high side outputs with digital feedback and 16 low side outputs with current controlled feedback.
- Two H-Bridges, one with current control and sensing and one for stepper motor control.

A third NVC project objective is to support new applications in Drive-by-Wire vehicle control and Energy Systems management technology. The increase in electrical systems on vehicles is significantly adding to the electrical power demand with a resulting trend toward 42V systems. To effectively address the applications, the NVC has implemented IO and CAN-bus interfaces that are protected for 42V capability.

5.4 Automotive Environment

Vehicle installations present a harsh environment for electronic control equipment. Automotive ECU's are subject to mechanical vibration; temperature ranges from -40°C to +85°C (excluding under-bonnet); electromagnetic fields that can exceed 200V/m; electrical transients, load dumps and potential mis-wiring of electrical systems causing short circuits to ground or positive supply; environments exposed water splash and dust [3]. It is desirable that an automotive development platform be as robust as possible to provide reliability and consistency throughout the development process.

The NVC has been designed specifically for use in an automotive environment where the main packaging features are shown in Figure 5. The hardware design approach is scalable, using separate PCB's such that the MPC565 and I/O circuitry could be used without the MPC8260 fast processor and Ethernet connections.

Fig. 5. Packaging details for the Network Vehicle Controller.

6 Potential Applications of NVC

The NVC approach has three main application areas:

- Applications that require very high processing power, for example developments in Camless Engines and Cylinder Pressure Engine Management Systems and for Kalman Filtering algorithms e.g. for determination of vehicle position relative to the road and use of that data for determination of slip angle.

- Applications that require fault tolerance, for example drive-by-wire and "integrated safety" applications.
- Rapid prototyping of systems by use of its flexible IO flexibility and auto-coding capabilities.

7 Conclusions

This paper has summarised the objectives for an advanced Network Vehicle Controller and described how this Ricardo project has addressed the issues of increasing electrical system complexity. In under 1 year a relatively small team of highly skilled engineers have created the prototype of what may become the standard network architecture and ECU performance.

References

[1] G, Leen, D. Heffernan, "Expanding Automotive Electronic Systems"; IEEE 0018-9162/02

[2] P. Jesty, K. Hobley, R. Evans, I. Kendall "Safety Analysis of Vehicle-based Systems", Safety Critical Systems Symposium 2000, 2000

[3] P. Miller, T. Dobedoe, G. Duncan, T. Pike, D. Sharred, P. Smout "Tables of parameter specifications for embedded system assemblies", http://www.surge-europe.com/guest/mtrx9711.html

[4] R. Griessbach, J. Berwanger, M. Peller, "Byteflight – neues Hochleistungs-Datenbussystem für sicherheitsrelevante Anwendungen, ATZ/MTZ"; Automotive Electronics, January 2000, Friedrich Vieweg & Sohn Verlagsgesellschaft mbH

[5] H. Kopetz, "Real Time Systems, design principles for distributed embedded applications"; Kluwer academic publishers, 1997

[6] B. Powel-Douglass, "Designing safety critical embedded systems"; pp34-47, Embedded Systems Programming Europe November 1999

[7] F. Redmill, T. Anderson, "Safety-critical systems; current issues, techniques and standards"; Chapman & Hall 1993

[8] J-L Camus, "Efficient development of safety-critical software"; IEE Electronics Systems and Software, February 2003 pp 38-43

Stephen Channon, Peter Miller
Ricardo Consulting Engineers Ltd.
Bedford Heights
Manton Lane
Bedford MK41 7PH
SAChannon@Ricardo.com
PMiller@Ricardo.com

Keywords: network vehicle control; fault tolerant; safety critical; high performance computing

OSGi – Service-Delivery-Platform for Car Telematics and Infotainment Systems

K. Hackbarth

Abstract

The importance of software in automobiles grow rapidly. Many components in cars are already controlled by software. This trend will continue with new telematic- and infotainment-systems, which are offering the driver new innovative comfort- and safety functions on request. Around 90% of all new innovations in the car will be electrical/electronical while 80% of these are software innovations (Source: A.T. Kearny - Global Automotive Practice; Software determines the future of Automotive Industry 2001).

In conclusion, the complexity of such a system will continously increase, however, software has at the same time a much shorter lifecycle than the car itself. An open, modular and scalable software-platform is necessarsy to handle this complexity and lifecycle management. This paper will focus on an OSGi-Alliance specified service-delivery-platform, which includes all mentioned characteristics.

1 OSGi-Alliance

The OSGi (Open Services Gateway Initiative) -alliance is an independent, non profit corporation that was established in 1999. More than 40 member companies, such as Alpine Electronics, BMW, Bombardier Transportation, Deutsche Telekom, Nokia, Robert Bosch GmbH, Philips and ProSyst, actively support the OSGi-Alliance.

The mission of the OSGi-Alliance is to specify a Java-based service-delivery-platform that allows service-providers, gateway operators, device- and car manufacturers to develop, deploy and manage network-based services in a standarized and cost effective manner.

As response to the high support of the automotive industry, the OSGi-Alliance has build a vehicle expert group, which specifies automotive relevant services, such as a Positioning API.

Global player commitment ensures that the standardization process will continue and will include both state-of-the-art and future standards, thus making it easier for service providers and car manufacturers to deliver solutions on time to market and to increase customer loyality.

2 Requirements of the Automotive Industry

Infotainment platforms are changing from simple audio devices to complex systems that in addition enable new generation of comfort- and safety-services, such as remote diagnostics, remote maintenance, entertainment services and many more. Following are some important requirements for such new infotainment platforms:

- Increasing complexity and number of integrated sub-systems.
- Parallel usage of different bus-systems demands intelligent connectivity.
- Shorter software lifecycles require faster development periods.
- Unpredictable technologies and applications require efficient extensibility.
- Secure communication.

In consideration to the mentioned requirements, the choice of a capable software-platform is very important and should have following characteristics:

- Open and standardized architecture.
- Performant and secure environment.
- Platform-independence in regards to the hardware.
- Reusage of software-modules.
- Dynamic exchangability of integrated functionalities and services.

3 OSGi as Key-Solution

The service platform which is specified by the OSGi-alliance meets all mentioned characteristics. The OSGi service platform Release 3 consists of three main entities:

A framework forms the core of the OSGi service platform specification. It provides a general-purpose, secure, managed Java framework that supports the deployment of extensible and downloadable service applications known as "bundles". In addition the framework manages the dependencies between bundles and services.

Bundles deployed in a Java ARchive (JAR) file are the only entities for deploying Java-based applications within the OSGi environment. A bundle comprises Java classes and other resources, such as an HTML page or images, which together can provide funtions to the end-user as well as to other bundles. By using the Java Native Interface (JNI), bundles can include Native Code as well. Installed bundles can register a number of services that can be shared with other bundles under strict control of the framework.

In the OSGi service platform, bundles are built around a set of cooperating services available from a shared service registry. The service object is owned by, and runs within, a bundle. This bundle must register the service object with the framework service registry so that the service´s funtionality is available to other bundles under the control of the framework.

Besides the framework the OSGi service platform release 3 includes several basic services as well as some vertical market specific services. Following are some of these services described:

- **User Admin Service Specification.** Bundles can use the user admin service to authenticate an inititator and represents his authentication as an Authorization object. Bundles that execute actions on behalf of this user can use the authorization object to verify that the user is authorized.

- **Configuration Admin Service Specification.** The configuration admin service is an important aspect of the deployment of an OSGi service platform. It allows an operator to set the configuration information of deployed bundles. Configuration is the process of defining configuration data of bundles and assuming that those bundles that data when they are active in the OSGi service platform.

- **Log Service Specification.** The log service provides a general purpose messages logger for the OSGi service platform. It consists of two services, one for logging the information and another for retrieving current or previously recorded log information.

- **Wire Admin Service Specification.** The wire admin service is an administrative service that is used to control a wiring topology in the OSGi service platform. It is intended to be used by user interfaces and management programs that control the wiring of services in an OSGi service platform. The purpose of wiring is services together is to allow configurable cooperation of bundles. For example, a temperature sensor can be connected to a heating module to provide a controlled system.

- **Metatype Specification.** The metatype specification defines interfaces that allow bundle developers to describe attribute types in a computer readable form using so-called metadata.

- **Measurement and State Specification.** The measurement class is a utility that provides a consistent way handling a diverse range of measurements for bundle developers. Its purpose is to simplify the correct handling of measurement in the OSGi service platform.

- **Positioning Specification.** The position class is a utility providing bundle developers with a consistent way of handling geographic positions in OSGi applications. The position class is intended to be used with the wire admin service but has wider applicability.

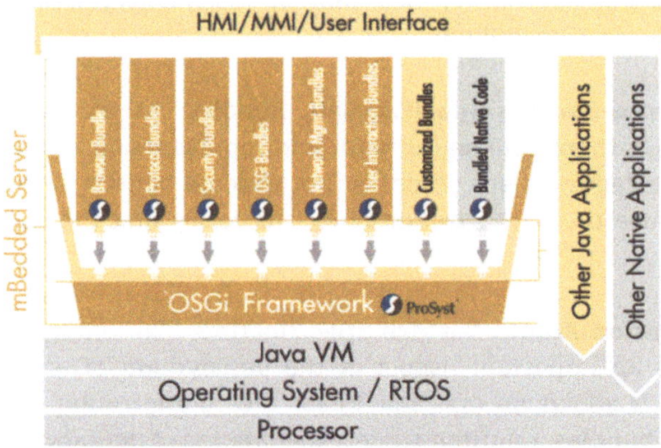

Fig. 1. Architecture of an OSGi -Framework.

When a service (bundle) is requested, an OSGi-compliant gateway, including such a framework and bundles, can both download and install bundles as well as uninstall them when they are not longer required. Due to Java´s dynamic class loading bundle installation and removal can be executed "on-the-fly" without restarting the entire system. The framework is an execution environment for applications that acts as a translator between different protocols and bus-systems.

Due to its modular structure, the OSGi specification does not compete with any other standard or protocol, but allows to support all existing and future standards. Such protocol bundles can be completely written in Java but can also use native code. Strong hardware dependencies of the used protocol can make it necessary to implement native code. Services and/or applications that run on such a framework can be connected to end-devices, such as a navigation system in a car, communicating with a GPS receiver or a security-service at home, communicating with sensors.

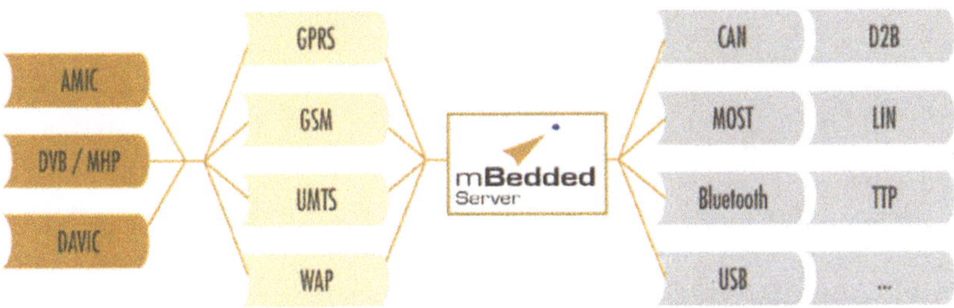

Fig. 2. OSGi does not compete to any other standard or protocol.

In the past car manufacturers and first-tier supplier had to cope with high costs of re-design a single native application due to platform changes, like the change of processors. Using the OSGi-concept car manufacturers can now reuse already developed service-bundles or applications.

In addition to the framework specification that enables local networking and connection between LAN and WAN, the OSGi-alliance designs crucial parts for a remote management system which are explained in the following

The OSGi specification offers interopability for all involved parties in different vertical market segments like the home automation and car infotainment/telematics areas. In order to use the advantages for real business scenarios, car manufacturers and service provider need remote administration and remote management capabilites. Responsible for the entire OSGi-network can be either the car manufacturer or a service aggregator. Such a gateway operator manages the registration of OSGi-frameworks located in the infotainment platform. This can be done either manually or automatically, using a bootstrapping mechnism. A gateway operator is able to remotly add new software components (bundles) and to update existing ones. Defective bundles can be stopped while ensuring the rest-functionality of the running framework. After the problem is solved the bundle can be restarted.

In general the OSGi specification provides only a mechnism but no management policies. The implementation of the policies – e.g. the location where a bundle is loaded – are provided by so called management agents. Management agents have permission to perfom administrative functions on the respective framework and provide the required policies. The OSGi specification does not define how an initial management agent is installed in the framework, because the installation process may differ for different deployment schemes. This pattern allows the vendor or operator to define the required policy in a simple and consistent way.

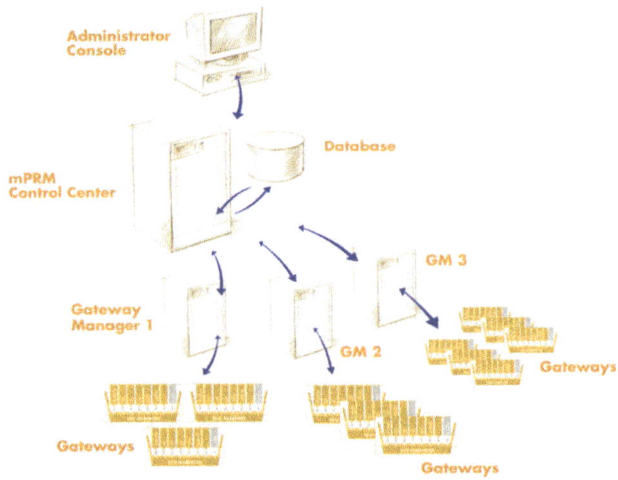

Fig. 3. Remote management of OSGi-gateways.

In addition to the gateway operator, service providers can use the network for offering new or additional functionalities although unable to execute the administrative the administrative actions as described above. Both parties can charge their customers for offered services. A customer either can be charged one time, using a service in defined time frame, or many time for single use of a service.

4 ProSyst's mBedded Server

ProSyst mBedded Server is a state-of-the-art product in the fast-growing world of embedded systems. It meets the requirements of the new generation of technology users, who need mobile access to a large number of services for communication and control.

mBedded Server is a reliable and secure platform with a small memory footprint, optimized for embedded devices. It is open to integration with any infotainment or telematics technology, making it perfectly suited for modern, high-technology vehicles.

ProSyst mBedded Server provides the solution for this paradigm. mBedded Server is a software platform which integrates all the diverse services in the vehicle, but still fits the limited resources available. It acts as a translator among various internal vehicle protocols and bus systems, and provides communication with the outside world. The service gateway provides the concept of error-free delivery, dynamic installation, and remote management of a wide range of telematics services.

mBedded Server is the technological basis for the external connection and the internal network, enabling new services such as remote problem diagnosis, maintenance, security, and entertainment. The end result is that Internet access is integrated into the telematics or infotainment system, while at the same time access to the vehicle and the important information therein is made available to the vehicle manufacturer.

ProSyst has added multiple, valuable capabilities to the platform, in order to utilize the OSGi specification in vehicle telematics and to provide an end-to-end solution. The ProSyst service platform sets new standards for flexibility, scalability and reliability.

5 ProSyst's mPower Remote Manager

ProSyst mPower Remote Manager is a powerful and efficient platform for remote management of a network composed of OSGi compatible gateways. Consequently, it can also manage a broad variety of intelligent devices. mPower Remote Manager has a flexible architecture, integrating the smooth and fast deployment of new services, the reliable maintenance of gateways and their components, as well as various retail and charging resources.

The remote management system enables the dynamic installation, configuration and update of a large number of services, the supervision of gateway runtime states, and the setting of security policies for the network and for end users.

Key Feature of the Power Remote Manager are:

- Provides storage and management of information about gateways, services and their configurations. The information about the gateways and their bundles is maintained in databases – either an LDAP directory server or an SQL RDBMS.
- Grouping of gateways according to various principles enables intelligent management on many-at-a-time basis. Provides life-cycle management – install, update, start, stop and uninstall for bundles and services. Enables control and configuration of the gateways, users, and installed services as well as their permissions.
- To enable large-scale deployment mPRM provides means to integrate a number of gateway manager subsystems, each controlling a separate subset of service gateways.
- Provides a highly optimised network communication protocol, which handles many concurrent client connections.
- The architecture is highly extensible and permits the deployment of custom back-end modules, thus facilitating the development of distributed services.
- The platform is modular, which allows flexible customization of the system behavior by adding and replacing back-end components.
- Provides secure communication and mutual certificate-based authentication. In this way, the system and the gateways are protected from unauthorized access.
- Web-based user interface provides easy to use access to the system functionality.
- Extensive maintenance of information on gateways allows framework state restoration after a possible failure. This also enables the use of platforms without persistent memory.
- Flexible registration and initial configuration of gateways in the management system.
- End user and service provider-oriented sub-system for service publishing and purchasing.
- Support for gateways and components designed to be compliant with OSGi 2.0 and OSGi 1.0 Specifications.
- The system is delivered with a variety of common interfaces and development resources for creation of a wide range of modules.

6 Current OSGi-Based Automotive Projects

6.1 3GT

The 3GT project is to help establish OSGi-based in-vehicle telematics platforms on the European mass market by ensuring interoperability between the products of different middleware providers, terminal manufacturers and service providers. This will be done by establishing common telematics interfaces for OSGi-based service delivery. More specifically, 3GT will develop OSGGi-based specifications for the interface between vehicles and control centers as well as for the interface between control centers and service providers. The 3GT project will test and validate these specifications at 5 European test sites in Gothenburg, Munich, Paris, Rüsselsheim and Torino. Project participants include: ERTICO- ITS Europe, Acunia, BMW, Robert Bosch, C.R.F., Ericsson Microwave Systems, Gatespace, Mizar Mediaservice, Adam Opel, Peugeot Citroen Automobiles (PCA), Netherlands Organisation For Applied Scientific Research (TNO), Vodafone Passo, Volvo Technological Development, Webraska Mobile Technologies, Wirelesscar Sweden.

6.2 ITEA EAST EEA

This automotive project is funded by the European Union and consists of major European automotive manufacturers, first-tier suppliers and research departments. The goal of EAST-EEA is to enable hardware and software interoperability of in-vehicle electronic control units through definition of an open, middleware based architecture. ProSyst's mBedded Server and mPower Remote Manager are being used as an enabling technology for implementing advanced software maintenance solutions for the lifecycle of a vehicle and delivering telematics, infotainment and diagnostics services.

6.3 Stadtinfokoeln – (www.stadtinfokoeln.de)

Stadtinfokoeln is a Cologne-based project, funded by the German Ministry of Education and Science (BMBF), focused on the delivery of parking services to the inhabitants or visitors of Cologne, Germany. The goal is to satisfy customer demand and to reduce traffic, which is related to the availability of parking space. An important role for this is the rapid development of the information- and Communication-Technology in the mobile environment. To fulfill all requirements the decision was made to use an OSGi service platform framework on the vehicle platform. This enables the rapid development and deployment of new services.

6.4 TOP-IQ

The Eureka-project TOP-IQ focuses on the development of a new generation of on-board computers (OBC) for luxury cars and trucks. Partners in the funded project are Groenveld Groep, ProSyst Software AG, Elemaster and esmertec. Based on the OSGi certified platform mBedded Server 5.1, the OBC facilitates telematics services for transport and logistics and offers the management, billing and delivery of new services such as online navigation, trailer management/tracking and tracing of cargo, etc. Due to the combination of ProSyst´s mBedded Server 5.1, esmertec´s RTOS/JVM jbed, Elemasters circuit motherboards and Groenvelds experience in developing systems for the automotive industry, the small-sized OBC requires only very small memory. The 2-year project will finish in April 2003. The prototype was already shown at the IAA Commercial Vehicles in September 2002, the product launch of the OBC is expected for the beginning of 2004.

Summary

As networks in homes, cars, and other environments continue to proliferate rapidly, OSGi is poised to capitalize on its adoption by key end-users, device developers, and service providers. The advantages of OSGi- hardware independence and portability, software modularity, openness and ability to integrate third party applications, seamless access to external services, development and maintenance cost reduction, to name a few- provide a compelling value proposition for the telematics industry. We expect OSGi to become much more visible in the value chain as telematics itself evolves.

References

[1] Kai Hackbarth; "Platform for end-to-end services for electronic devices" ECE-Magazine October 2002.
[2] OSGi Service Platform Release 3, http://www.osgi.org, April 2003.
[3] OSGI-Alliance – Implementation and Deployment Fact Sheet, http://www.osgi.org, March 2003.

Kai Hackbarth
Duerener Str. 405
50858 Cologne
Germany
k.hackbarth@prosyst.com

Keywords: OSGi-alliance, open services gateway initiative, service-delivery-platform, lifecycle management

Automotive Sensor Integration

Dr. G. Teepe, Motorola GmbH
T. Görnig, Conti Temic microelectronic GmbH

Abstract

Next generation sensor systems for automotive applications must operate in an increasingly networked environment. The advent of the FlexRay networking system brings a new view to in-vehicle networking, where system integration, diagnostics and serviceability is given priority over the more traditional view on networks, characterised by transmission bandwidth and latency time. In fact new requirements such as diagnostics, real-time behaviour and fault-tolerance are becoming primary networking criteria. Sensor integration into the larger car electronic system must cope with these features with seamless integration into the software and hardware architecture of the vehicle. New automotive electronics design will deviate from function based system partitioning and move towards a software centric system architecture [1].

1 Electronic Architecture

As a reference architecture, figure 1 shows the network of a high-end passenger car. The powertrain bus is a high-speed CAN bus with up to 1Mbit/s datarate, the multimedia communication is assumed through the MOST-bus and the Body-network is built by the fault-tolerant low-speed CAN. In this architecture the LIN-bus acts as a sub-bus from the high-level network system underneath the body, communications or the powertrain network.

In this architecture sensors are part of the functional subsystems like engine control, seat-control or navigation, just to name a few. In this reference architecture sensor data is available on the different busses as a "pass-through" from the different electronic function blocks. As an example the speed information is extracted from the four wheelspeed sensors and captured and consensed by the ABS-system. We assume here that we want only one "car-speed" information signal be distributed, rather that the speeds of each individual wheel. This speed information is then made available to the powertrain bus participants and through the gateway to all the other busses as well. The speed information is then consumed by e.g. the radio, which needs this information for volume adaptation, the rain-sensor for adaptive wiping or the motormanagement for injection-parameter settings. There is a huge amount of consumers for the speed information in the car, and it becomes clear, that there is a great value in having a network distribute this information, as otherwise it would have to be captured for each different application over and over again. This principle applies to many other signals in the vehicle.

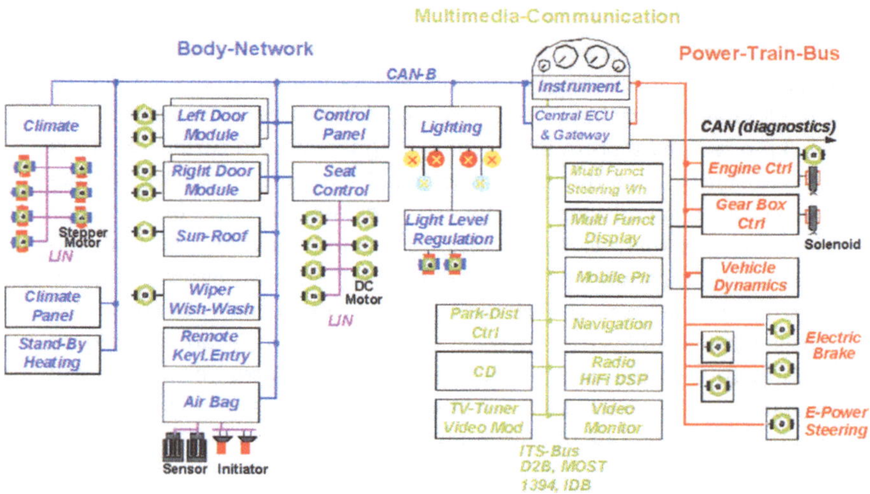

Fig. 1. Vehicle reference architecture.

As architecture development continues, we are seeing major shifts in development:

- The shift towards time-triggered architectures based on LIN and FlexRay for ease of system integration in a real-time environment [2].
- Fault tolerant systems with higher system availability especially for x-by-wire applications.
- Decomposition of functional system blocks into smaller entities, governed by the mechatronics concept of distribution and standard interfaces.
- Concentration of computing performance in a few powerful nodes for a magnitude leap in computing performance increase in the car.

Figure 2 shows a "backbone" concept, where a few fundamental distinctions are made. Especially the separation of input capture and Actuation is a fundamental deviation of the traditional reference architecture of figure 1. The backbone bus, here marked with "FlexRay" carries the complete internal data information of signal variables and actuation commands.

At this point in time this architecture is still a concept, rather than a full implementation, and demonstrates the advantages:

- Sensors are becoming standard parts with separate standard interfaces electrically as well as mechanically.
- Actuators are reduced to its primary function (e.g. valve control, motor-control, electric switch) however with the intelligence necessary to run the bus protocol and the minimum control loops required to perform its local function.
- Computing elements are tailored to the required performance. I.e. if a 200MIPS-machine is required, this computer is hooked on the "backbone" and is easily upgraded if more computing power is required.
- Redundancy is automatically present when several computing engines take part in the network and are configured such that these can mutually replace each others tasks.
- The HMI (Human-Machine-Interface) is a separate sensor-input which senses the driver's wish. By this means the driver input to go faster could be an electronic gas pedal or a side-stick sensing hand movement against a back-force.

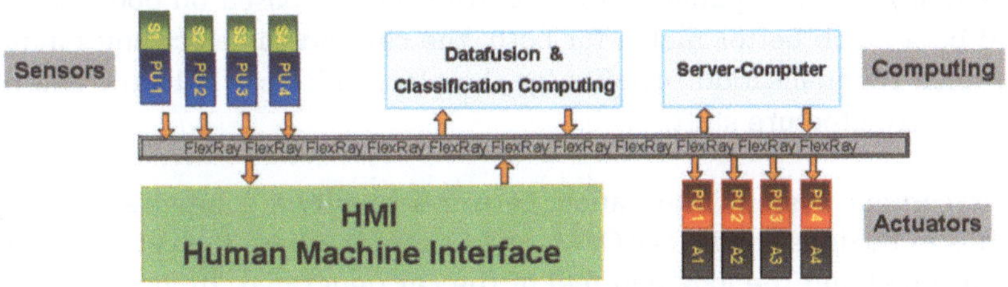

Fig. 2. Fundamental backbone architecture concept.

The advantages are clear: the flexibility of such a system is much higher than a traditionally configured system according to figure 1, which is centred around function groups. In the new architecture of figure 2, the function is a software task on one of the central computers taking some of the captured signals and commanding certain actuators actions. In total the benefits are:

- System integration becomes a software driven process on a few large computers, analog to the traditional computer systems. As an added benefit the well known methods for software integration in real-time systems can be applied with a wealth of industrial expertise available [3].

- Components can become standard parts, much discoupled from the semiconductor growth pattern. This would solve the problem of lifetime discrepancy between electronics and automobile. The rapid electronic progress makes semiconductor components obsolete earlier than the lasting car's lifetime, leading to spare parts supply issues in the current context of rigid production flow and lengthy qualification time. However, once these components are intelligent parts with bus interfaces and local intelligence, adhering to the mechatronics concept, those problems are swiftly solved through backwards interface compatibility [2, 7].
- Newest technology can be applied faster: as the components have simpler boundary definitions, new components can be incorporated with the next cost effective semiconductor technology without lengthy qualification cycles. This leads to faster technology introductions, much like in the computer and network industry [6].
- A strong discoupling between the different semiconductor technologies can be achieved: Sensor capture technology is separate from power-drivers at the actuators and again different from the high performance computing nodes. As such this architecture ideally matches the semiconductor industry's technology split.

In summary the new generic architecture concept, based on open standards would be a much better match for both, the car and the electronics industry. Long term rationalisation benefits for the electronics car industry will result from this architecture shift.

In this context system integration becomes a software integration charter, which is to be done by the car OEM. In fact this is where the car functionality is determined and the look and feel of the car generated. In view to enable a flexible system, the software system must allow for swift software block integration. The approaches in discussion today are centred around a middleware concept described in figure 3.

In essence middleware can be considered as a software bus or transport mechanism where the particular server functions can be hooked into. The beauty of the concept is that the client task can be switched from one machine to the other easily, allowing for spatial and temporal redundancy. The middleware data repository is in essence a real-time data-base regularly updated with the newest data from where the function-tasks draw its data.

Sensor Integration in this concept is achieved as an extraction from the software functions and placed into the middleware repository. Physically this is achieved through a software routine running on the functional machine and placing this data at disposition to the other tasks.

In essence a sensor always needs computing resources to place the data at disposition of other functions. In one way the sensor itself can hold some computing resources able to join data to the repository. This can be achieved through a functional computer to which the sensor is connected to, or in the most complex form, as a sensor which possesses itself of the computing and network resources to pass the digital data forward.

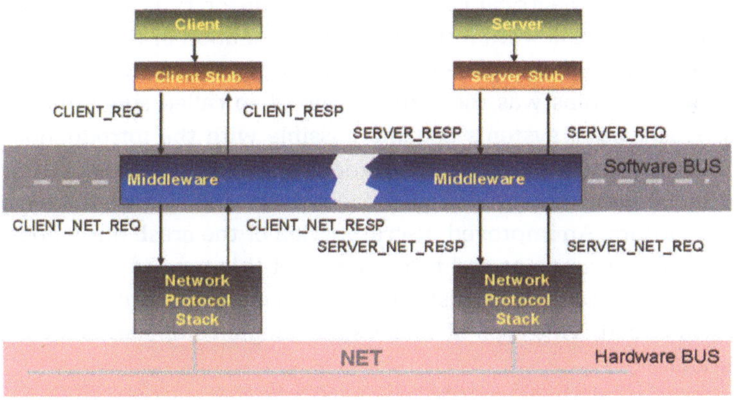

Fig. 3. Middleware Software Bus.

In its most abstract form a sensor can be a Kalman filter, deriving a specific sensor variable from other captured data. Data-Fusion makes use of this method to refine and solidify certain captured data. An example of data fusion is the match of radar and vision data in a data-fusion unit. Here the radar-measured distance between the car and the next obstacle is matched with vision data captured by a CMOS-sensor. The result is more precise and of higher reliability than a single measurement. An other example for a calculated sensor signal is the calculation of tyre pressure from suspension and wheel speed data, which has been tried for some time, but it seems that the required precision is only reached through direct measurement, especially as the tyre temperature can be taken at the same time. In essence a trend towards more calculation can be noted which leads to a reduction of data capturing.

Through the shift from mechanical to semiconductor technology, microsystems have become pervasive in the car, resulting into more signals captured in the vehicle. Over the last decade this has led to double-digit growth rates of semiconductor sensor sales.

2 Sensor Technologies

The development of the sensor technologies can be seen in the development of automotive passive safety systems. As early airbag systems used mechanical crash sensors that have been mounted on structural elements in the front of the vehicle. The electrical contacts of this sensor type were closed on an impact and the airbags were inflated. A discrimination of the crash pulse was not possible with this type of sensor. The next big step in the development of airbag control systems was the introduction of so called single point sensing systems. This type of systems became possible with the introduction of electronic crash sensors. The early generations have been piezoelectric sensors, followed by bulk micromachined resistive sensors and surface micromachined capacitive sensors. An improved discrimination of the crash pulse became possible with the development and introduction of this type of sensors. The standard design of this sensors consists of a sensing element, signal processing and an interface circuit. This type of sensors had single components for the different functions.

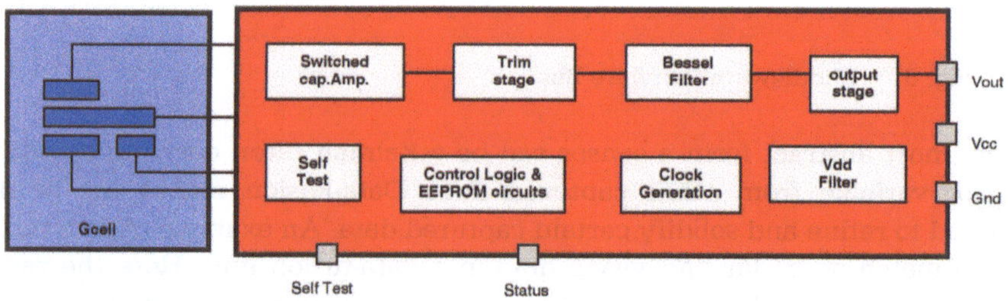

Fig. 4. Sensor block diagram.

Fig. 5. Sensing element with signal processing and interface ASIC.

The next generation of airbag control modules use sensor microsystems (MEMS/MST) [13, 14]. This type of sensors are the result of a consequent integration of the up to date separated functions into two or even single chip designs (figure 4). Several types of this sensor are already in production. A typical assembly can be seen in figure 5, the sensing element in figure 6. This technological step does of course not only apply to accelerometers, it is also to be seen on the various pressure sensors that are used in the vehicle, e.g. engine management, airbag side sensors and also yaw rate sensors that are used at ESP systems or roll over protection systems.

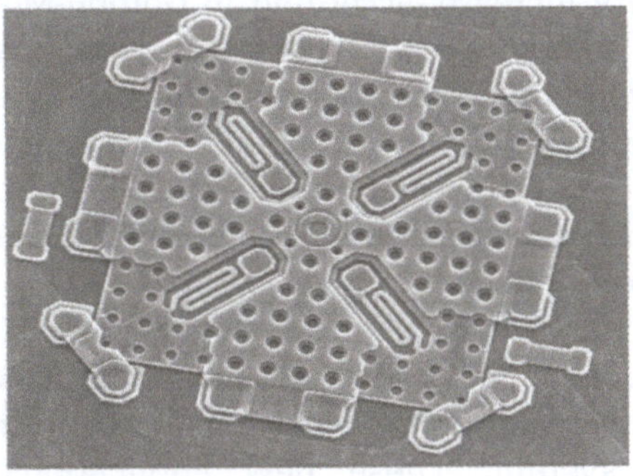

Fig. 6. REM picture of sensing element.

2 Summary and Outlook

The trend towards active/passive safety puts new demands to the design of sensor systems. Due to the increased number of sensors, the necessity to build up sensor clusters is obvious. Another important issue is the wiring effort for this type of systems. It has to be reduced for costs, assembly, weight, quality and reliability reasons.

Future sensor technologies will also extend to complete sensor clusters, which are necessary to provide an optimum on traffic safety, as well as to supply all relevant information from this sensor elements to other control modules in the vehicle. These are comfort systems, e.g. information and navigation, driver assistance and comfort systems, such as lane keeping systems and crash avoidance and protection systems.

Bus systems will be used for the connection and transmission of information between the different modules. There will be different types of bus systems, depending on the different applications that we can find in future vehicles, there will be passive safety system related bus systems like the Bosch/Siemens/Temic Sensor and Deployment Bus [11, 12] and the related development of bus compatible igniters for the actuators of the passive safety system like it is done in the cooperation project IBA [15]. On the Vehicle level the FlexRay bus system [10] for fault-tolerant x by wire applications is in development, complementing the existing LIN [9] and CAN [8] networks.

In summary we have to recognise that the semiconductor microsystem technologies have led to a massive increase of sensor data captured in the vehicle, making cars safer, more reliable and of higher use-value. This trend will continue as long as the sensor technology can profit from the tremendous semiconductor productivity increase, according to Gordon-Moore, which we continue to observe. As an additional element of growth, the sensor data distribution will reach beyond the network of the vehicle. Progress in wireless technologies will ease capturing of data around the car, sourced from other cars, from sensors on the road infrastructure and consensed from a central server. This can be information about the actual road conditions due to weather influences, traffic situations or detours.

With cheap communication technology available, mobile ad-hoc networks will form on the road and between traffic participants to transmit pervasive sensor data through the system. These new applications called "Ambient Intelligence" [4] will open the next chapter of exponential growth for microsystems, semiconductors, automotive electronics and will propel the whole car industry. It will lead us into an unprecedented decade of accident free driving, combined with unparalleled information and service functions for driver and passenger.

References

[1] Gerd Teepe: "Sensors in the next Generation Automotive Networks" Editor: Sven Krüger, Wolfgang Gessner, AMAA-Berlin, March 2002, Publication: Advanced Microsystems for Automotive Applications Yearbook 2002, pp276, ISBN 3-540-43232-9, Springer Verlag Berlin, Heidelberg, New York, 2002

[2] Gerd Teepe: "Elektronik im Lebenszyklus eines Kraftfahrzeugs" VDA-Technischer Kongress 20-21 März 2002, Stuttgart Liederhalle, Tagungsband:VDA, Verband der Automobilindustrie, Westendstraße 61, D60325 Frankfurt, Germany

[3] Christopher Temple, Gerd Teepe: "Composable Architectures" Elektronik Automotive, Magazin für Entwicklungen in der Kfz-Elektronik und Telematik, April 2002, WEKA Fachzeitschriften Verlag GmbH, Gruber Strasse 46a, 85586 Poing, Germany ISSN 0013-5658

[4] Gerd Teepe, Thomas Böhm: „Ambient Intelligence - Ein Konzept für den unfallfreien Verkehr der Zukunft" Automotive Electronics Sonderdruck II/2002

[5] Gerd Teepe, Don Remboski, Richard Baker: „Towards Information Centric Automotive System Architectures" Paper Number SAE 2002-21-0057, Proceedings of the 2002 International Congress on Transportation Electronics, October 21-23, 2002 Cobo-Center, Detroit Michigan;ISBN 0-7680-1113-2, ISSN 98-86269, SAE/P-381 SAE Customer Service email: CustomerService@sae.org

[6] Gerd Teepe: „Schnelle Innovationszyklen in der Halbleiterindustrie im Kontrast zu langlebigen Produktstrategien der Anwenderindustrien" 6. Handelsblatt-Jahrestagung HALBLEITER-INDUSTRIE 2002 18-19 September 2002, Kempinski Hotel Atlantic, Hamburg

[7] Gerd Teepe: „Open Assembly Architectures for flexible Plug&Play Car Solutions" MI-Congress 2002, „Automobile Electronics, From Systems to Vehicle Networks, from Semiconductors to Software" Liederhalle Stuttgart, Germany, November 26-27, 2002 Verlag Moderne Industrie, 56895 Landsberg am Lech

[8] Can in Automation (CIA) webpage on: www.can-cia.org

[9] LIN-Consortium; "LIN-Specification Version 1.3" webpage on: www.lin-subbus.org

[10] FlexRay Consortium webpage on: www.flexray-group.org

[11] Knut Balzer, Christian Zelger, Thomas Goernig, "BST Deployment and Sensor Bus", Airbag 2000+, 5th International Symposium on Sophisticated Car Occupant Safety Systems, Karlsruhe, Germany, Dec. 4-6, 2000. Publication: ICT Airbag 2000+ Yearbook, V12, ISSN 0722-4087, DWS, Karlsruhe, 2000

[12] Common Bosch Siemens Temic Bus Description Rev.: 3.0, October 2002, http://www.temic.com/

[13] Thomas Goernig: "True 360° Sensing Using Multiple Systems" Editor: Sven Krüger, Wolfgang Gessner, AMAA-Berlin, March 2002, Publication: Advanced Microsystems for Automotive Applications Yearbook 2002, pp106, ISBN 3-540-43232-9, Springer Verlag Berlin, Heidelberg, New York, 2002

[14] Thomas Goernig, "Advanced Sensors for Future Restraint Systems", Airbag 2002, 6th International Symposium on Sophisticated Car Occupant Safety Systems, Karlsruhe, Germany, Dec. 2-4, 2002. Publication: ICT Airbag 2000+ Yearbook, V10, ISSN 0722-4087, DWS, Karlsruhe, 2002

[15] Thomas Goernig: "IBA - Integrated Bus Compatible Igniters" Editor: Sven Krüger, Wolfgang Gessner, AMAA-Berlin, March 2002, Publication: Advanced Microsystems for Automotive Applications Yearbook 2002, pp325, ISBN 3-540-43232-9, Springer Verlag Berlin, Heidelberg, New York, 2002

Dr. Gerd Teepe
Motorola GmbH
Schatzbogen 7
81829 München
Germany
gerd.teepe@motorola.com

Thomas Görnig
Conti Temic microelectronic GmbH
Ringlerstrasse 17
85057 Ingolstadt
Germany
thomas.goernig@temic.com

Keywords: Sensor Integration, LIN, CAN, FlexRay, Automotive Networks, Multiplexing, BST Sensor and Deployment Bus, IBA, Automotive electronic Architectures, Microsystems, Ambient Intelligence, Sensor Fusion, Mobile ad-hoc Networks

Automotive Real Time Cost Effective Multi-Media Network Today

H. A. Parmar, Communication and Control Electronics

Abstract

Networked infotainment systems have become common in premium brands such as Mercedes and Jaguar, and as the mass market car brands look for differentiation for their products they are searching for cost effective means to implement similar networks in their vehicles. This paper discusses C&CE's experience gained by implementing real time digital optical systems both at DaimlerChrysler and Jaguar. C&CE have gone for an evolutionary approach to improve performance and reduce cost. The new SMARTwireX –D2B system retains all the advantages of D2B and provides an easy upgrade route from an optical to electrical system while increasing the available bandwidth by 4x. The paper discusses how the SMARTwireX-D2B meets the present and future data capacity requirements of car makers with data rates and bandwidth efficiencies well in excess of legacy networks such as MOST. The technology offers a full system solution to both car makers and set makers with a flexible 'Network Management software module' that will allow set makers to integrate their Applications in a seamless manner, using a clearly defined API.

1 Introduction

This paper describes the experiences that have been gained by Communication and Control Electronics (C&CE) in the last 10 years through implementing real time multimedia data buses in production vehicles. The paper highlights the experience gained by the company to solve the challenges of providing multimedia network solutions e.g. connecting CD Changer, Navigation, Phone, MMI interfaces, etc. It also demonstrates how SMARTwireX-D2B solves many of the challenges in terms of hardware and software solutions for both the set makers and car makers.

2 Implementation Examples: D2B Optical Systems at DaimlerChrysler and Jaguar

D2B Optical is a multimedia system architecture that has been used across the entire platform range at DaimlerChrysler (Mercedes) from 1998 onwards. One of the primary drivers for this architecture was the requirement to link together multiple functions – ranging from HU Tuner/Navigation/TV/CD-Changer/Amplifier, phone module and voice IO to provide hands free control of the complete multimedia functionality. Another driver was the fact that these functions were available from a variety of set makers making it important to enable an interoperable network that could support this diversity of supplier. DaimlerChrysler was one of the first car makers to benefit from a distributed network and the ability to offer their customers the flexibility to pick and choose the features that they wanted.

The Multimedia network is also linked on to a CAN bus, which is used to provide diagnostic information about the multimedia system to determine its status during production, at dealer configuration or if a fault occurs after sale.

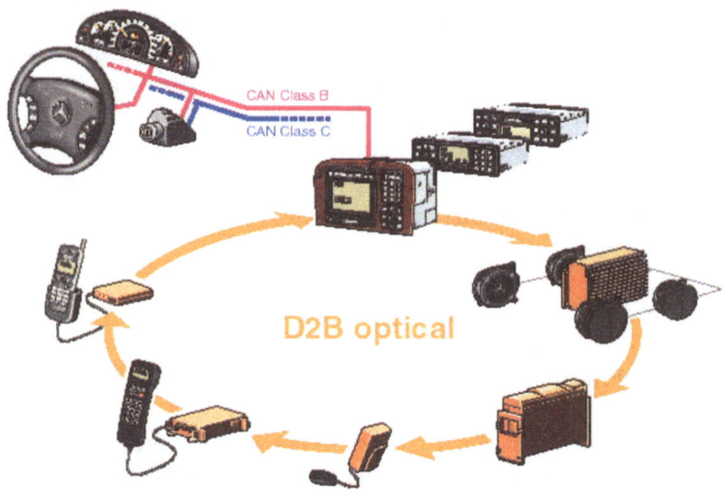

Fig. 1. D2B optical system as fitted into DaimlerChrysler vehicles from 1998 onwards.

C&CE and Jaguar were able to extend the capability of the integrated multimedia communication system, as supplied as a "cross platform" solution at Jaguar which has been incorporated in to the X-Type, S-Type and, with all the features shown below, in Jaguar's XJ model from 2003.

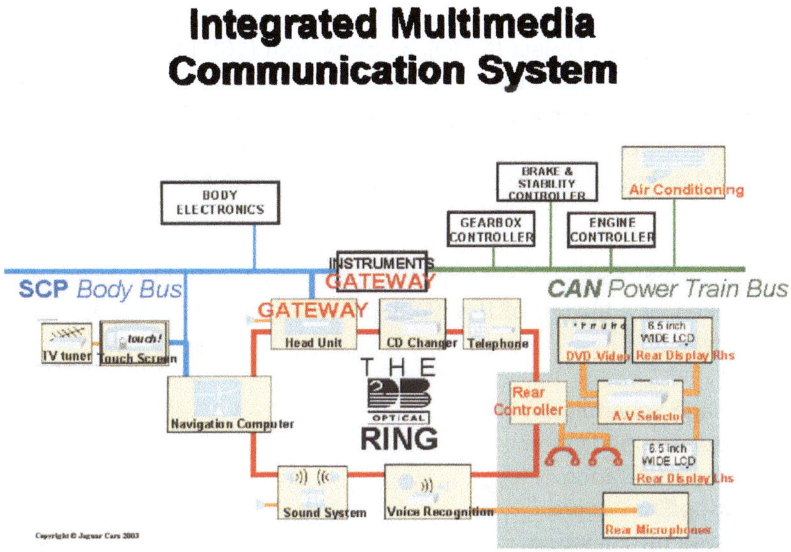

Fig. 2. D2B optical system as fitted to Jaguar vehicles from 2000.

The system uses modules from a variety of specialist providers including Alpine, Visteon, Denso and Motorola. With this approach Jaguar and their dealers can offer customers the flexibility to choose from a low specification fit, consisting of just a HU (Tuner/Tape/AMP in one module), to a high specification multimedia system consisting of variety of Audio/Video devices. This level of dealer fit options can be offered if the basic network structure is already fitted to the vehicle. This then provides car makers and dealers with a flexible and cost effective solution.

3 Lessons Learnt from Working with Carmakers and Set Makers

During the implementation phase of multimedia D2B optical networked systems, C&CE, the car makers and set makers learned several lessons. For example, it was realized that asking several set makers to implement the network management core software for allowing the nodes in the network to start up, shut down, raise alarm, etc from a paper specification is a process that can introduce interoperability errors that needed to be found and fixed during the validation phase of the program.

To resolve these issues it became clear that C&CE should define and produce a structured and precise Network Management software object that would work with a defined API link both at Application level and at the Low-level communication level. C&CE's approach to this requirement was to develop the Network Management Object (NMO), a lightweight (16K ROM and 2K RAM for a slave) and flexible solution that forces a small number of constraints on the application developer. Subsequent implementations of D2B Optical nodes such as the Delphi's SDARS unit for Mercedes have been developed using the C&CE's NMO linked to the Application code in a clean manner using a predefined API.

The D2B Network Management object is written in the 'C' high-level language and is supplied as a library of functions providing the following key software functionality:

- ▶ Message Scheduler.
- ▶ Bandwidth Allocation.
- ▶ Slave Initialisation.
- ▶ Start-up/Shut-down.
- ▶ Address Initialisation.
- ▶ Source Data Routing.
- ▶ Alarm & Fault Monitoring.
- ▶ Retry Mechanisms.
- ▶ Normal Operation.

This product is provided under license as an object code library. Use of a standard code block ensures compatibility of networked components greatly reducing implementation risk for all customers.

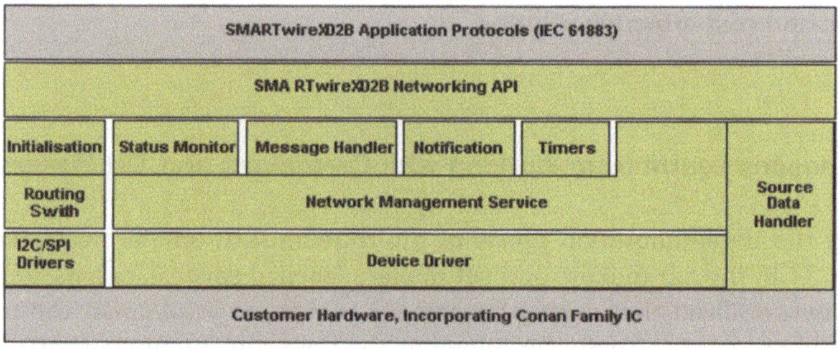

Fig. 3. Network management SW object module concept.

C&CE manages the development and maintenance of the Source Code and the production of libraries for a variety of target microprocessors, including:

- Hitachi H8.
- Intel 8051.
- Infineon 167.
- NEC V850.

Target hardware integration is achieved via a simple device driver layer provided as 'C' Source Code, which can be adapted by the customer as necessary.

4 Real Project Lifecycle Experiences of Implementing 'Electronic Systems in Vehicles'

Typical multimedia networking projects can take 18 to 24 months to implement from the start of system definition through to production. In addition there will be 1 to 2 interior "facelifts" that occur in a platform's 5 to 6 year lifetime. What is more difficult for car makers and their suppliers when defining the multimedia system is how they can future proof the system. The rapid advances in consumer electronics need to be integrated into the platform quickly in such a way that enables them to provide leading edge features to their customers throughout the lifetime of the platform.

Experience has shown that system requirements can change right up to the launch date, as marketing departments want to deliver the latest features in their vehicles. This is bar far the most compelling reason to adopt a network. Integration of last minute changes can be done with low risk if the methodology adopted to define requirements is well controlled and managed. One such method of managing change that is suited to the automotive industry is how to translate:

New requirements → software change → new feature change presented to customer

Using a central port to upload software that can target any of the multimedia devices will be quick and flexible. To do this in an efficient way set makers and network providers need to have the capability to share the onboard processor and memory resources.

The ability to test a new feature implemented, making sure that it does not impact or interfere with a different node on the same network is an important aspect to consider. An important point is that validation testing both at node level and at system level should be started prior to starting any integration activity. The earlier in the development cycle that problems are discovered the less it costs to fix them, so testing early and frequently is an important philosophy for such a complex system.

5 An Evolutionary Approach to Improving Performance and Reduce Overall System Cost

Whereas 5 years ago the premium car brands were looking towards digitally networked infotainment systems and have generally selected systems based on optical technology running over plastic optical fibre (POF) such as D2B Optical, today the volume car brands are also seriously considering the adoption of networking, but with a very different set of constraints.

Is a need to combine functions and features such as:

- Phone function.
- High Fidelity Audio features.
- Navigation Feature.
- Digital TV.
- DVD Video.
- Voice control.

In volume vehicle production a more radical approach is needed compared to the traditional "one box in the 1DIN slot" method. Such diverse functions are difficult to source from a single supplier, but could be achieved by working with a variety of set makers. This is leading towards the volume carmakers specifying a much more distributed system.

Going through a process of implementing a digital network in a vehicle will enable the vehicle to benefit from the strengths of a variety of set makers, which can lead to significant improvements in the features offered to an end customer.

Such a radical change of architecture can be challenging for the automotive industry, so it is important to utilise series production proven technology where possible.

Volume car makers are clear that they are excited by the benefits that a networked infotainment system will offer but it must be production proven and it must be cost effective.

The challenge that C&CE has taken up is to enable the production proven D2B system to be transitioned to moving to a cost effective solution using a lower cost physical layer compared to optical systems, while achieving a 4x bandwidth improvement to support leading edge applications such as digital video.

Moving D2B from an optical physical layer (PHY) using POF to an electrical PHY over unshielded twisted pair (UTP) is the route that C&CE has undertaken. An electrical PHY layer eliminates the need for expensive Fibre Optic Transceiver (FOT) and reduces the cost of the multimedia components by 60%. With over 50 individual D2B networked component types in production today. A proven low risk route exists to upgrade to SMARTwireX-D2B. Several million D2B optical nodes have been shipped to date.

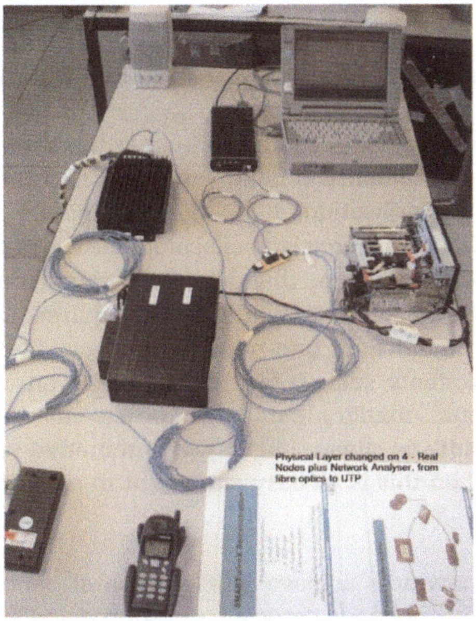

Fig. 4. Physical layer change from fibre optics to UTP.

A proof of concept of the low risk route that this technology offers has been achieved by converting the physical layer from an optical system to a UTP based solution, without having to make any software changes to the nodes that have been designed for the optical solution. Such an example is shown in figure 4 and figure 5, where C&CE has taken actual production units from DaimlerChrysler and converted them with a small daughter board consisting of the C&CE's new UTP physical layer controller (UTP IC) and a suitable automotive grade connector.

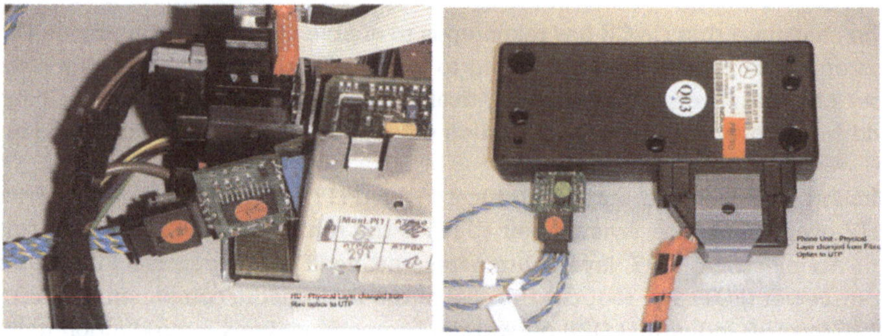

Fig. 5. Detailed picture of the small PCB used on the HU and a phone module – conversion from fibre optics to UTP.

6 Cost Effective and Efficient use of Available Bandwidth

Cost effective use of bandwidth can be defined as the ability to efficiently control and manage the availability of bandwidth on a network at a given time, in order to maximise the utilisation of this bandwidth resource. An example of managing bandwidth as resource is to allocate bandwidth between source and destination when selected rather than allocating source data resource permanently. This would be wasted bandwidth, if not required at the same time and as a result require higher data rate network, and would require networks to provide faster components, such as faster fibre optic transceivers (FOTs), faster microprocessor, faster memory etc. In an office environment it might be argued that bandwidth is cheap, but in the automotive world where EMC is critical, as is cost, then this is an easy statement to make but a very hard one to justify.

SMARTwireX-D2B networks achieve efficient use of bandwidth through the ability to make fast connections between any source and any destination. This enables systems designers to make maximum use of the available 25Mbps bandwidth, which can be delivered in a cost effective way over low cost UTP cables.

7 SMARTwireX-D2B Testing Carried out to Meet Carmakers Needs

EMC

For anybody embarking on a microelectronic implementation in a vehicle, one of the major hurdles that needs to be overcome is meeting the stringent EMC requirements of car makers. The importance of EMC is continuing to increase as more and more electronic systems and networks are installed within vehicles. For all these new systems to work reliably together car makers need to ensure that one electronic system does not interfere in any way with another electronics system. Whilst developing the PLC device for the SMARTwireX technology C&CE have carried out many of these stringent tests with car makers either at their premises or in independent labs. Over a period of 3 years C&CE has amassed a great deal of knowledge in the field of EMC/EMI behavior in automotive electronic systems. This testing has been performed in Europe and Japan.

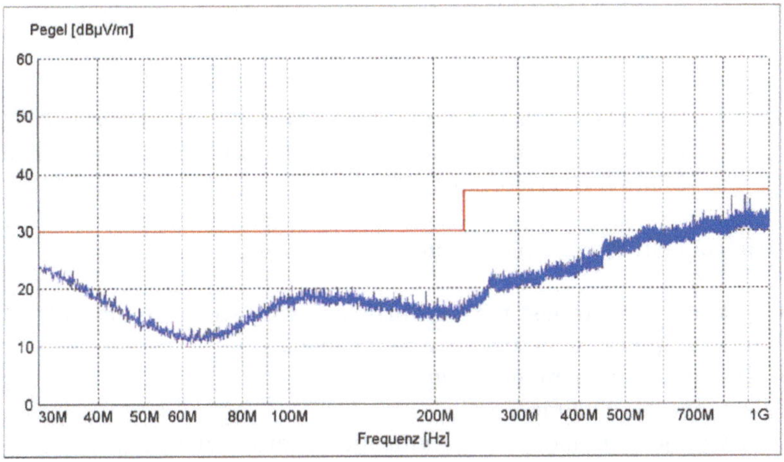

Fig. 6. Point-to-Point configuration vertical biconic antenna tests (MIRA).

Protocols Tests

C&CE have carried out testing in collaboration with a large French OEM to prove the SMARTwireX-D2B protocol to the OEM's satisfaction. This work has involved creating a modular approach to both hardware and software modules that can be transferred to product development. The concept involved creating a Media Access Controller (MAC) Board consisting of C&CE's network transceiver device IC and a Physical Layer Controller (PLC). The MAC Board communicates to a microprocessor board that could be flashed to become a network node (master or slave) via a serial link. We were able to download C&CE's Network Management Object code (NMO) and test code, which was applied to these boards and monitored using a network analyzer and oscilloscope. Figure 8 shows the concept adapted to testing the SMARTwireX system.

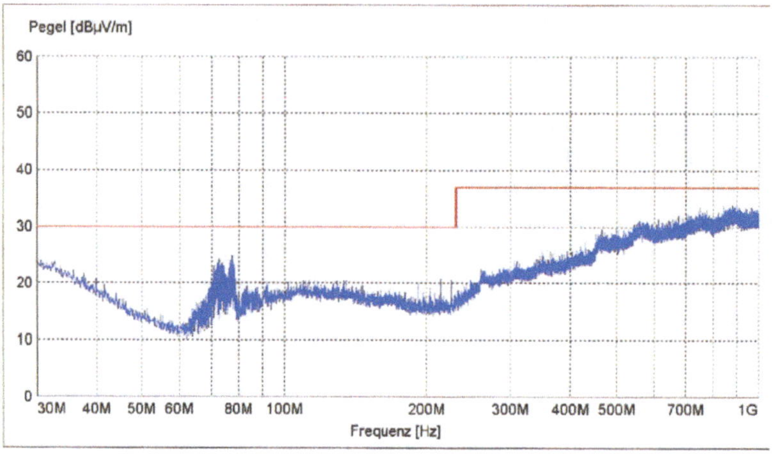

Fig. 7. Vertical emission test result from 1.5M of UTP cable between 2 PLC boards in a point-to-point transmission conducted at a German OEM's R&D laboratories.

The Protocol testing was split up into 5 categories:

- Physical layer robust tests criteria involving short-circuiting UTP leads to 12V applied to actual TX and RX leads.
- Network Management robustness.
- Min and Max voltage.
  Error injection and Alarm notification over the network.
- Power degradation faults applied and recovery mechanisms.
- Power consumption measurements.
- Digital Source data delay over the network testing.
- Control messages and source data transfer reliably over the SMARTwireX-D2B bus.

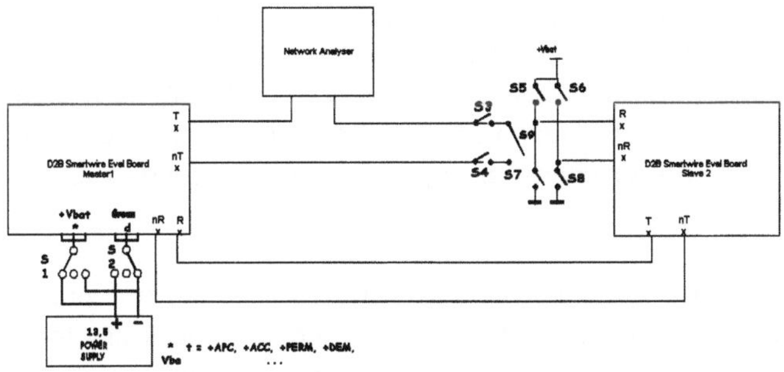

Fig. 8. SMARTwireX-D2B – protocol test set up used for OEM testing.

8 Future Requirements – How to Meet Them

Future requirements are to have commercially available devices such as DVD, DTV, PDAs, Portable PC, etc. plugged into a network, which will allow multiple users to use these devices at any time. One method that allows multiple users to use these devices is to provide a 'convince port' that will allow customers to plug into a cost effective multimedia network, such as SMARTwireX-D2B and use these devices in conjunction with existing items already fitted in the network.

9 Linking other Networks to Increase Functionality and Features of Vehicles

Vehicles in the last 5-10 years have become system platforms rather than just a body, engine and 4 wheels. Car makers have needed additional skills sets in electronic hardware and software design, and fundamentally the ability to test and validate the new systems that are fitted into to their vehicles.

To test these new functions and features, it is vital that during manufacture and dealer fit, a central access port must be provided. This determines what is fitted, what works, what does not, find faults, configure and update software to upgrade new additional features.

Experience gained by C&CE working closely with DaimlerChrysler and Jaguar during the last 5 years has shown that a need to link multimedia data bus such as D2B Optical to other networks in the vehicle such as SPC/CAN/J1850. There are two vital reasons.

With the voice recognition unit on the multimedia D2B Optical system, incorporated by Jaguar, it was possible with simple voice commands to easily control the air-conditioning, which is networked to the CAN network. The gateway function within the head unit allowed 2-way communication between the voice unit and the air-conditioning unit, hence extended and enhanced the control offered by Jaguar.

Central diagnostic information gathering from a central port. As servicing information gathering unit is vital. It has become necessary for all networked devices to provide information that can be transferred to this central diagnostic unit. In many examples means a gateway unit that sits on two or more networks.

10 Digital Video Requirements in Vehicles

In recent years challenges that have been laid down to all car makers and set makers are how to achieve realistic, cost effective solution within a vehicle to support multi-channel video.

Several solutions have been brought forward all providing varying degree of success:

1. Total Analogue solution – provided for many years and tends to be dedicated with a large amounts of wiring. This has heavy cost penalty limited in flexibility and does not tend to offer the control to all users.
2. Hybrid solution to provide analogue video over RGB cables but controlled using a control bus by the users. This offers a middle ground solution by enabling more flexible control and routing of video information, but falls short of the complete flexibility offered by a full digital system. This is the kind of approach that Jaguar have taken as employed in their XJ launched in 2003 (see figure 2).
3. An all-digital multimedia approach is being sought by many car makers for video, audio and flexible control for all the users in the vehicle. For this approach to succeed technology providers, set makers and car makers need to work together to develop the right hardware and software solutions that meet the bandwidth, cost and copyright protection requirements of the market.

C&CE is currently cooperating with other technology providers to develop cost effective digital video solutions over the SMARTwireX-D2B network.

11 Conclusion

Car makers are paying increasing attention to the differentiation that they can achieve for their brand by providing their customers with leading edge Infotainment functions that have clear consumer value and can be accessed intuitively. This trend started with the premium brands, and has rapidly established itself in the mainstream as well.

A multimedia network provides the car maker with a flexible approach to enabling this increased functionality, whether through factory fitted standard specification modules, factory fitted options or dealer fitted options. However, the optical networks that have been used in the premium market, have been a significant barrier to adoption in the volume segment which has so far been unable to benefit from this opportunity.

With the development of SMARTwireX-D2B, C&CE is aiming to remove this barrier and give the mainstream of the car industry the opportunity to install leading edge features while still meeting stringent cost constraints.

References

[1] D2B Optical Basic protocols, C&CE.
[2] Application Protocols, C&CE.
[3] EMC Report, C&CE.
[4] Permission from Jaguar Cars.
[5] Permission from DaimlerChrysler.

H . A. Parmar
Communication and Control Electronics
Occam Court
The Surrey Research Park
Guilford, Surrey, GU2 7YQ
UK.
hari.parmar@candc.co.uk

Keywords: real time automotive network, multimedia application, UTP cost effective solution

11 Conclusion

Advanced Sensorless Distributed Seat Concept

R. Jiménez, J. López, A. Ferré, Lear Automotive

Abstract

Electrical complexity in products yielded from Automotive Industry goes on the increase. During the last years, new functionality and cost-effective strategies have been introduced, for reducing for instance: weight, packaging and assembly labours of components and devices. Several concepts and approaches have been put forward and developed in literature. Among these concepts and solutions, those regarding the chance of improvement as technology integrators are most accepted. As mentioned supplier, Lear is working on advanced designs for reaching this market's common aiming. Involved in a wider dual voltage concept design, that ensure scalable integration and platform's differentiation, a ripple counting strategy, by means of application and developing of an intelligent switch module that plays the role of a CAN-gateway in a automotive seat has been carried out. This new module connects the seat to the rest of the car driving and storing the position of the different loads reducing the number of wires involved in integration. As major that allows cost-effective integration, an embedded hybrid HW/SW implementation has been selected under 8-bit microcontroller programming.

1 Introduction

Current high-end car includes different electronics systems inside the seats. It could be mentioned, as common ones: position adjustment by motors, which is considered a must, being the number of movements a variable provided from manufacturer strategies (car level and market segment) and seat heating, that is also provided for "cold" countries. The last one is usually performed by a single wire resistor, which follows a pattern under the upholstery to cover as much surface as possible. Another usual comfort system integrated is lumbar adjustment, being this device responsible for the feeling of the lower part of the backrest, creating (or not) a hard support zone by height adjustment. Moreover ventilation, massage, etc. are unusual devices that may appear in an electric seat.

Each system has its own interconnecting methodology and interface. For example, devices that incorporate movements managed by motors usually include four wires per connection interface (two for powering, two for a Hall-Effect position sensor). Heating mats, both cushion and backrest are connected in series with a single 2-way connector. Lumbar adjustment may include up to four Electro-valves and an air pump, referring a few features. Along what seen, for each configuration, a specific control unit is designed, to match all required power consumption features, specific wire harnesses and connectors constraints.

When an OEM wants to offer several versions of an electric seat, all these multiple approaches suppose geometrical increase of complexity by means of assembling chain, design concept and so on. A system design focused to the highest-class version would be too expensive for the lower ones, which are usually produced at higher ratios. By other hand, a design of a module for each version may lead expensive developments for small series. Furthermore, different versions may be desired for different markets strategies.

Nevertheless, a consideration should be in mind regarding harness constraint provided by control unit definition. Each control unit version and seat configuration will lead to a special wire harness, so that, as mentioned in paragraph before, wide diversity of harness, connectors and assembly constraints would have to be well managed yielding high level of complexity and an overcost focused mainly at manufacturing. Solutions for cost-effective strategies are called to be a must. In this paper, an open architecture based on a hierarchical network, where a master control device associated with the keyboard module, controls a group of slaves (one for each load) is proposed.

Particularly, the technique for motion control implemented on the actuators allowing the modularity of the system is detailed.

2 The Distributed Architecture Concept

The structure defined before, is conceived by means of a master module which is used both, as a switch controller and gateway between the CAN net and the LIN sub-network. This master module does not include in any case power devices for load control, getting simplified wiring concept and reducing the complexity of the board layout.

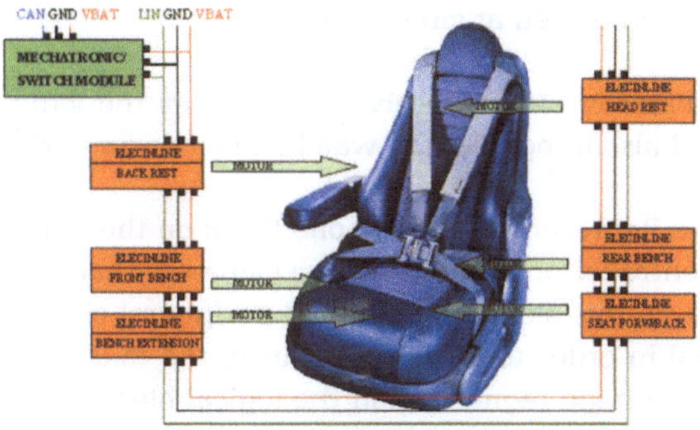

Fig. 1. Distributed architecture system.

Sub-elements connected to the LIN network, also called smart connectors due to its load proximity, are designed under a high degree of versatility, for covering the maximum variety of loads that could appear in a seat system. At the same time, these smart connectors are sized and optimised, looking for highest component integration.

Leaded by a cost-effective design, an improved control has been introduced in selected architecture, fitting compressed algorithm in an 8-bits structure.

A block diagram of one node is represented on figure 2:

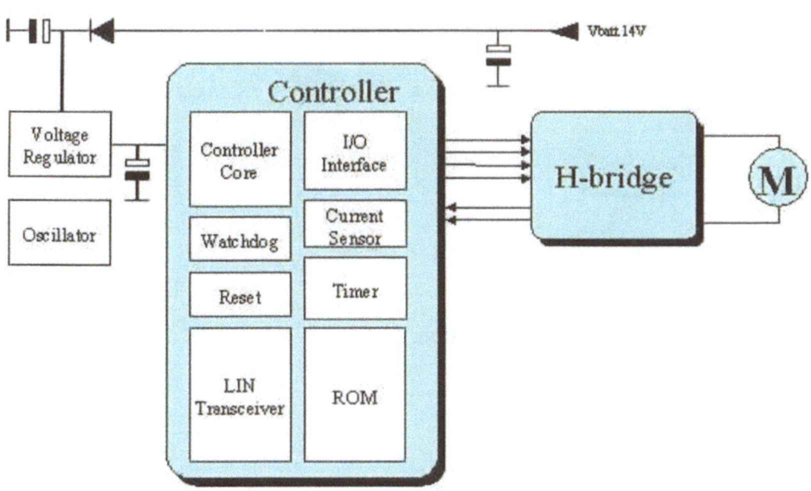

Fig. 2. Distributed smart connection block diagram.

Three main blocks can be seen at circuit sketch:

- Voltage regulator & LIN-transceiver: It supplies the logical module on the board and also integrates the watchdog circuitry and the LIN transceiver.
- Control block: Represented by only one block on the diagram. Actually, both microcontroller and passive-active elements that act as peripheral circuits build up this element. Regarded peripheral circuits prepare the current signal in order to be easily dealt by the micro. Due to the versatility of the voltage regulator and the logical block on the automotive environment, it could be considered the integration in an ASIC.
- Power block: As seen, an integrated H-Bridge topology power block is used. It allows the maximum flexibility on the node application.

3 Ripple Counting Technique

The position control of different motor groups is realised by current sensing technique. It allows control without including a dedicated sensor such as a usual magnetic one.

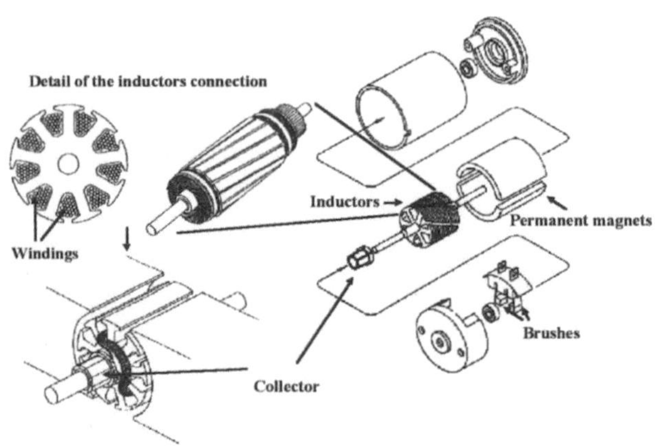

Fig. 3. DC brush motor description.

As well known, Ripple Counting technique is based on effect produced by the own DC-brush motor basis. A DC motor is made of variable number of inductors turned around a core and connected serial like. Two fixed contacts (brush) will carry out bias voltage connection to inductors and thus will characterise the one that is touched by. Each inductor contacts periodically to the brush within a short period of time that corresponds to certain angle of the motor's rotor.

The alternating contact between the serial inductors produce a ripple signal superposed to the nominal current that feeds the motor. This variation is caused by the different value of the inductor group (windings) that contribute, at certain time-window, to the signal performance according to the position of the brushes over the inductor collector.

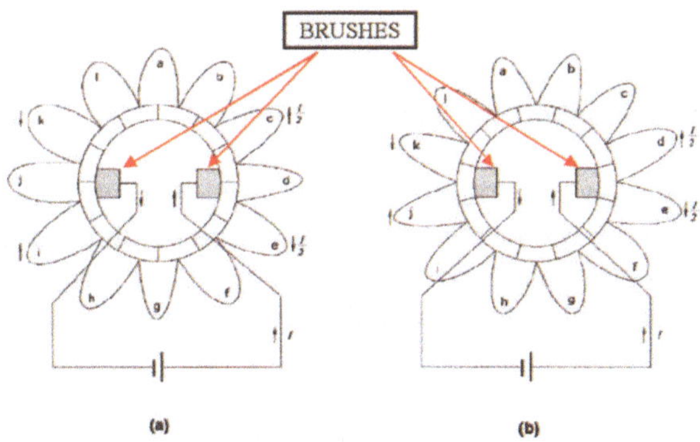

Fig. 4. Different brush-contact possibilities.

As can be seen in the figure 4, there could be two different situations observed. The one on the left, where contribution is reduced up to ten of the twelve inductors that composed the motor and brushes are short-circuiting two opposite poles of the motor, therefore, the global inductance decreases (a). Or, on the other hand (b), all the inductors are contributing to the motor energising.

A characteristic current's behaviour captured from motor sensing is shown in the figure. As often observed, it presents notable initial current-peak, corresponding to the initial torque, and a forthcoming current steady state, more or less reached after the first turn of the motor. Figure 5 clearly shows the mentioned ripple signal superposed to the typical nominal current and fixed around 5 amps. This classic well-like figure is very regular in normal conditions, allowing likelihood for pattern recognition identifying the shape of the different windings where the brush is displacing on.

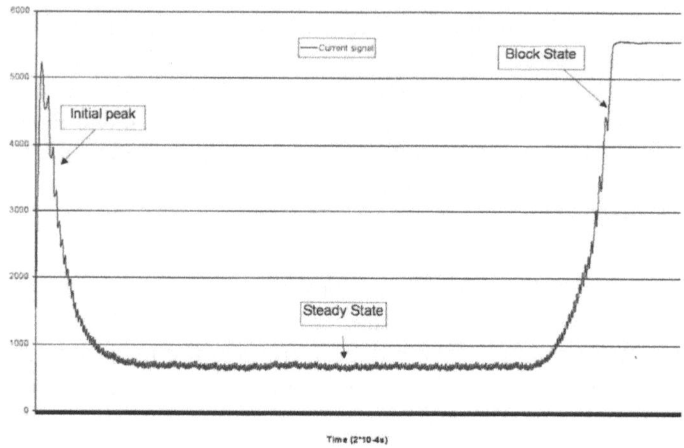

Fig. 5. Current curve from a start-action capture.

3.1 Advantages in Front of Hall Sensor

Classic control method for a DC-brush motor is the introduction of a Hall magnetic sensor inside the motor chassis. It counts the revolutions around the axis by means of magnetic field produced at axis soldered magnet device. Such a way measuring system implies manipulation of the motor and extra wiring for these elements.

Introducing ripple counting suppression of Hall sensor can be achieved; ripple technique extracts the rotational speed information from the current sense, and adds two interesting major effects, which are:

- Accuracy increase: Information from any turn is deeper than from magnetic sensor as changes at inductors are directly reported and therefore detected.
- High Time-Reaction performance: because of detecting events before the arrival to the end of track. Moreover measurable is directly proportional to motor torque while Hall sensor detects the blocked motor as a consequence of it so that averaged reaction time is higher, and avoids errors due to the mechanical effect of the compression.

3.2 Developed Ripple-Counting Technique

The main feature of ripple counting technique is the evaluation of the motor average current comparing against the real one. For this purpose, leaded by a fully covered software solution, a strategy implemented by means of microcontroller provided of an well sized A/D converter capable for sampling analogue signals at Niquist's ratio around (>5kHz), and with at least a 10-bit resolution, should been selected. This parameters are critical constraints for microcontroller selection, and thus to well manage such a proposed system.

Focused to this cost-effective sensor-less motor control system development, an implementation of the current ripple detection over a separate hardware and one supported on software approach, has been disposed. For testing, both were managed by 8-bit microcontroller with an 8-bit A/D converter for sampling the level of current at low speed (500Hz). The ripple signal was read using only an inlet comparing port. Figure 6 shows the two structures analysed.

4 Evaluation and Results

Tests, listed below, were carried out intending position regaining analysis and examination, for both of them, two seating positions were stored (10% = Pos. 1 and 90% = Pos. 2):

A) **Test 1:**
An adjustment cycle covers a sequential starting of the positions 1-2-1. After achieving each position, a short hand-operated control is made (each case) for both directions (< 10 % of the adjusting path), without coming to a block run. All five cycles is a block run into the nearest mechanical impact to be made, 10 flows (= 50 cycles) with the load statuses, no-load operation and nominal load are to be executed. Subsequently, the deviation is determined in relation to the initial value.

B) **Test 2** (specially to verify the stop phase algorithm):
B.1) Start in position 1 and perform 20 start-stop sequences in same direction without reaching any block position. Return automatically to position 1 in one movement. Subsequently, the deviation is determined in relation to the initial value.
B.2) Start in position 2 and perform 20 start-stop sequences in same direction without reaching any block position. Return automatically to position 2 in one movement. Subsequently, the deviation is determined in relation to the initial value.

All tests have been done with the longitudinal axis. After the test, the maximum error tolerance allowed for both test methods is around ±2,5mm, which was completely fulfilled by te algorithm. Preliminary tests, with an 8-bit microcontroller running at clock frequencies of 8MHz showed feasibility for implementation, and a suitable communication support, reasoned on an acceptable bandwidth ratio.

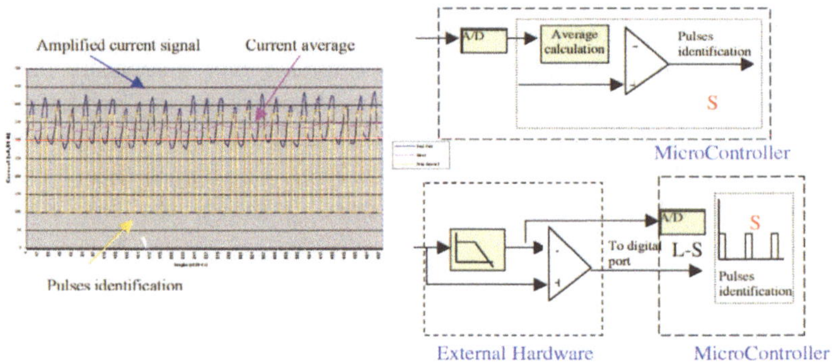

Fig. 6. Capture and processing strategies: full software (top) and mixed solution (bottom).

At pictures showed on figure 6 can be observed that, in the hardware version, digital pulses are introduced to the microcontroller directly, particulary the in-port can be an interruption and thus knowing the speed of the motor only by computing the time between pulses. On the other hand, calculus is more complicated and computing efforts requirements higher. These situation lead to a malfunctioning behavior due to delays caused by cascade computing algorithm.

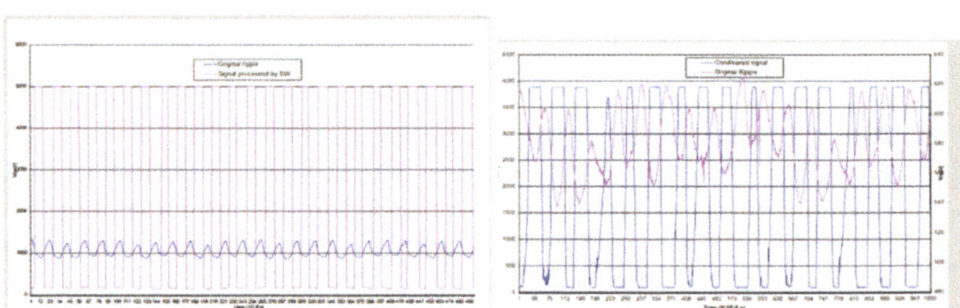

Fig. 7. Ripple detection results: software solution (left side) and hardware. (right side).

During the process, the microcontroller executes a patented software algorithm based on the prediction of the pole sequence combined with a self-learning function to accept the detected pulse as shown in the figure 7.

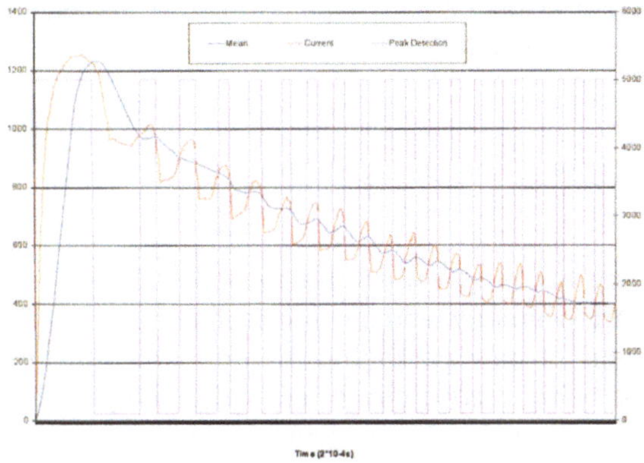

Fig. 8. Initial inrush current peak captured with detection response.

When the motor enters on the steady state, detection is easily done, and the microcontroller can solve the whole group of peaks, including those yielded when the motor is blocked due to a mechanical stop.

At H-bridge opening, the motor is moving at a certain speed and is also supporting an effort directly related to the current intensity that crosses it. At stop case, the motor still goes moving up to certain distance, depending of the inertial speed, the effort done by the own motor and the electromotive force that appears from movement-to-voltage transformation.

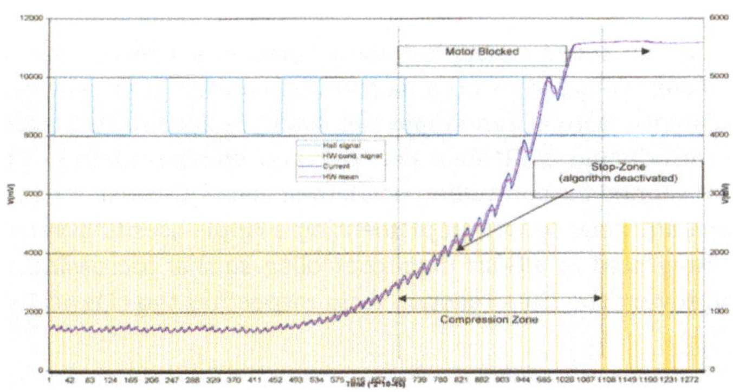

Fig. 9. Stop situation managing performance; hall sensor and ripple one.

On these assumptions, computing and monitoring the displacement of the seat from these variables is possible. Results show that displacement depends mainly on the speed of the seat, and presents variable behaviour when the seat moves towards ahead or backwards. This is because the asymmetries in the electromechanical and mechanical elements.

5 Next Steps

Although detailed system has put forward the goodness of such ripple counting approach, several improvements and new performance of the above structure is ongoing by means of new packaging features and ASIC integration. New concept of nodes is being carried out for achieving a higher level of electronic integration. For scaling developed concept, an integrated circuit in ywhich will be include: microcontroller, voltage regulator with the communication physical layer and some annexed circuitry, that build up the peripheral device circuitry, will be considered and implemented. Proposed integration will improve, by cost-effective approach, the whole system and, specially, by a better placement of the nodes next to the loads.

Yielded first test results involving 42V loads are being evaluated. As assumed, control algorithm well managed the ripple counting technique with the implied current reduction. Such goodness managing 42V loads allows for applying on other controlled motor devices as power windows, sunroofs, etc.

6 Conclusions

In this work, an advanced sensor-less distributed seat concept is presented. The seat has been designed using a distributed concept that provides a high degree of scalability, both at functional and electrical levels. This scalability is achieved by introducing a LIN-bus network and small actuators. The small actuators are designed for controlling position of each DC-brush motor without the need of an additional sensor, implementing a ripple counting variant capable of being integrated in a 8-bit microcontroller, so that increasing the resolution and improving the time-reaction ratio respecting the classical solution.

References

[1] S. J. Chapman, Electric Machinery Fundamentals, McGraw Hill, 1999, (Third Edition).

[2] CRSM1 Current Ripple DC Motor Speed Module Datasheet, available at http://www.cdasys.com/CRSM1.pdf.

[3] US Patent, No 6150785.

[4] US Patent, No 6086177.

[5] US Patent, No 6078153.

[6] US Patent, No 5668451.

Rafael Jiménez, J. López, Antoni Ferré
Lear Automotive (EEDS) Spain, SL
European Technological Center
Applied Research Dpt.
43800 Valls
Spain
rjimenez03@lear.com
jlopez15@lear.com
aferre@lear.com

Keywords: automotive, memory seats ripple counting, sensor-less

Appendix A

List of Contributors

List of Contributors

Appendix B

List of Keywords

List of Keywords

MIX
Papier aus verantwortungsvollen Quellen
Paper from responsible sources
FSC® C105338

If you have any concerns about our products,
you can contact us on
ProductSafety@springernature.com

In case Publisher is established outside the EU,
the EU authorized representative is:
Springer Nature Customer Service Center GmbH
Europaplatz 3, 69115 Heidelberg, Germany

Printed by Libri Plureos GmbH
in Hamburg, Germany